Lecture Notes in Computer Scie___ _   ₁ᵤ₅₄

Edited by G. Goos, J. Hartmanis and J. van Leeuwen

Springer

Berlin
Heidelberg
New York
Barcelona
Hong Kong
London
Milan
Paris
Singapore
Tokyo

Edwin R. Hancock   Marcello Pelillo (Eds.)

Energy Minimization Methods in Computer Vision and Pattern Recognition

Second International Workshop, EMMCVPR'99
York, UK, July 26-29, 1999
Proceedings

Springer

Series Editors

Gerhard Goos, Karlsruhe University, Germany
Juris Hartmanis, Cornell University, NY, USA
Jan van Leeuwen, Utrecht University, The Netherlands

Volume Editors

Edwin R. Hancock
University of York, Department of Computer Science
York YO1 5DD, UK
E-mail: erh@cs.york.ac.uk

Marcello Pelillo
Università Ca' Foscari di Venezia, Dipartimento di Informatica
Via Torino 155, I-30172 Venezia Mestre, Italy
E-mail: pelillo@dsi.unive.it

Cataloging-in-Publication data applied for

Die Deutsche Bibliothek - CIP-Einheitsaufnahme

**Energy minimization methods in computer vision and pattern
recognition** : second international workshop ; proceedings /
EMMCVPR '99, York, UK, July 26 - 29, 1999 / Edwin R. Hancock ;
Marcello Pelillo (ed.). - Berlin ; Heidelberg ; New York ; Barcelona ;
Hong Kong ; London ; Milan ; Paris ; Singapore ; Tokyo : Springer,
1999
  (Lecture notes in computer science ; Vol. 1654)
  ISBN 3-540-66294-4

CR Subject Classification (1998): I.4, I.5, I.2.10, I.3.5, F.2.2, F.1.1

ISSN 0302-9743
ISBN 3-540-66294-4 Springer-Verlag Berlin Heidelberg New York

© Springer-Verlag Berlin Heidelberg 1999
Printed in Germany

Typesetting: Camera-ready by author
SPIN: 10704012   06/3142 – 5 4 3 2 1 0   Printed on acid-free paper

Preface

Collected in this volume are the papers presented at the 2nd International Workshop on Energy Minimization Methods in Computer Vision and Pattern Recognition (EMMCVPR'99), held at the University of York, England, from July 26 through July 29, 1999. The workshop is the second in what we hope will become a series. The first meeting was held in Venice in May 1997. The motivation in starting this series of meetings was the feeling that energy minimization methods, a topic which has roots in various disciplines such as physics, statistics, and biomathematics, represent a fundamental methodology in computer vision and pattern recognition. Although the subject is traditionally well represented in major international conferences in the fields of computer vision, pattern recognition, and neural networks, our primary motivation in organizing this workshop series was to offer researchers the chance to report their work in a focussed forum that allows for intensive informal discussion.

We received 35 submissions for this workshop. Each paper was reviewed by three committee members who were asked to comment on the technical quality of the submissions and provide suggestions for possible improvement. Based on the comments of the reviewers as well as on time and space constraints we selected five papers to be delivered as long oral presentations and 17 papers for regular oral presentation. We make no distinction between these two types of papers in this book. The book is organized into seven sections on shape, minimum description length, Markov random fields, contours, search and consistent labeling, tracking and video, and biomedical applications. We believe that this topical coverage represents a good snapshot of the state of the art in the subject.

Finally, we must offer thanks to those who have helped us in bringing reality to the idea of holding this workshop. Firstly, we thank the program committee for reviewing the papers and providing insightful comments to their authors. We also gratefully acknowledge the work of the following people who helped in the review process: J. Clark, H. Deng, N. Duta, F. Ferrie, G. Guo, E. Mémin, P. Pérez, and L. Wang. Although the workshop was intended to be small we hope that this book will reach a larger audience. In this respect we are extremely grateful to Alfred Hofmann at Springer-Verlag who responded positively to our proposal to publish this volume in the Lecture Notes in Computer Science series. At York most of the hard work of assembling the reviews has been very professionally executed by Sara-Jayne Farmer. We also warmly acknowledge the help of Massimo Bartoli in assembling the final proceedings volume.

May 1999

Edwin Hancock
Marcello Pelillo

Program Co-Chairs

Edwin R. Hancock
Marcello Pelillo

University of York, UK
University of Venice, Italy

Program Committee

Patrick Bouthemy
Joachim Buhmann
Mario Figueiredo
Davi Geiger
Anil K. Jain
Josef Kittler
Stan Z. Li
Maria Petrou
Anand Rangarajan
Kaleem Siddiqi
Shimon Ullman
Lance Williams
Alan L. Yuille
Josiane Zerubia
Steven W. Zucker

INRIA, France
University of Bonn, Germany
Instituto Superior Tecnico, Portugal
New York University, USA
Michigan State University, USA
University of Surrey, UK
Nanyang Technological University, Singapore
University of Surrey, UK
Yale University, USA
McGill University, Canada
The Weizmann Institute of Science, Israel
University of New Mexico, USA
Smith-Kettlewell Eye Research Institute, USA
INRIA, France
Yale University, USA

Table of Contents

Shape

Minimum Description Length

Markov Random Fields

Biomedical Applications

A Hamiltonian Approach to the Eikonal Equation

Kaleem Siddiqi[1], Allen Tannenbaum[2], and Steven W. Zucker[3]

[1]School of Computer Science & Center for Intelligent Machines
McGill University, 3480 University Street, Montréal, PQ, Canada H3A 2A7.

[2]Department of Electrical and Computer Engineering
University of Minnesota, 200 Union Street S. E. Minneapolis, MN 55455.

[3]Department of Computer Science & Center for Computational Vision & Control
Yale University, New Haven, CT 06520-8285.

Abstract. The eikonal equation and variants of it are of significant interest for problems in computer vision and image processing. It is the basis for continuous versions of mathematical morphology, stereo, shape-from-shading and for recent dynamic theories of shape. Its numerical simulation can be delicate, owing to the formation of singularities in the evolving front, and is typically based on level set methods introduced by Osher and Sethian. However, there are more classical approaches rooted in Hamiltonian physics, which have received little consideration in the computer vision literature. Here the front is interpreted as minimizing a particular action functional. In this context, we introduce a new algorithm for simulating the eikonal equation, which offers a number of computational advantages over the earlier methods. In particular, the locus of shocks is computed in a robust and efficient manner. We illustrate the approach with several numerical examples.

1 Introduction

Variational principles emerged naturally from considerations of energy minimization in mechanics [11]. We consider these in the context of the eikonal equation, which arises in geometrical optics and, recently, which has become of great interest for problems in computer vision [4]. It is the basis for continuous versions of mathematical morphology [3, 16, 24, 25], as well as for Blum's grassfire transform [2] and new dynamic theories of shape representation including [9, 22, 21]. It has also been widely used for applications in image processing and analysis [17, 5], shape-from-shading [10] and stereo [8].

The numerical simulation of this equation is non-trivial, because it is a hyperbolic partial differential equation for which a smooth initial front may develop singularities or *shocks* as it propagates. At such points, classical concepts such as the normal to a curve, and its curvature, are not defined. Nevertheless, it is precisely these points that are important for the above applications in computer vision since, e.g., it is they which denote the skeleton (see Figures 4 and 5). To continue the evolution while preserving shocks, the technology of level set methods introduced by Osher and Sethian [14], has proved to be extremely powerful. The approach relies on the notion of a weak solution, developed in viscosity theory [6, 12], and the introduction of an appropriate entropy condition to select it. Care must be taken to use an upwind scheme to compute derivatives, so that information is not blurred across singularities. The representation of the evolving front as a level set of a hypersurface allows topological changes to be handled in a natural way, and robust, efficient implementations have recently been developed [18].

Level set methods are Eulerian in nature because computations are restricted to grid points whose locations are fixed. For such methods, the question of computing the locus of shocks for dynamically changing systems remains of crucial importance, i.e., the methods are shock *preserving* but do not explicitly *detect* shocks. One approach, such as that taken in [20], is to rely on one-sided interpolation of the underlying hypersurface between grid points, to provide sub-pixel estimates of the singularities. Such methods suffer the disadvantage that the interpolation step is computationally very expensive, and introduces numerical thresholds for shock detection. Hence, in order to obtain satisfactory results, high order accurate numerical schemes must be used to simulate the evolving front [13].

On the other hand, there are more classical methods rooted in Hamiltonian physics, which can also be used to study shock theory. To the best of our knowledge, these have not been considered in the computer vision literature. The purpose of this paper is to introduce these methods and a straightforward algorithm for simulating the eikonal equation. The approach offers a number of computational advantages, in particular, the locus of shocks is computed in a robust and efficient manner. The proposed algorithm is Lagrangian in nature, i.e., the front is explicitly represented as a sequence of marker particles. The motion of these particles is then governed by an underlying Hamiltonian system. Such systems are of course fundamental in classical physics, and the technique we elucidate for shock tracking therefore has a natural physical interpretation based on elementary Hamiltonian and Lagrangian mechanics.

2 The Eikonal Equation

We begin by showing the connection between a monotonically advancing front, and the well known eikonal equation. Consider the curve evolution equation

$$\frac{\partial \mathcal{C}}{\partial t} = F\mathcal{N}, \tag{1}$$

where $\mathcal{C}$ is the vector of curve coordinates, $\mathcal{N}$ is the unit inward normal, and $F = F(x, y)$ is the speed of the front at each point in the plane, with $F \geq 0$ (the case $F \leq 0$ is also allowed). Let $T(x, y)$ be a graph of the solution surface, obtained by superimposing all the evolved curves in time (see Figure 1). In other

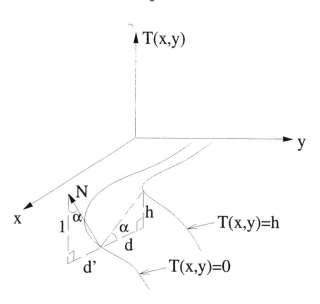

Fig. 1. A geometric view of a monotonically advancing front (Eq. 1). $T(x,y)$ is a graph of the 'solution' surface, the level sets of which are the evolved curves.

words, $T(x,y)$ is the time at which the curve crosses a point (x,y) in the plane. Referring to the figure, the speed of the front is given by

$$F(x,y) = \frac{d}{h} = \frac{1}{\tan(\alpha)} = \frac{1}{d'} = \frac{1}{\|\nabla T\|}.$$

Hence, $T(x,y)$ satisfies the *eikonal equation*

$$\|\nabla T\| \, F = 1. \tag{2}$$

A number of algorithms have been recently developed to solve a quadratic form of this equation, i.e., $\|\nabla T\|^2 = \frac{1}{F^2}$. These include Sethian's fast marching method [18], which relies on an interpretation of Huygens's principle to efficiently propagate the solution from the initial curve, and Rouy and Tourin's viscosity solutions approach [15]. However, neither of these methods address the issue of shock detection explicitly, and more work has to be done to track shocks.

A different approach, which is related to the solution surface $T(x,y)$ viewed as a graph, has been proposed by Shah *et al* [19, 22]. Here the key idea is to use an edge strength functional v in place of the surface $T(x,y)$, computed by a linear diffusion equation. The framework provides an approximation to the reaction-diffusion space introduced in [9]. However, it does not extend to the extreme cases, i.e., morphological erosion by a disc structuring element (reaction) or motion by curvature (diffusion). Hence, points of maximum (local) curvature are interpreted as skeletal points, and the framework provides a type of regularized skeleton. Its relation to the classical skeleton, obtained from the eikonal equation with $F = 1$, is as yet unclear. For example, the curvature maxima based skeleton

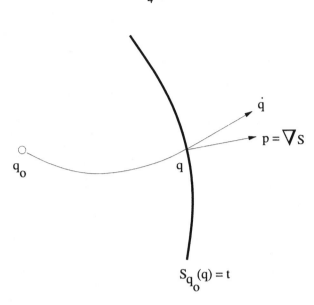

Fig. 2. Direction of a ray $\dot{q}$ and the direction of motion of the wave front $\mathbf{p}$. From [1].

may not be connected (see the examples in [19, 22]). Nevertheless, the framework is computationally very efficient since the governing equation is linear and can be implemented using finite central differences. Furthermore, it can be applied directly to greyscale images as well as to curves with triple point junctions.

In the next section, we shall consider an alternate framework for solving the eikonal equation, which is based on the canonical equations of Hamilton. The technique is widely used in classical mechanics, and rests on the use of a Legendre transformation (see [1] for the precise definition) which takes a system of n second-order differential equations to a (mathematically equivalent) system of $2n$ first-order differential equations. We believe that for a number of vision problems involving shock tracking and skeletonization, this represents a natural way of implementing the eikonal equation.

3 Hamilton's Canonical Equations

Following Arnold [1, pp. 248–258], we shall use Huygens' principle to show the connection between the eikonal equation and the Hamilton-Jacobi equation. For every point $\mathbf{q_0}$, define the function $S_{\mathbf{q_0}}(\mathbf{q})$ as the optical length of the path from $\mathbf{q_0}$ to $\mathbf{q}$ (see Figure 2). The wave front at time t is given by $\{\mathbf{q} : S_{\mathbf{q_0}}(\mathbf{q}) = t\}$. The vector $\mathbf{p} = \frac{\partial S}{\partial \mathbf{q}}$ is called the *vector of normal slowness of the front*. By Huygens' principle the direction of the ray $\dot{\mathbf{q}}$ is conjugate to the direction of motion of the front, i.e., $\mathbf{p} \cdot \dot{\mathbf{q}} = 1$. Note that these directions do not coincide in an anisotropic medium.

Let us specialize to the case of a monotonically advancing front in an inhomogeneous but isotropic medium (Eq. 1). Here the speed $F(x, y)$ depends only on position (not on direction), and the directions of $\mathbf{p}$ and $\dot{\mathbf{q}}$ coincide. The action function minimized, $S(\mathbf{q}, t)$, is defined as

$$S_{\mathbf{q}_0, t_0}(\mathbf{q}, t) = \int_\gamma L \, dt,$$

along the extremal curve γ connecting the points $(\mathbf{q}_0, t_0)$ and $(\mathbf{q}, t)$. Here the Lagrangian

$$L = \frac{1}{F(x, y)} \, \|\partial \gamma / \partial t\|$$

is a conformal (infinitesimal) length element, and we have assumed that the extremals emanating from the point $(\mathbf{q}_0, t_0)$ do not intersect elsewhere, i.e., they form a *central field of extremals*. Note that for an isotropic medium the extremals are straight lines, and that for the special case $F(x, y) = 1$, the action function becomes Euclidean length.

It can be shown that the vector of normal slowness, $\mathbf{p} = \frac{\partial S}{\partial \mathbf{q}}$, is not arbitrary but satisfies the Hamilton-Jacobi equation

$$\frac{\partial S}{\partial t} = -H\left(\frac{\partial S}{\partial \mathbf{q}}, \mathbf{q}\right), \tag{3}$$

where the Hamiltonian function $H(\mathbf{p}, \mathbf{q})$ is the Legendre transformation with respect to $\dot{\mathbf{q}}$ of the Lagrangian function $L(\mathbf{q}, \dot{\mathbf{q}})$. Rather than solve the nonlinear Hamilton-Jacobi equation for the action function S (which will give the solution surface $T(x, y)$ to Eq. 2), it is much more convenient to look at the evolution of the phase space $(\mathbf{p}, \mathbf{q})$ under the equivalent Hamiltonian system

$$\dot{\mathbf{p}} = -\frac{\partial H}{\partial \mathbf{q}}, \qquad \dot{\mathbf{q}} = \frac{\partial H}{\partial \mathbf{p}}.$$

This offers a number of advantages, the most significant being that the equations become linear, and hence trivial to simulate numerically. In the following we shall derive this system of equations for the special case of a front advancing with speed $F(x, y) = 1$.

4 The Hamilton-Jacobi Skeleton Flow

For the case of a front moving with constant speed, recall that the action function being minimized is Euclidean length, and hence S can be viewed as a Euclidean distance function from the initial curve $\mathcal{C}_0$. Furthermore, the magnitude of its gradient, $\|\nabla S\|$, is identical to 1 in its smooth regime, which is precisely where the assumption of a central field of extremals is valid.

With $\mathbf{q} = (x, y)$, $\mathbf{p} = (S_x, S_y)$, associate to the evolving plane curve $\mathcal{C} \subset \mathbf{R}^2$ the surface $\tilde{C} \subset \mathbf{R}^4$ given by

$$\tilde{C} := \{(x, y, S_x, S_y) : (x, y) \in C, \ S_x^2 + S_y^2 = 1, \ \mathbf{p} \cdot \dot{\mathbf{q}} = 1\}.$$

The Hamiltonian function obtained by applying a Legendre transformation to the Lagrangian $L = \|\dot{\mathbf{q}}\|$ is given by

$$H = \mathbf{p} \cdot \dot{\mathbf{q}} - L = 1 - (S_x^2 + S_y^2)^{\frac{1}{2}}.$$

The associated Hamiltonian system is:

$$\dot{\mathbf{p}} = -\frac{\partial H}{\partial \mathbf{q}} = (0,0), \qquad \dot{\mathbf{q}} = \frac{\partial H}{\partial \mathbf{p}} = -(S_x, S_y). \tag{4}$$

Evolve $\tilde{C}$ under this system of equations and let $\tilde{C}(t) \subset \mathbf{R}^4$ denote the resulting (contact) surface. Now project $\tilde{C}(t)$ to $\mathbf{R}^2$ to get the parallel evolution of C at time t, $C(t)$.

5 Numerical Simulations

Fig. 3. The original binary shapes used in our experiments range in size from 128x128 to 168x168 pixels^2.

In this section we apply the above theory to formulate an efficient algorithm for simulating the eikonal equation, while tracking the shocks which form. Recall that since the approach is a Lagrangian one, marker particles will have to first be placed along the initial curve, which in our simulations is assumed to be a simple closed curve in the plane.[1] The evolution of marker particles is then governed

---

[1] The method also extends naturally to a set of open curves by interpreting S as an outward distance function from the collection of curve segments. The initial marker particles are placed on the boundaries of an infinitesimal dilation of each open curve, and are then evolved in an outward direction.

by Eq. 4. With $\mathbf{q} = (x, y)$, $\mathbf{p} = (S_x, S_y) = \nabla S$, the system of equations

$$\dot{S}_x = 0, \dot{S}_y = 0; \qquad \dot{x} = -S_x, \dot{y} = -S_y$$

gives a gradient dynamical system. The second equation indicates that the trajectory of the marker particles will be governed by the vector field obtained from the gradient of the Euclidean distance function S, and the first indicates that this vector field does not change with time, and can be computed once at the beginning of the simulation. Projecting this 4D system onto the (x, y) plane for each instance of time t will give the evolved curve $C(t)$.

In order to obtain accurate results, three numerical issues need to be addressed. First, in order to obtain a dense sequence of marker particles, a continuous representation of the initial shape's boundary ($T(x, y) = 0$, see Figure 1) is needed. Second, it is possible for marker particles to drift apart in rarefaction regions, i.e., concave portions of the curve may fan out. Hence, new marker particles must be interpolated when necessary. Third, whereas finite central differences are adequate for estimating the gradient of the Euclidean distance function in its smooth regime, such estimates will lead to errors near singularities, where S is not differentiable. Hence, we use ENO interpolants for estimating derivatives [13]; the key idea is to obtain information from the "smooth" side, in the vicinity of a singularity. The algorithm may now be stated as follows:

1. Take as the initial curve $T(x(s), y(s)) = 0$, the given boundary of an object, assumed to be a simple closed curve in the plane.
2. Create an ordered sequence of marker particles at positions Δs apart along the boundary.
3. Compute a Euclidean distance transform, where each grid point in the interior of the boundary is assigned its Euclidean distance to the closest marker particle.
4. For each grid point in the interior of the boundary compute and store the components of the vector field ∇S, using ENO interpolants.

5. Do for **step** from **0** to **TOTALSTEPS** {
   Do for **particle** from **0** to **NPARTICLES** {
      • Update the particle's position based on
         ∇S at the closest grid point:
         $x(step + 1) = x(step) - \Delta t \times S_x,$
         $y(step + 1) = y(step) - \Delta t \times S_y$
      • **if** (Distance(particle,next_particle) $> a\Delta s$){
         interpolate a new particle in between.
      }
   }
}

The original binary shapes used in our experiments are depicted in Figure 3. The simulations are based on a piecewise circular arc representation of the boundary, obtained using the contour tracer developed in [20] on the signed distance transform of the original shape. Prior to obtaining the contour, the

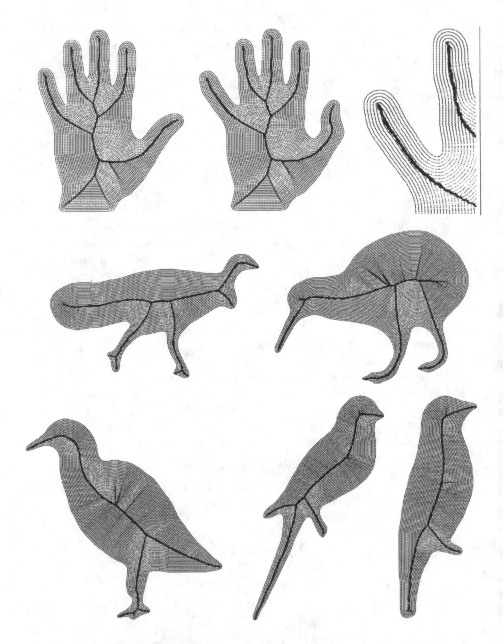

Fig. 4. The evolution of marker particles under the Hamiltonian system. The initial particles are placed on the boundary, and iterations of the process are superimposed. These correspond to level sets of the solution surface $T(x, y)$ in Figure 1. Individual marker particles are more clearly visible in the zoom-in on the fingers of the hand (top right). See Section 5 for a discussion.

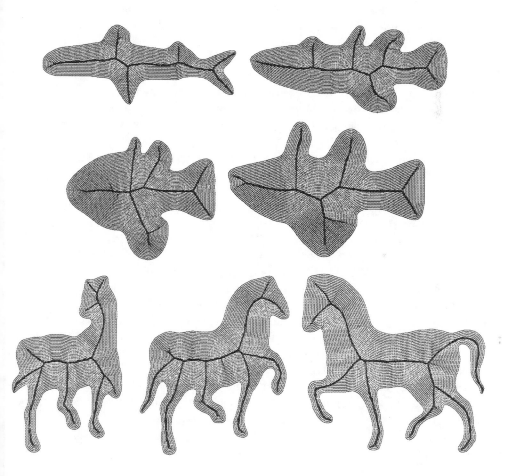

Fig. 5. The evolution of marker particles under the Hamiltonian system. The initial particles are placed on the boundary, and iterations of the process are superimposed. Each iteration gives a level set of the solution surface $T(x, y)$ in Figure 1. See Section 5 for a discussion.

distance transform is Gaussian blurred very slightly ($\sigma = 0.5$ pixels) to combat discretization. The birth of new marker particles (step 5) is also based on circular arc interpolation. Figures 4 and 5 depict the evolution of marker particles, with speed $F = 1$, for several different shapes. For all simulations, the spacing Δs of initial marker particles is 0.25 pixels, the spacing criterion for interpolating a new particle in the course of the evolution is $a\Delta s = 0.75$ pixels, and the resolution of the Euclidean distance transform S is the same as that of the original binary image. The timestep Δt is 0.5 pixels, and results for every second iteration are saved. The superposition of all the level curves gives the solution surface $T(x, y)$ in Figure 1. It is important to note that in principle higher order interpolants can be used for the placement of marker particles, and the resolution of the exact distance transform is not limited by that of the original binary shape.

The results are comparable to those obtained using higher order ENO implementations, although the algorithm is computationally more efficient (linear in the number of marker particles). Informal timing experiments indicate that the efficiency of the algorithm exceeds that of level set methods, except under the "fast marching" implementation, with which it compares favorably. However, when shock detection is included, the Hamiltonian approach has important conceptual and computational advantages. In particular, in contrast with level set approaches, topological splits are not explicitly handled, but shocks (collisions of marker particles) are. In effect, the marker particles are jittered back and forth along the crest lines of the distance function S, leading to thick traces.

The above simulation of the eikonal equation has a variety of applications in computer vision [4, 9, 22, 21, 10, 8], mathematical morphology [3, 16, 24, 25], and image processing and analysis [17, 5]. If desired, it is also possible to formulate an explicit stopping condition for the marker particles. The key idea is to consider the net outward flux per unit volume of the vector field underlying the Hamiltonian system, and to detect locations where energy is lost. As a bi-product of this analysis, which will be described in future work, the skeleton can be robustly and efficiently obtained using only local parallel computations. Figure 6 illustrates the potential of this method, on the same set of shapes. These results may be interpreted as a "stopping potential"; as marker particles enter the regime of negative flux (shown in white) they can be extinguished.

6 Conclusions

In this paper we have introduced a new algorithm for simulating the eikonal equation. The method is rooted in Hamiltonian physics and offers a number of computational advantages when it comes to shock tracking. In future work we plan to extend our results to 3D, where the underlying Hamiltonian system will have the same structure, and the same divergence analysis can be carried out. However, the placement and interpolation of marker particles on the propagating surface will be more delicate. In closing, we note that in related recent work, vector fields rooted in magneto-statics have been used for extracting symmetry and edge lines in greyscale images [7], and that a wave propagation framework

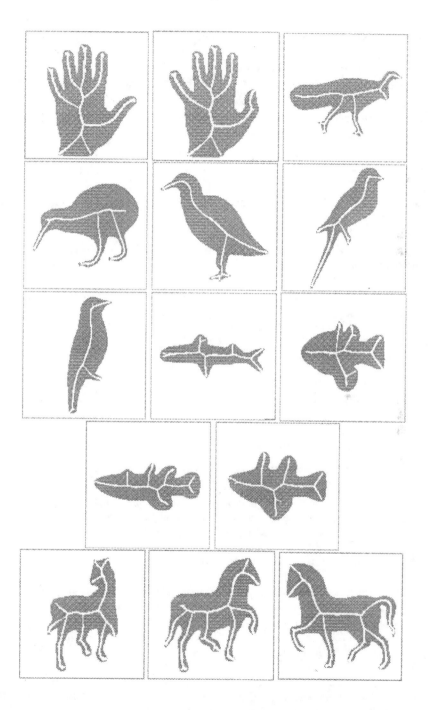

Fig. 6. A divergence-based skeleton, superimposed in white on the original binary shapes (shown in grey). Comparing with the Hamiltonian system based flows in Figures 4 and 5, these maps can be used to formulate a explicit stopping condition for the individual marker particles.

on a discrete grid has been proposed for curve evolution and mathematical morphology [23].

Acknowledgements This work was supported by the Natural Sciences and Engineering Research Council of Canada, the National Science Foundation, the Air Force Office of Scientific Research, the Army Research Office and by a Multi University Research Initiative grant.

References

1. V. Arnold. *Mathematical Methods of Classical Mechanics, Second Edition.* Springer-Verlag, 1989.
2. H. Blum. Biological shape and visual science. *J. Theor. Biol.*, 38:205–287, 1973.
3. R. Brockett and P. Maragos. Evolution equations for continuous-scale morphology. In *Proceedings of the IEEE Conference on Acoustics, Speech and Signal Processing*, San Francisco, CA, March 1992.
4. A. R. Bruss. The eikonal equation: Some results applicable to computer vision. In B. K. P. Horn and M. J. Brooks, editors, *Shape From Shading*, pages 69–87, Cambridge, MA, 1989. MIT Press.
5. V. Caselles, J.-M. Morel, G. Sapiro, and A. Tannenbaum, editors. *IEEE Transactions on Image Processing, Special Issue on PDEs and and Geometry-Driven Diffusion in Image Processing and Analysis*, 1998.
6. M. G. Crandall, H. Ishii, and P.-L. Lions. User's guide to viscosity solutions of second order partial differential equations. *Bulletin of the American Mathematical Society*, 27(1):1–67, 1992.
7. A. D. J. Cross and E. R. Hancock. Scale-space vector fields for feature analysis. In *Conference on Computer Vision and Pattern Recognition*, pages 738–743, June 1997.
8. O. Faugeras and R. Keriven. Complete dense stereovision using level set methods. In *Fifth European Conference on Computer Vision*, volume 1, pages 379–393, 1998.
9. B. B. Kimia, A. Tannenbaum, and S. W. Zucker. Shape, shocks, and deformations I: The components of two-dimensional shape and the reaction-diffusion space. *International Journal of Computer Vision*, 15:189–224, 1995.
10. R. Kimmel, K. Siddiqi, B. B. Kimia, and A. Bruckstein. Shape from shading: Level set propagation and viscosity solutions. *International Journal of Computer Vision*, 16(2):107–133, 1995.
11. C. Lanczos. *The Variational Principles of Mechanics.* Dover, 1986.
12. R. J. LeVeque. *Numerical Methods for Conservation Laws.* Birkhauser Verlag, 1992.
13. S. Osher and C.-W. Shu. High-order essentially non-oscillatory schemes for Hamilton-Jacobi equations. *SIAM Journal of Numerical Analysis*, 28:907–922, 1991.
14. S. J. Osher and J. A. Sethian. Fronts propagating with curvature dependent speed: Algorithms based on hamilton-jacobi formulations. *Journal of Computational Physics*, 79:12–49, 1988.
15. E. Rouy and A. Tourin. A viscosity solutions approach to shape-from-shading. *SIAM. J. Numer. Analy.*, 29(3):867–884, June 1992.
16. G. Sapiro, B. B. Kimia, R. Kimmel, D. Shaked, and A. Bruckstein. Implementing continuous-scale morphology. *Pattern Recognition*, 26(9), 1992.

17. J. Sethian. *Level Set Methods: evolving interfaces in geometry, fluid mechanics, computer vision, and materials science.* Cambridge University Press, Cambridge, 1996.
18. J. A. Sethian. A fast marching level set method for monotonically advancing fronts. *Proc. Natl. Acad. Sci. USA*, 93:1591–1595, February 1996.
19. J. Shah. A common framework for curve evolution, segmentation and anisotropic diffusion. In *Conference on Computer Vision and Pattern Recognition*, pages 136–142, June 1996.
20. K. Siddiqi, B. B. Kimia, and C. Shu. Geometric shock-capturing eno schemes for subpixel interpolation, computation and curve evolution. *Graphical Models and Image Processing*, 59(5):278–301, September 1997.
21. K. Siddiqi, A. Shokoufandeh, S. J. Dickinson, and S. W. Zucker. Shock graphs and shape matching. *International Journal of Computer Vision*, to appear, 1999.
22. Z. S. G. Tari, J. Shah, and H. Pien. Extraction of shape skeletons from grayscale images. *Computer Vision and Image Understanding*, 66:133–146, May 1997.
23. H. Tek and B. B. Kimia. Curve evolution, wave propagation and mathematical morphology. In *Fourth International Symposium on Mathematical Morphology*, June 1998.
24. R. van den Boomgaard. Mathematical morphology: extensions towards computer vision. Ph.D. dissertation, University of Amsterdam, March 1992.
25. R. van den Boomgaard and A. Smeulders. The morphological structure of images: The differential equations of morphological scale-space. *IEEE Transactions on Pattern Analysis and Machine Intelligence*, 16(11):1101–1113, November 1994.

Topographic Surface Structure from 2D Images Using Shape-from-Shading

Philip L. Worthington   Edwin R. Hancock

Department of Computer Science, University of York, UK
{plw,erh}@minster.cs.york.ac.uk

Abstract. This paper demonstrates how a new shape form shading scheme can be used to extract topographic information from 2D intensity imagery. The shape-from-shading scheme has two novel ingredients. Firstly, it uses a geometric update procedure which allows the image irradiance equation to be satisfied as a hard-constraint. This not only improves the data-closeness of the recovered needle-map, but also removes the necessity for extensive parameter tuning. Secondly, we use curvature information to impose topographic constraints on the recovered needle-map. The topographic information is captured using the shape-index of Koenderink and van Doorn [14] and consistency is imposed using a robust error function. We show that the new shape-from-shading scheme leads to a meaningful topographic labelling of 3D surface structures.

1   Introduction

Marr identified shape-from-shading (SFS) as providing one of the key routes to understanding 3D surface structure via the $2\frac{1}{2}$D sketch [17]. The process has been been an active area of research for over two decades. It is concerned with recovering 3D surface-shape from shading patterns. The subject has been tackled in a variety of ways since the pioneering work of Horn and his co-worker's in the 1970's [8,13].

The classical approach to shape-from-shading is couched as an energy minimisation process using the apparatus of variational calculus [13,9]. Here the aim is to iteratively recover a needle-map representing local surface orientation by minimising an error-functional. The functional contains a data-closeness term, and a regularizing term that controls the smoothness of the recovered needle-map. Since the recovery of the needle-map is under-constrained, the variational equations must be augmented with boundary constraints.

Despite considerable progress in the recovery of needle-maps using shape-from-shading [2,4,21], there are few examples of the use of the method for 3D surface analysis and recognition from 2D imagery [1]. One of the reasons for this is the lack of surface detail contained in the needle-map. This can be attributed to the fact that most shape from shading schemes over-smooth the recovered needle map. This is a disappointing omission since there is psychophysical evidence that shading information is a useful shape cue [15].

We have recently embarked on a programme of work aimed at using shape-from-shading as a means of 3D surface analysis and recognition. The contributions to date have been two-fold. We have commenced by bringing the apparatus of robust statistics to bear on the problem of needle map recovery [22, 23]. By using robust error kernels to model the smoothness of the needle map, we have limited some of the problems of over smoothing of surface detail. Our second contribution has been to show how the extracted needle map information can be used for view-based object recognition [24]. Here shape-from-shading has been shown to deliver usable information for a simple histogram-based recognition scheme.

Encouraged by these first results, we are currently investigating how more sophisticated surface representations can be elicited from 2D intensity imagery using shape-from-shading. In particular, we would like to capture the differential or topographic structure of surfaces. Although this is a routine procedure in range imagery [5], there has been little effort directed at extracting topographic structure using shape-from-shading. One notable exception is the work of Lagarde and Ferrie [6] . Here the curvature consistency process of Sander and Zucker [20] is applied to the needle map as a post-processing step so as to improve the organisation of the field of principal curvature directions. There is no attempt to exploit curvature consistency constraints in the recovery of needle maps via the image irradiance equation.

To meet the goal of recovering topographic information, we present a new shape-from-shading algorithm. The algorithm is based on a geometric interpretation of the ambiguity structure of the image irradiance equation in the under-constrained conditions that apply in shape-form-shading. . At each image location, the available intensity information together with the physics of the image irradiance equation mean that the recovered surface normal must fall on a cone of ambiguity. The axis of the cone points in the light source direction. We can impose organisation on neighbouring surface normals by allowing them to rotate on their respective cones of ambiguity so as to satisfy consistency constraints. Here we impose the neighbourhood organisation constraints in such a way as to encourage curvature consistency. Our modelling of curvature consistency is based around Koenderink and Van Doorn's shape index [14]. This is a scale-invariant measure which captures the different topographic classes using a continuous angular variable. Using the shape index allows surfaces to be segmented into meaningful topographic structures such as ridges or valleys, saddle points or lines, and, domes or cups. These structures can be further organised into simply connected elliptical or hyperbolic regions which are separated from one-another by parabolic lines. Our consistency model uses robust error kernels to model acceptable local variations in the shape-index. The model encourages parabolic structures (i.e. ridges and ravines) to be thin and contour-like. Hyperbolic and elliptical structures (domes, cups etc.) are encouraged to form contiguous regions.

2 The Variational Approach to SFS

Central to shape-from-shading is the idea that local regions in an image $E(x,y)$ correspond to illuminated patches of a piecewise continuous surface, $z(x,y)$. The measured brightness $E(x,y)$ will vary depending on the material properties of the surface (whether matte or specular), the orientation of the surface at the co-ordinates (x,y), and the direction of illumination.

The *reflectance map*, $R(p,q)$ characterises these properties, and provides an explicit connection between the image and the surface orientation. The surface orientation is characterised by the components of the surface gradient in the x and y direction, i.e. $p = \frac{\partial z}{\partial x}$ and $q = \frac{\partial z}{\partial y}$. The shape from shading problem is to recover the surface $z(x,y)$ from the image $E(x,y)$. As an intermediate step, we may attempt to recover a set of surface normals or *needle-map*, describing the orientations of surface patches which locally approximate $z(x,y)$.

To simplify the problem, most research has concentrated on recovering ideal Lambertian surfaces illuminated by a single point source located at infinity [3]. A Lambertian surface has a matte appearance and reflects incident light uniformly in all directions. Hence, the light reflected by a surface patch in the direction of the viewer is simply proportional to the orientation of the patch relative to the light source direction. If $\mathbf{n} = (-p, -q, 1)^T$ is the local unit surface normal, and $\mathbf{s} = (-p_l, -q_l, 1)^T$ the global light source direction, then the reflectance function is given by $R(p,q) = \mathbf{n} \cdot \mathbf{s}$.

The image irradiance equation states that the measured brightness of the image is proportional to the radiance at the corresponding point on the surface, which is $R(p,q)$. Normalising both image intensity and reflectance map, the constant of proportionality becomes unity, and the image irradiance equation is simply $E(x,y) = R(p,q)$.

2.1 Horn and Brooks Algorithm

This equation succinctly describes the mapping between the x,y co-ordinate space of the image and the the p,q gradient-space of the surface, but provides insufficient constraints for the unique recovery of the needle-map. Additional constraints, based on assumptions about the structure of the recovered surface, must be utilised. Invariably, it is smoothness of the needle-map that is assumed. Hence, the goal is to recover the smoothest surface satisfying the image irradiance equation. This is posed as a variational problem in which a global error-functional is minimized through the iterative adjustment of the needle map. Here we consider the formulation of Brooks and Horn [3], which is couched in terms of unit surface normals. The Horn and Brooks error functional is defined to be

$$I = \int\int \underbrace{\left(E(x,y) - \mathbf{n} \cdot \mathbf{s}\right)^2}_{Brightness Error} + \lambda \underbrace{\left(\left\|\frac{\partial \mathbf{n}}{\partial x}\right\|^2 + \left\|\frac{\partial \mathbf{n}}{\partial y}\right\|^2\right)}_{Regularizing Term} + \underbrace{\mu\left(\|\mathbf{n}\|^2 - 1\right)}_{Normalizing Term} \, dxdy$$

$$(1)$$

The functional has three distinct terms. Firstly, the brightness error encourages data-closeness of the measured image intensity and the reflectance function. It is the only term which directly exploits shading information. The *regularizing term* imposes the smoothness constraint on the recovered surface normals; it penalises large local changes in surface orientation, measured by the magnitudes of the partial derivatives of the surface normals in the x and y directions. The final term imposes normalization constraints on the recovered normals. The constants μ and λ are Lagrangian multipliers.

The functional is minimized by applying variational calculus and solving the Euler equation:

$$\left(E - \mathbf{n} \cdot \mathbf{s}\right)\mathbf{s} + \lambda\nabla^2\mathbf{n} - \mu\mathbf{n} = 0 \tag{2}$$

To obtain a numerical scheme for recovering the needle-map we must discretise this variational equation to the pixel lattice by indexing the surface normals $\mathbf{n}_{i,j}$ according to their co-ordinates (i,j) on the pixel-lattice. With this notation, the discrete numerical approximation to the Laplacian is

$$\left\{\nabla^2\mathbf{n}\right\}_{i,j} \approx \frac{4}{\epsilon^2}\left(\bar{\mathbf{n}}_{i,j} - \mathbf{n}_{i,j}\right) \tag{3}$$

where

$$\bar{\mathbf{n}}_{i,j} = \frac{1}{4\epsilon}\left(\mathbf{n}_{i,j+1} + \mathbf{n}_{i,j-1} + \mathbf{n}_{i+1,j} + \mathbf{n}_{i-1,j}\right) \tag{4}$$

is the average normal over the local 4-neighbourhood and ϵ is the spacing of pixel-sites on the lattice. Upon substitution, the Euler equation becomes

$$\left(E_{i,j} - \mathbf{n}_{i,j} \cdot \mathbf{s}\right)\mathbf{s} + \frac{4\lambda}{\epsilon^2}\left(\bar{\mathbf{n}}_{i,j} - \mathbf{n}_{i,j}\right) - \mu_{i,j}\mathbf{n}_{i,j} = 0 \tag{5}$$

Rearranging this equation to isolate $\mathbf{n}_{i,j}$ yields the following fixed-point iterative scheme for updating the estimated normal at the surface point corresponding to image pixel (i,j), at epoch $k + 1$, using the previously available estimate from epoch k:

$$\mathbf{n}_{i,j}^{(k+1)} = \frac{1}{1 + \mu_{i,j}\left(\frac{\epsilon^2}{4\lambda}\right)} \times \left(\bar{\mathbf{n}}_{i,j}^{(k)} + \frac{\epsilon^2}{4\lambda}\left(E_{i,j} - \mathbf{n}_{i,j}^{(k)} \cdot \mathbf{s}\right)\mathbf{s}\right) \tag{6}$$

At first-sight, it appears necessary to solve for the Lagrangian multiplier, $\mu_{i,j}$ on a pixel-by-pixel basis. However, it is important to note that $\mu_{i,j}$ only enters the update equation as a multiplying factor which does not effect the direction of update, so we can replace this factor by a normalization step. Finally, we comment on the geometry of the Horn and Brooks needle-map update equation. It is clear that there are two distinct components. The first of these is in the direction of the average neighbourhood normal $\bar{\mathbf{n}}_{i,j}^{(k)}$. This component has a local smoothing effect. The second component is in the direction of the light-source direction $\mathbf{s}$. This can be viewed as responding to the physics of the image irradiance equation and has step-size proportional to $E_{i,j} - \mathbf{n}_{i,j}^{(k)} \cdot \mathbf{s}$. The relative step-sizes are controlled by the Lagrange multiplier.

The principal criticism of the Horn and Brooks algorithm and similar approaches, is the tendency to over-smooth the recovered needle-map. Specifically, the smoothness term dominates the data term. Since the smoothness constraint is formulated in terms of the directional derivatives of the needle-map, it is trivially minimised by a flat surface. Thus, the conflict between the data and the model leads to a strongly smoothed needle-map and the loss of fine-detail. The problem is exacerbated by the need to select a conservative value for the Lagrange multiplier in order to ensure numerical stability [11].

Horn [11] attempts to reduce the model dominance problem by annealing the Lagrange multiplier as a final solution is approached. Meanwhile, we have used the apparatus of robust statistics to moderate the penalization of discontinuities [22].

3 A Novel Framework for SFS

The idea underpinning our new framework for shape-from-shading is to guarantee data-closeness by treating the IIR as a hard constraint. In other words we aim to recover a valid needle-map which satisfies the IIR at every iteration. Subject to this data-closeness constraint, the task of shape-from-shading becomes that of iteratively improving the needle-map estimate. Here, we do this using curvature consistency constraints.

Our approach is a geometric one. We view the IIR as defining a cone of ambiguity about the light source direction for each surface normal. The individual surface normals which constitute the needle-map can only assume directions the fall on this cone. At each iteration the updated normal is free to move away from the cone under the action of the local consistency constraints. However, it is subsequently mapped back onto the closest normal residing on the cone. By applying this constraint, we gain dual advantages in terms of both numerical stability and obviating the need for a Lagrange multiplier. More importantly, the needle-map evolves via a series of intermediate states which are each solutions of the IIR.

3.1 Hard Constraints

The new framework requires us to minimize the constraint functional

$$I_C = \int \int \psi \left(\mathbf{n}(x,y), \mathbf{N}(x,y) \right) dx dy \qquad (7)$$

whilst satisfying the hard constraint, imposed by the image irradiance equation

$$\int \int (E - \mathbf{n} \cdot \mathbf{s}) \, dx dy = 0 \qquad (8)$$

Here $\mathbf{N}(x,y)$ is the set of local neighbourhood vectors about location (x,y). For example, in terms of lattice coordinates i, j, the 4-neighbourhood of $\mathbf{n}_{i,j}$ is defined as

$$\mathbf{N} = \{ \mathbf{n}_{i+1,j}, \mathbf{n}_{i-1,j}, \mathbf{n}_{i,j+1}, \mathbf{n}_{i,j-1} \} \qquad (9)$$

The function $\psi\left(\mathbf{n}(x,y),\mathbf{N}(x,y)\right)$ is a localized function of the current surface normal estimates. The size of the neighbourhood may be varied according to the nature of ψ. Clearly, it is possible to incorporate the hard data-closeness constraint directly into ψ, but this needlessly complicates the mathematics. Instead, we choose to impose the constraint after each iteration by mapping the updated normals back to the most similar normal lying on the cone.

The resulting update equation for the surface normals can be written as

$$\mathbf{n}_{i,j}^{k+1} = \Theta\hat{\mathbf{n}}_{i,j}^k \tag{10}$$

where $\hat{\mathbf{n}}_{i,j}^k$ is the surface normal that minimises the constraint functional I_C. The hard image irradiance constraint is imposed by the rotation matrix Θ which maps the updated normal to the closest normal lying on the cone of ambiguity. Another way to look at this is that we allow the smoothness constraint to select the direction of the normal estimate in the image plane only, whilst fixing the angle between the normal estimate and the light source direction.

To achieve the rotation, we define an axis perpendicular to the intermediate update normal, $\bar{\mathbf{n}}_{i,j}^k$, and the light source direction. The axis of rotation is found by taking the cross-product of the intermediate update with the light source direction $(u,v,w)^T = \bar{\mathbf{n}}_{i,j}^k \times \mathbf{s}$. The angle of rotation is the difference between the angle subtended by the intermediate update and the light source, and the apex angle of the cone of ambiguity. Since the image is normalized, the latter angle is simply $\cos^{-1} E$, giving a rotation angle of

$$\theta = -\cos^{-1}\left(\frac{\bar{\mathbf{n}}_{i,j}^k \cdot \mathbf{s}}{\|\bar{\mathbf{n}}_{i,j}^k\|\,\|\mathbf{s}\|}\right) + \cos^{-1} E \tag{11}$$

Hence, the rotation matrix is given by

$$\Theta = \begin{pmatrix} c + u^2 c' & -ws + uvc' & vs + uwc' \\ ws + uvc' & c + v^2 c' & -us + vwc' \\ -vs + uwc' & us + vwc' & c + w^2 c' \end{pmatrix}$$

where $c = \cos(\theta)$, $c' = 1 - c$, and $s = \sin(\theta)$. Figure 1 illustrates the update process, and compares it with the Horn and Brooks update equation.

3.2 Initialization

The new framework requires an initialization which ensures that the image irradiance equation is satisfied. This differs from the Horn and Brooks algorithm, which is usually initialized by estimating the occluding boundary normals, with all other normals set to point in the light source direction.

To satisfy the image irradiance equation, we must choose a normal direction from the infinite possibilities defined by the cone of ambiguity.

We choose to initialize each normal such that its projection onto the image plane lies in the opposite direction to the image gradient direction, as shown

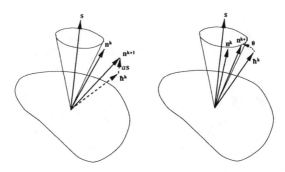

Fig. 1. Comparison of update process for Horn and Brooks (left) and the new framework with the same smoothness constraint (right). Horn and Brooks allows the updated normal to move away from the cone of ambiguity, sacrificing brightness error for smoothness. The movement in the light source direction is $a$sin the left-hand diagram, where $a = \frac{\epsilon^2}{2\lambda}(E - \mathbf{n} \cdot \mathbf{s})$ from Equation 10. In contrast, the new framework forces the brightness error to be satisfied by using the rotation matrix Θ to map the smoothness update to the closest point on the cone.

in Figure 2. This results in an initialization with an implicit bias towards convex rather than concave surfaces. In other words, bright regions are assumed to correspond to peaks, and the image gradient direction points towards these peaks.

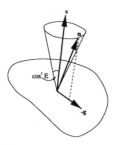

Fig. 2. The set of surface normals at a point which satisfy the Image Irradiance equation define a cone such that E-n.s = 0. A normal from this set is chosen such that the direction of its projection to the image plane is opposite to the maximum intensity gradient direction, **g**.

We have also applied this initialization to the Horn and Brooks algorithm in place of the traditional occluding boundary initialization. We find that this initialization produces significantly better and faster results for the Horn and Brooks algorithm. However, in common with the Horn and Brooks algorithm, our schemes are sensitive to initialization. It is impossible to say whether using the image gradient as described above is the best possible initialization, but it does have stable properties and is, intuitively, a reasonable method of estimating initial normal direction.

4   Needle Map Smoothness Constraints

Before we describe our modelling of curvature consistency, we pause to consider how needle-map smoothness can be incorporated into our new framework for shape-from-shading. In a recent paper [23], we showed how needle-map smoothness could be modelled using robust error kernels. Here the adopted framework was based on a regularised energy-function similar to that underpinning the Horn and Brooks algorithm. However, rather than using a quadratic smoothness prior, we used a continuous variant of the Huber robust error kernel. In this section we show how this smoothness model can be used in conjunction with our geometric needle-map update process In essence, we consider that the recovered surface should be smooth, except where there is a high probability that a discontinuity is present, in which case the smoothing is reduced.

We define the robust regularizer constraint function as

$$\psi\left(\mathbf{n}, \mathbf{N}\right) = \rho_\sigma\left(\left\|\frac{\partial \mathbf{n}}{\partial x}\right\|\right) + \rho_\sigma\left(\left\|\frac{\partial \mathbf{n}}{\partial y}\right\|\right) \tag{12}$$

where $\rho_\sigma(\eta)$ is a robust kernel defined on the residual η and with width parameter σ. Applying the calculus of variations to the resulting constraint function I_C yields the general update equation

$$\mathbf{n}_{i,j}^{(k+1)} = \Theta\left(\frac{\partial}{\partial x}\rho_\sigma'\left(\left\|\frac{\partial \mathbf{n}}{\partial x}\right\|\right) + \frac{\partial}{\partial y}\rho_\sigma'\left(\left\|\frac{\partial \mathbf{n}}{\partial y}\right\|\right)\right) \tag{13}$$

where

$$\rho_\sigma'\left(\left\|\frac{\partial \mathbf{n}}{\partial x}\right\|\right) = \frac{\partial}{\partial \mathbf{n}_x}\left(\rho_\sigma\left(\left\|\frac{\partial \mathbf{n}}{\partial x}\right\|\right)\right) \qquad \rho_\sigma'\left(\left\|\frac{\partial \mathbf{n}}{\partial y}\right\|\right) = \frac{\partial}{\partial \mathbf{n}_y}\left(\rho_\sigma\left(\left\|\frac{\partial \mathbf{n}}{\partial y}\right\|\right)\right) \tag{14}$$

In [23] we experimented with several robust error kernels, including Li's Adaptive Potential Functions [16], and the Tukey [7] and Huber [12] estimators. However, the sigmoidal-derivative M-estimator, a continuous version of Huber's estimator, proved to possess the best properties for handling surface discontinuities, and is defined to be

$$\rho_\sigma(\eta) = \frac{\sigma}{\pi}\log\cosh\left(\frac{\pi\eta}{\sigma}\right) \tag{15}$$

Substituting Equation 15 into Equation 13 yields the update equation

$$\mathbf{n}_{i,j}^{(k+1)} = \Theta\left(\left\|\frac{\partial \mathbf{n}_{i,j}^{(k)}}{\partial x}\right\|^{-1}\tanh\left(\frac{\pi}{\sigma}\left\|\frac{\partial \mathbf{n}_{i,j}^{(k)}}{\partial x}\right\|\right)\left(\mathbf{n}_{i+1,j}^{(k)} + \mathbf{n}_{i-1,j}^{(k)}\right)\right)$$

$$+ \left(\frac{\pi}{\sigma} \left\| \frac{\partial \mathbf{n}_{i,j}^{(k)}}{\partial x} \right\|^{-2} \operatorname{sech}^2 \left(\frac{\pi}{\sigma} \left\| \frac{\partial \mathbf{n}_{i,j}^{(k)}}{\partial x} \right\| \right) - \right.$$

$$\left. \left\| \frac{\partial \mathbf{n}_{i,j}^{(k)}}{\partial x} \right\|^{-3} \tanh \left(\frac{\pi}{\sigma} \left\| \frac{\partial \mathbf{n}_{i,j}^{(k)}}{\partial x} \right\| \right) \right) \left(\frac{\partial \mathbf{n}_{i,j}^{(k)}}{\partial x} \cdot \frac{\partial^2 \mathbf{n}_{i,j}^{(k)}}{\partial x^2} \right) \frac{\partial \mathbf{n}_{i,j}^{(k)}}{\partial x}$$

$$+ \left\| \frac{\partial \mathbf{n}_{i,j}^{(k)}}{\partial y} \right\|^{-1} \tanh \left(\frac{\pi}{\sigma} \left\| \frac{\partial \mathbf{n}_{i,j}^{(k)}}{\partial y} \right\| \right) \left(\mathbf{n}_{i,j+1}^{(k)} + \mathbf{n}_{i,j-1}^{(k)} \right)$$

$$+ \left(\frac{\pi}{\sigma} \left\| \frac{\partial \mathbf{n}_{i,j}^{(k)}}{\partial y} \right\|^{-2} \operatorname{sech}^2 \left(\frac{\pi}{\sigma} \left\| \frac{\partial \mathbf{n}_{i,j}^{(k)}}{\partial y} \right\| \right) - \right.$$

$$\left. \left\| \frac{\partial \mathbf{n}_{i,j}^{(k)}}{\partial y} \right\|^{-3} \tanh \left(\frac{\pi}{\sigma} \left\| \frac{\partial \mathbf{n}_{i,j}^{(k)}}{\partial y} \right\| \right) \right) \left(\frac{\partial \mathbf{n}_{i,j}^{(k)}}{\partial y} \cdot \frac{\partial^2 \mathbf{n}_{i,j}^{(k)}}{\partial y^2} \right) \frac{\partial \mathbf{n}_{i,j}^{(k)}}{\partial y} \right\} \quad (16)$$

It is illuminating to consider the behaviour of this update equation for small and large smoothness errors. Firstly, the averaging of the neighbourhood normals is moderated by a function of the form $\frac{1}{\|\eta\|} \tanh \left[\frac{\pi}{\sigma} \eta \right]$. This averaging effect is most pronounced when the smoothness error is small i.e. when the surface is already approximately flat. The other contribution to the smoothness process is of the form $\eta^2 (\frac{1}{\|\eta\|^2} \operatorname{sech}^2 \eta - \frac{1}{\|\eta\|^3} \tanh \eta)$. This term vanishes at the origin and tends towards zero for large values of η, only kicking-in at intermediate error conditions. Thus, there is no smoothing pressure from either process upon strong discontinuities.

The robust regularizer approach provides significantly improved results over the simple Horn and Brooks smoothness constraint, but at the expense of introducing the parameter, σ.

5 Curvature Consistency

Needle-map smoothness appears to be an over-strong and inappropriate constraint for shape from shading. This is primarily because real surfaces are more likely to be *piecewise* smooth; in other words, formed of smooth regions separated by sharp discontinuities in depth or orientation. The over-smoothing problem is exacerbated by the difficulty of formulating the continuous concept of smoothness on a discrete pixel lattice, as clearly illustrated by the fact that the Horn and Brooks smoothness constraint is trivially minimized by a needle-map corresponding to a planar surface.

Here we take a different tack by using curvature consistency. Although the curvature classes either side of a depth discontinuity may be completely unrelated, this is not the case for an orientation discontinuity. Orientation discontinuities usually correspond to ruts or ridges. Furthermore, the curvature classes for locations either side of a rut or a ridge should be the most similar classes, either trough or saddle rut for a rut, or dome or saddle ridge for a ridge. This property of smooth variation in class suggests that curvature consistency may be a more appropriate constraint for SFS than smoothness, which strongly penalises legitimate orientation discontinuities.

The use of a curvature consistency measure was introduced to SFS by Ferrie and Lagarde [6]. They use global consistency of principal curvatures [18] to refine the surface estimate returned by local shading analysis. Curvature consistency is formulated in terms of rotating the local Darboux frame to ensure that the principal curvature directions are locally consistent.

An alternative method of representing curvature information is to use $H - K$ labels, but these require us to set 4 thresholds to define the classes in terms of the mean and Gaussian curvatures. However, we propose to use curvature consistency based upon the shape index of Koenderink and van Doorn [14]. This is a continuous measure which encodes the same curvature class information as $H - K$ labels in an angular representation, and has the further advantage of not requiring any thresholds.

5.1   The Shape Index

We reformulate the definition of the shape index in terms of the needle-map. This allows us to use the needle-map directly, rather than needing to reconstruct the surface.

The differential structure of a surface is captured by the Hessian matrix which may be approximated in terms of surface normals by

$$\mathcal{H} = \begin{pmatrix} \left(\frac{\partial \mathbf{n}}{\partial x}\right)_x & \left(\frac{\partial \mathbf{n}}{\partial x}\right)_y \\ \left(\frac{\partial \mathbf{n}}{\partial y}\right)_x & \left(\frac{\partial \mathbf{n}}{\partial y}\right)_y \end{pmatrix} \tag{17}$$

where $(\cdots)_x$ and $(\cdots)_y$ denote the x and y components of the parenthesized vector respectively.

The eigenvalues κ_1 and κ_2 of the Hessian matrix, found by solving the eigenvector equation $|\mathcal{H} - \kappa \mathbf{I}| = 0$, are the principal curvatures of the surface. Koenderink and van Doorn [14] used the two eignevalues to define the shape index

$$\phi = \frac{2}{\pi} \arctan \frac{\kappa_2 + \kappa_1}{\kappa_2 - \kappa_1} \qquad \kappa_1 \geq \kappa_2 \tag{18}$$

This may be expressed in terms of surface normals thus

$$\phi = \frac{2}{\pi} \arctan \frac{\left(\frac{\partial \mathbf{n}}{\partial x}\right)_x + \left(\frac{\partial \mathbf{n}}{\partial y}\right)_y}{\sqrt{\left(\left(\frac{\partial \mathbf{n}}{\partial x}\right)_x - \left(\frac{\partial \mathbf{n}}{\partial y}\right)_y\right)^2 + 4\left(\frac{\partial \mathbf{n}}{\partial x}\right)_y \left(\frac{\partial \mathbf{n}}{\partial y}\right)_x}} \tag{19}$$

Figure 3 shows the range of shape index values, the type of curvature which they represent, and the grey-levels used to display different shape-index values.

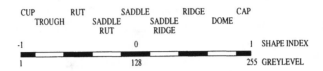

Fig. 3. The shape index scale ranges from -1 to 1 as shown. The shape index values are encoded as a continuous range of grey-level values between 1 and 255, with grey-level 0 being reserved for background and flat regions (for which the shape index is undefined).

Table 1 shows the relationship between the shape-index and the mean and Gaussian curvature classes. It is important to note that there are adjacency constraints applying to the topographic classes In particular, the the cup (C) and dome (D) surface types may not appear adjacent to each other on a surface. Moreover, elliptic regions on the surface (those for which K is positive) must be separated from hyperbolic regions (those for which K is negative) by a parabolic line (where $K=0$). Parabolic lines are effectively zero crossings of the mean or Gaussian curvatures. In other words, domes and cups are enclosed by ridge or valley-lines.

Class	Symbol	H	K	Region-type	Shape Index
Dome	D	-	+	Elliptic	$\left[\frac{5}{8}, 1\right)$
Ridge	R	-	0	Parabolic	$\left[\frac{3}{8}, \frac{5}{8}\right)$
Saddle ridge	SR	-	-	Hyperbolic	$\left[\frac{1}{8}, \frac{3}{8}\right)$
Plane	P	0	0	Hyperbolic	Undefined
Saddle-point	S	0	-	Hyperbolic	$\left[-\frac{1}{8}, \frac{1}{8}\right)$
Cup	C	+	+	Elliptic	$\left[-\frac{5}{8}, -1\right)$
Rut	V	+	0	Parabolic	$\left[-\frac{5}{8}, -\frac{3}{8}\right)$
Saddle-rut	SV	+	-	Hyperbolic	$\left[-\frac{3}{8}, -\frac{1}{8}\right)$

Table 1. Topographic classes.

5.2 Adaptive Robust Regularizer Using Curvature Consistency

As stated above, since the shape index is an angular, physical measure, we expect it to vary gradually over a smooth surface. For instance, with reference to Figure 3, we would not expect the shape index at adjacent pixels to differ by more than one curvature class unless they lie on opposite sides of a surface discontinuity. Since the over-smoothing effect of the quadratic smoothness constraint stems directly from the indiscriminate averaging of normals lying across a discontinuity, we anticipate that weighting according to curvature consistency will reduce the problem in a physically-principled manner.

To meet these goals we use curvature consistency to control the robust weighting kernel applied to the variation in the needle-map direction. The idea is a simple one. We use the variance of the shape-index in the neighbourhood $\mathbf{N}$ to control the width σ of the robust error-kernel applied to the directional derivatives of the needle map. The kernel width determines the level of smoothing applied to the surface normals in the neighbourhood. If the variance of the shape index is large i.e. the neighbourhood contains a lot of topographic structure, then we choose a small kernel width. This limits the local smoothing and allows significant local variation in the local needle-map direction. From a topographic viewpoint, we can see the rationale for this choice by considering the behaviour of the needle-map and the shape-index at ridges and ravines. For such features, the direction of the needle-map changes rapidly in a particular direction. These two structures are parabolic lines which intercede between elliptic and hyperbolic regions. As a result there is a rapid local variation in shape index. Turning our attention to the case where the shape-index variance is small, then the kernel width is large. This is the case when we are situated in a hyperbolic or elliptic region. Here the shape-index is locally uniform and the needle-map direction varies slowly.

Once again, we use the robust error-kernel of Equation 12 to model needle-map smoothness. However, instead of using a fixed kernel of width, σ, we adapt the width. The variance dependance of the kernel is controlled using the exponential function

$$\sigma = \sigma_0 \exp\left(-\left(\frac{1}{N}\sum_{l\in\mathbf{N}}\frac{(\phi_l-\phi_c)^2}{\Delta\phi_d^2}\right)^{\frac{1}{2}}\right) \qquad (20)$$

Here ϕ_c is the shape index associated with the central normal of the neighbourhood, $\mathbf{n}_{i,j}$, ϕ_l is one of the neighbouring shape-index values and $\Delta\phi_d$ is the difference in shape index between the centre values of adjacent curvature classes listed in Table 1. The number of neighbourhood normals used in calculating the finite difference approximations to $\frac{\partial\mathbf{n}}{\partial x}$ and $\frac{\partial\mathbf{n}}{\partial y}$ is denoted N, and σ_0 is a reference kernel width which we set to unity. Using the scale of Figure 3, $\Delta\phi_d = \frac{1}{8}$. To summarise, if the shape index varies significantly over the neighbourhood, a small value of σ results, and the robust regularizer saturates to produce a heavy smoothing effect. In contrast, when the shape-index values are already similar, the kernel is widened so that little smoothing occurs. When this model is used, the needle-map update equation is identical to that of Equation 13. However, now the error kernels adapt locally in line with the shape-index variance.

6   Experiments

In this section we provide some experimental evaluation of the new shape-from-shading technique. The evaluation focuses on the quality of the shape-index information extracted from the intensity images used in our experiments. The images used in our study are taken from the Columbia University COIL database. We furnish some comparison between the use of curvature consistency constraints and the simple needle-map smoothness constraint.

We commence in Figure 4 by showing a sequence which shows the shape index evolving with iteration number. In each case, the left-hand column shows the results obtained using the curvature consistency constraint developed in the paper, whilst the right-hand column shows the results of using a quadratic smoothness constraint within the same framework.

Using the curvature consistency scheme, the elliptic and hyperbolic region classes become more connected while the parabolic lines (i.e. ridges and ravines) become thinner and more continuous. In contrast, using the smoothness constraint leads to loss of the ridge and ravine structure. The regions are noisy and exhibit poor connectivity.

The final shape-index images contain some features which merit special mention. In particular the ravines defining the boundaries of the wing of the duck are well segmented. In addition, the slot in the top of the piggy-bank is correctly identified as ravine.

Figure 5 demonstrates that the curvature classes recovered using shape-from-shading are stable to viewpoint changes. Here we show six views of a toy-duck. Notice how the valley lines around the beak and the wing are well recovered at each viewing angle. Also notice how the shape of the addle structure below the wing is maintained.

Finally, we provide an initial comparison of the schemes using a simple synthetic image. The object used in this study is a hemisphere on a plane. The ideal shape index histogram for the object would consist of a small peak corresponding to the boundary where the hemisphere meets the plane at -0.5 (rut), and a large peak at 1.0 (dome). Figure 6 shows the shape index histograms recovered using curvature consistency (left) and needle-map smoothness (right).

7 Conclusions

This paper has presented a new shape-from-shading algorithm which uses curvature consistency to extract topographic information from 2D intensity images. The approach has two novel ingredients. Firstly, we provide a geometric framework for needle-map recovery. This process iterates between two steps. Firstly, we modify the local surface normal direction to satisfy local consistency constraints. Secondly, we back-project the updated normals onto a cone of shading ambiguity so as to satisfy the image irradiance equation as a hard constraint. The second novel contribution resides in the modelling of curvature consistency. Here we use the Koenderink and Van Doorn shape-index as a measure of surface topography. We use the variance of the shape-index to control the width of a robust error kernel which controls needle-map smoothness errors. In this way we ensure that the local smoothness of the needle-map responds to the variability of the local surface topography. We illustrate the comparative advantages of the new method on real-world imagery from the COIL object data-base. Here it proves to be reliable in delivering both smooth elliptic and hyperbolic regions together with thin and continuous parabolic lines.

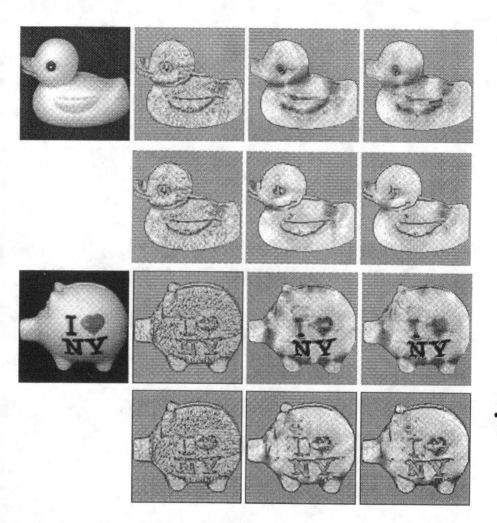

Fig. 4. Evolution of Shape Index classes with iterations of the SFS schemes. Top row, Smoothness constraint. Second row, Curvature consistency constraint. Third row, Smoothness constraint. Bottom row, Curvature consistency constraint.

Our future plans revolve around exploiting the topographic information delivered by the new shape-from-shading scheme for 3D object recognition from 2D imagery.

References

1. Belhumeur, P.N. and Kriegman, D.J. (1996) What is the Set of Images of an Object Under All Possible Lighting Conditions? *Proc. IEEE Conference on Computer Vision and Pattern Recognition*, pp. 270-277.
2. Bichsel, M. and Pentland, A.P. (1992) A Simple Algorithm for Shape from Shading, *Proc. IEEE Conference on Computer Vision and Pattern Recognition*, pp. 459-465.

Fig. 5. Shape index labels remain fairly stable with changes of object viewpoint, even for complex, non-Lambertian objects.

3. Brooks, M.J. and Horn, B.K.P. (1986) Shape and Source from Shading, *IJCAI*, pp. 932-936, 1985.
4. Bruckstein, A.M. (1988) On Shape from Shading, *CVGIP*, Vol. 44, pp. 139-154.
5. Dorai C.and Jain A.K., "Shape Spectrum based View Grouping and Matching of 3D free-from objects", *IEEE PAMI*, **19**, pp. 1139–1146, 1997.
6. Ferrie, F.P. and Lagarde, J. (1990)Curvature Consistency Improves Local Shading Analysis, *Proc. IEEE International Conference on Pattern Recognition*, Vol. I, pp. 70-76.
7. Hoaglin, D.C., Mosteller, F. and Tukey, J.W. (eds.) (1983) *Understanding robust and exploratory data analysis*, Wiley, New York.
8. Horn, B.K.P. (1975) Obtaining Shape from Shading Information. In Winston, P.H. (ed.), *The Psychology of Computer Vision*, McGraw Hill, NY, pp.115-155.

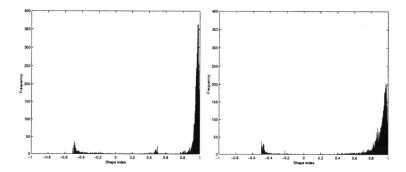

Fig. 6. Distributions of shape index values resulting from the curvature consistency scheme (left) and using the smoothness constraint (right) applied to a synthetic image of a hemisphere on a plane. Ground-truth for this object would result in a distribution with a tall delta function at shape index value 1, and a smaller delta at -0.5 corresponding to the perceived rut at the boundary of the sphere where it meets the plane.

9. Horn, B.K.P. and Brooks, M.J. (1986) The Variational Approach to Shape from Shading, *CVGIP*, Vol. 33, No. 2, pp. 174-208.
10. Horn, B.K.P. and Brooks, M.J. (eds.), *Shape from Shading*, MIT Press, Cambridge, MA, 1989.
11. Horn, B.K.P. (1990) Height and Gradient from Shading, *IJCV*, Vol. 5, No. 1, pp. 37-75.
12. Huber, P. (1981) *Robust Statistics*, Wiley, Chichester.
13. Ikeuchi, K. and Horn, B.K.P. (1981) Numerical Shape from Shading and Occluding Boundaries, *Artificial Intelligence*, Vol. 17, No. 3, pp. 141-184.
14. Koenderink, J.J. and van Doorn, A.J. (1992) Surface Shape and Curvature Scales, *Image and Vision Computing*, Vol. 10, pp. 557-565.
15. Koenderink, J.J., van Doorn, A.J., Christou, C. and Lappin, J.S. (1996) Perturbation study of shading in pictures, *Perception*, Vol. 25, No. 9, pp. 1009-1026.
16. Li, S.Z. (1995) Discontinuous MRF Prior and Robust Statistics: a Comparative Study, *Image and Vision Computing*, Vol. 13, No. 3, pp. 227-233.
17. Marr, D.C. (1982), *Vision*, Freeman, San Francisco.
18. Parent, P. and Zucker, S.W. (1985) Curvature Consistency and Curve Detection, *J. Opt. Soc. America A*, Vol. 2, No. 13, p. 5.
19. Pentland, A.P. (1984) Local Shading Analysis, *IEEE PAMI*, Vol. 6, pp. 170-187.
20. Sander P. and Zucker S.W., "Inferring Surface Structure and Differential Structure from 3D Images", *IEEE PAMI*, **12**, 833-854, 1990.
21. Szeliski, R. (1991) Fast Shape from Shading, *CVGIP: Image Understanding*, Vol. 53, pp. 129-153.
22. Worthington, P.L. and Hancock, E.R. (1997) Needle Map Recovery using Robust Regularizers, *Proc. British Machine Vision Conference*, BMVA Press, Vol. I, pp. 31-40.
23. Worthington, P.L. and Hancock, E.R. (1998) Needle Map Recovery using Robust Regularizers, *Image and Vision Computing*, to appear.
24. Worthington P.L., Huet B. and Hancock E.R., "Appearence based object recognition using shape-from-shading", Proceedings of the 14th International Conference on Pattern Recongition, pp. 412–416, 1998.

Harmonic Shape Images: A Representation for 3D Free-Form Surfaces Based on Energy Minimization

Dongmei Zhang and Martial Hebert

The Robotics Institute, Carnegie Mellon University
Pittsburgh, PA 15213
{dzhang, hebert}@ri.cmu.edu

Abstract. A new representation called Harmonic Shape Images for 3D free-form surfaces is proposed in this paper. This representation is based on the well-studied theory of Harmonic Maps which studies the mapping between different metric manifolds from the energy-minimization point of view. The basic idea of Harmonic Shape Images is to map a 3D surface patch with disc topology to a 2D domain and encode the shape information of the surface patch into the 2D image. Due to the application of harmonic maps in generating Harmonic Shape Images, Harmonic Shape Images have the following advantages: they preserve both the shape and the continuity of the underlying surfaces, they are robust to occlusion and they are independent of surface sampling strategy. The proposed representation is applied to solve the surface-registration problem. Experiments have been conducted on real data and results are presented in the paper.

1 Introduction

Shape representation is a fundamental issue in computer vision. In the past few decades, the objects under study have evolved from 2D objects to 3D polygonal objects, parametric surfaces and free-form surfaces. In this paper, the objects that are studied are 3D free-form surfaces represented by polygonal meshes.

A large amount of research has been done on surface registration([17]-[21]) and surface representation([5]-[16]). According to the way which objects are represented, existing representations of 3D free-form surfaces can be regarded as either global or local. Examples of global representations include algebraic polynomials[15], spherical representations such as EGI[5], SAI[6][7] and COSMOS[11], triangles and crease angle histograms[12], and HOT curves[13]. Although global representations can describe the overall shape of an object, they usually have difficulties handling occlusion and clutter.

Among the local surface representations proposed so far, the one that uses local shape signature[16] is among those that have been quite successful in real applications. This representation is independent of surface topology and easy to compute. The matching performance using this representation decreases gracefully as the occlusion and clutter increase. However, the major limitation of this representation is that it provides only sets of individual point correspondences that have to be grouped into sets of mutually consistent correspondences. This limitation stems from the fact that the local signatures only partially capture the shape of the surface.

In this paper, as part of our continuing effort to develop the data-level representations for surface matching, we investigate the surface-matching problem using a mathematical tool called harmonic maps with the goal of addressing the limitation of the representation in[16]. The harmonic map theory studies mappings between different metric manifolds from an energy-minimization point of view. With the application of harmonic maps, a surface representation called Harmonic Shape Images is created and used for surface matching. Furthermore, owing to the properties of harmonic maps, Harmonic Shape Images are able to provide all the point correspondences once two regions are matched. This will be shown in detail in the paper.

The basic idea of Harmonic Shape Images is to map a 3D surface patch with disc topology to a 2D domain and to encode the shape information of the surface patch into the 2D image. This simplifies the surface-matching problem to a 2D image-matching problem. When constructing Harmonic Shape Images, harmonic maps provide a mathematical solution to the mapping problem between a 3D surface patch with disc topology and a 2D domain.

This paper is organized as follows: the concept of Harmonic Shape Images is explained in Section 2; in section 3, the generation of Harmonic Shape Images is discussed in detail; the properties of Harmonic Shape Images are discussed in section 4; in Section 5, as an application example, Harmonic Shape Images are applied to solve the surface-matching problem. Conclusions and future work are presented at the end of the paper.

2 The Concept of Harmonic Shape Images

Studied in this paper, the 3D free-form surfaces are represented by polygonal meshes. According to [18], a free-form surface S is defined to be a smooth surface such that the surface normal is well defined and continuous almost everywhere, except at vertices, edges and cusps. Comparing two such meshes directly is difficult due to the following reasons: the topology may be different for different objects; the sampling may be different even for the same object; the surfaces may not be complete because of occlusion and clutter in the scene.

In order to address the above difficulties, it is advantageous for the shape representation to have the following characteristics: it is local; it is defined on a simple domain; and it fully captures the shape of the original surface. Local representations have the power to deal with occlusion and clutter. This has already been demonstrated by some of the existing representations[16]. Defining the shape representation on a simple domain, e.g., a 2D image, facilitates the comparison of the shape representations. In other words, the problem of surface matching in 3D can be reduced to the problem of image matching in 2D. If the shape representation fully captures the shape of the original surface, e.g., curvature and continuity, then correspondences between the two surfaces can be established once their shape representations match.

The development of Harmonic Shape Images was motivated by the above requirements. Given a 3D surface S as shown in Fig. 1(a), let v denote an arbitrary vertex in S. Let $D(v, R)$ denote the surface patch which has the central vertex v and radius R and has disc topology. $D(v, R)$ is connected and consists of all the vertices in S whose surface distance is less than, or equal to, R. The overlaid region in Fig. 1(a) is an example of $D(v, R)$. Its amplified version is shown in Fig. 1(b). If the unit disc is selected to be the 2D domain and $D(v, R)$ is mapped onto the domain using certain strategy, then the resultant image $HI(D(v, R))$ is called harmonic image as shown in Fig. 1(c). The harmonic image preserves the shape and continuity of $D(v, R)$. This will be explained further in the next section. Because correspondences are established between the vertices in $D(v, R)$ and the vertices in $HI(D(v, R))$, the Harmonic Shape Image of $D(v, R)$, $HSI(D(v, R))$, can be obtained by associating shape attributes, e.g., curvature, at each vertex of $D(v, R)$ with the corresponding vertex in $HI(D(v, R))$. Fig. 1(d) shows the Harmonic Shape Image of the surface patch in Fig. 1(b). The curvature values are gray-scale coded.

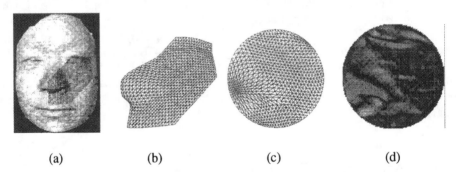

(a)   (b)   (c)   (d)

Fig. 1. (a) A surface patch (overlaid region) $D(v, R)$ on a given surface; (b) Amplified version of $D(v, R)$ in (a); (c) The harmonic image of $D(v, R)$; (d) The Harmonic Shape Image of $D(v, R)$.

Harmonic Shape Images can be generated for any vertex on a given surface as long as there exists a valid surface patch at that vertex. Here, a valid surface patch means a connected surface patch with disc topology. The generation of Harmonic Shape Images will be discussed in detail in the next section.

3 The Generation of Harmonic Shape Images Based on Energy Minimization

Given a surface patch $D(v, R)$ and a selected 2D domain M, the generation of its Harmonic Shape Image can be summarized as two steps: at the first step, a mapping between $D(v, R)$ and M is constructed, which results in the harmonic image $HI(D(v, R))$; at the second step, shape attributes are stored at each vertex of the harmonic image, which results in the Harmonic Shape Image $HSI(D(v, R))$. In comparison with the second step, the first step is much more difficult and complicated. The mathematical tool, harmonic maps, is employed to construct the mapping between *D(v, R)* and *M*. In this section, we first briefly introduce harmonic maps and then discuss in detail how to use harmonic maps to generate the harmonic image *HI(D(v, R))*. At last, we will talk about how to obtain Harmonic Shape Image *HSI(D(v, R))* from harmonic image *HI(D(v, R))*.

3.1 Harmonic Maps

According to [2], the concept of harmonic maps is closely related to the concept of geodesics. Geodesics are the shortest connection between two points in a metric continuum, e.g., a Riemannian manifold. Geodesics are critical points of the following length integral

$$L(c) = \int_0^1 \left| \frac{\partial}{\partial t} c \right|^2 dt \tag{1}$$

where $c : [0,1] \to N$ is the parameterization of the curve proportional to arc length. The generalization of the energy integral in (1) for maps between Riemannian manifolds leads to the concept of harmonic maps. Harmonic maps are critical points of the corresponding integral where energy density is defined in terms intrinsic to the geometry of the source and target manifolds and the map between them.

Formally, harmonic maps are defined as follows[1][2]. Let *(M, g)* and *(N, h)* be two smooth manifolds of dimensions *m* and *n*, respectively and let $\phi : (M, g) \to (N, h)$ be a smooth map. Let *(x^i)*, *i = 1, ..., m* and *(y^α)*, *α = 1, ..., n* be local coordinates around *x* and *φ(x)*, respectively. Take *(x^i)* and *(y^α)* of *M* and *N* at corresponding points under the map *φ* whose tangent vectors of the coordinate curves are $\partial / \partial x^i$ and $\partial / \partial y^\alpha$, respectively. Then the energy density of *φ* is defined as

$$e(\phi) = \frac{1}{2} \sum_{i,j=1}^m g^{ij} \sum_{\alpha,\beta=1}^n \frac{\partial f^\alpha}{\partial x^i} \frac{\partial f^\beta}{\partial x^j} h_{\alpha\beta} \tag{2}$$

in which g^{ij} and $h^{\alpha\beta}$ ($h_{\alpha\beta}$ is the inverse of $h^{\alpha\beta}$) are the components of the metric tensors in the local coordinates on *M* and *N*. The energy of *φ* in local coordinates is given by the number

$$E(\phi) = \int_M e(\phi) dM \tag{3}$$

If *φ* is of class C^2, $E(\phi) < \infty$, and *φ* is an extremum of the energy, then *φ* is called a harmonic map and satisfies the corresponding Euler-Lagrange equation.

The above is a general definition of harmonic maps. Now let us look at a special case in which both the source manifold and the target manifold are surfaces in the 3D Euclidean space. To be more specific, let *D* be a surface of disc topology and *P* be a planar region. According to the results in the theory of harmonic maps[2], the following problem has a unique solution: given a homeomorphism *b* between the boundary of D and the boundary of P, there exists a unique harmonic map $\phi : D \rightarrow P$ that agrees with b on the boundary of D and minimizes the energy functional of D.

Furthermore, the harmonic map ϕ has the following properties[2]: it always exists; it is unique and continuous; it is one-to-one and onto and it is intrinsic to D and P. All the above properties show that the harmonic map ϕ is a well-behaved mapping.

Recall that the mapping to be constructed from the surface patch $D(v, R)$ to the 2D domain M; it can be seen that harmonic maps provide a good solution to this problem.

3.2 Approximation of Harmonic Maps

As we have already seen in the previous section, harmonic maps are solutions of partial differential equations. Due to the expensive computational cost in solving partial differential equations and the discrete nature of surfaces we deal with in practice, it is natural to look for an approximation of harmonic maps when mapping $D(v, R)$ to M.

Eck et al. proposed an approximation approach to harmonic maps in [4]. Eck's approximation consists of two steps. At the first step, the boundary of the 3D surface patch is mapped onto the boundary of an equilateral triangle that is selected to be the 2D target domain. At the second step, the interior of the surface patch is mapped onto the interior of the equilateral triangle with the boundary mapping as a constraint. Our approach uses the same interior mapping strategy as that of Eck's approach but a different target domain and a different boundary mapping strategy. In this section, we will first discuss the interior mapping with a given boundary mapping and then discuss our boundary mapping in detail.

3.2.1 Interior Mapping

Let D be a 3D surface patch with disc topology and P be a unit disc in 2D. We use $D(v, R)$ to denote that the central vertex of D is v and the radius of D is R which is surface distance. Let ∂D and ∂P be the boundary of D and P, respectively. Let v_i^i, $i=1,...,n^i$, be the interior vertices of D. The interior mapping ϕ maps v_i^i, $i=1,...,n^i$, onto the interior of the unit disc P with a given boundary mapping $b : \partial D \rightarrow \partial P$. ϕ is obtained by minimizing the following energy functional:

$$E(\phi) = \tfrac{1}{2} \sum_{\{i,j\} \in Edges(D)} k_{ij} \|\phi(i) - \phi(j)\|^2 \tag{4}$$

In (4), for simplicity of the notation, we use $\phi(i)$ and $\phi(j)$ to denote $\phi(v_i^i)$ and $\phi(v_j^i)$, which are the images of the vertices v_i^i and v_j^i on P under the mapping ϕ. The values of $\phi(i)$ define the mapping ϕ. k_{ij} serve as spring constants with the definition in (5)

$$k_{ij} = ctg\,\theta(e_{mi}, e_{mj}) + ctg\,\theta(e_{li}, e_{lj}) \tag{5}$$

in which $\theta(e_{mi}, e_{mj})$ and $\theta(e_{li}, e_{lj})$ are defined in Fig2.. If e_{ij} is associated only with one triangle, then there will be only one term on the right of (5).

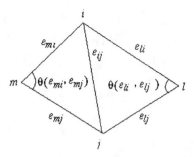

Fig. 2. Definition of spring constants.

An instance of the functional $E(\phi)$ can be interpreted as the energy of a spring system by associating each edge in D with a spring. Then the mapping problem from D to P can be viewed as adjusting the lengths of those springs when squeezing them down onto P. If the energy of D is zero, then the energy increases when the springs are squeezed down to P because all the springs are compressed. Different ways of adjusting the spring lengths correspond to different mappings ϕ. The best ϕ minimizes the energy functional $E(\phi)$. Using the definition of the spring constants in (5), the best ϕ best preserves the ratios of edge lengths in D; therefore, the shape of D, under the boundary mapping b.

The minimum of the energy functional $E(\phi)$ can be found by solving a sparse linear least-square system of (4) for the values $\phi(i)$. This results in a set of linear equations that can be written in matrix form

$$A_{n\times n} X_{n\times 2} = b_{n\times 2} \qquad (6)$$

In (6), $X_{n\times 2} = [\phi(1),...,\phi(n)]^T$, $\phi(i) = [\phi_x(i), \phi_y(i)]$. $X_{n\times 2}$ denotes the unknown coordinates of the interior vertices of D when mapped onto P under ϕ. $A_{n\times n}$ is a sparse matrix which has the following structure:

$$
A_{n\times n} =
\begin{array}{c}
\begin{array}{ccccccccccc}
1 & .. & .. & .. & m & .. & i & .. & j & .. & .. & . & n
\end{array}\\
\left[
\begin{array}{ccccccccccc}
* & . & & & & & & & & & \\
 & & * & & . & & * & & & & \\
 & & * & & & . & * & & . & * & \\
 & & * & & & & * & & . & & \\
 & & & & & & & & * & &
\end{array}
\right]
\begin{array}{c}
1\\ m\\ i\\ j\\ n
\end{array}
\end{array}
$$

If D is considered to be a non-directional graph, then $A_{n\times n}$ can be interpreted as an adjacency matrix of the graph. All the diagonal entries of $A_{n\times n}$ are non-zero. For an arbitrary row i in $A_{n\times n}$, if vertex v_i^i is connected to vertices v_m^i and v_j^i, then only the mth and jth entries in row i are non-zero. Similarly, the ith entry in row m and row j are also non-zero. Therefore, in addition to the usual properties that matrix $A_{n\times n}$ has in least-square problems, $A_{n\times n}$ is sparse in this particular case. The boundary condition is accommodated in the matrix $b_{n\times 2}$. In $b_{n\times 2}$, if a vertex v_i^i is connected to boundary vertices, then its corresponding entry i in $b_{n\times 2}$ is weighted by the coordinates of those boundary vertices. Otherwise, the entries in $b_{n\times 2}$ are zero.

After ϕ is solved from (6), the vertices in D are mapped onto the domain P. The 2D image of D after being mapped onto P is named the harmonic image $HS(D)$ of D. Examples of D and $HS(D)$ are shown in Fig. 3(a) and (b) respectively.

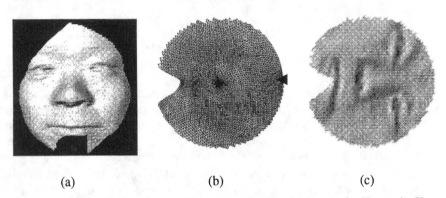

(a)   (b)   (c)

Fig. 3. (a) An example of surface patch; (b) Its harmonic image; (c) Its Harmonic Shape Image. (c) is the same as (b) except that it has a shape attribute gray-scale coded at each vertex. It will be introduced later in the paper.

3.2.2 Boundary Mapping

The construction of the boundary mapping $b : \partial D \rightarrow \partial P$ is illustrated in Fig. 4.

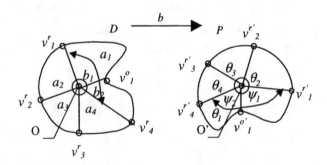

Fig. 4. Illustration of the boundary mapping between the surface patch and the 2D domain.

First of all, let us define the vertices and vectors in Fig. 4. O is the central vertex of D and O' is the center of P. v_i, $i=1,...,5$ are the boundary vertices of D. D is said to have radius R when the surface distance from any vertex in D to the central vertex O is less than, or equal to, R. For some boundary vertices, e.g., v_i^r, $i=1,...,4$, the surface distance between any of them and the central vertex O is equal to R; for other boundary vertices, e.g., v_1^o, the surface distance is less than R. The vertices in the former case are called radius boundary vertices and the vertices in the later case are called occluded boundary vertices. Radius boundary vertices are determined by the size of the surface patch, while occluded boundary vertices are determined by occlusion (either self occlusion or occlusion by other objects). The vector from the central vertex O to a radius vertex v_i^r is called a radius vector, while the vector from O to an occluded boundary vertex v_j^o is called an occlusion vector.

Now let us define the angles in Fig. 4. Angles a_i, $i=1,...,4$ are the angles between two adjacent radius vectors $\overrightarrow{v_i^r O}$ and $\overrightarrow{v_{i+1}^r O}$. Angles b_j, $j=1, 2$, are the angles between two adjacent occlusion vectors, or one occlusion vector and one adjacent radius vector, in an occlusion range. An occlusion range is a consecutive sequence of occlusion boundary vertices except for the first and last ones. For example, (v_4^r, v_1^o, v_1^r) is an occlusion range. The sum of b_j over an occlusion range is the angle a_i formed by the first and last radius vectors in that range. For example, the sum of b_j over (v_4^r, v_1^o, v_1^r) is a_1.

The construction of the boundary mapping consists of two steps. At the first step, the radius boundary vertices are mapped onto the boundary of the unit disc P, which is a unit circle. In Fig. 4, v_i^r, $i=1,...,4$, are mapped to $v_i^{r'}$, $i=1,...,4$, respectively. It can be seen that once the angles θ_i are determined, the positions of $v_i^{r'}$ are determined. θ_i is computed as follows:

$$\theta_i = \frac{a^i}{\sum_{k=1}^{n} a_k} 2\pi \tag{7}$$

At the second step, the occlusion boundary vertices in each occlusion range are mapped onto the interior of the unit disc P. For example, in Fig. 4, v_1^o, which is in the occlusion rangle (v_4^r, v_1^o, v_1^r), is mapped onto $v_1^{o'}$. Once the angles ψ_j and the radii r_j are determined, the position of v_1^o is determined. ψ_j are computed as follows.

$$\psi_j = \frac{bj}{\sum_{m=1}^{n} b_m} a_i \tag{8}$$

in which n is the number of angles within the occlusion range and a_i is the angle corresponding to the occlusion rangle. r_j is defined to be

$$r_j = \frac{dist(r_j, O)}{R} \tag{9}$$

in which $dist(r_j, O)$ is the surface distance between the occlusion boundary vertex v_j^o and the central vertex O. R is the radius of the surface patch D.

Two issues need to be mentioned regarding the above boundary mapping. The first one is that the boundary vertices of D need to be ordered in either a clock-wise or counter-clock-wise manner before constructing the boundary mapping. Similarly, when mapping vertices onto the boundary of P, either clockwise or counter-clockwise order needs to be determined. The two orders must remain consistent for all surface patches for the convenience of matching.

The second issue is how to select the starting vertex among the boundary vertices of D. If the starting vertex is always mapped to the same vertex on the boundary of the unit disc, then different starting vertices will result in different boundary mappings. This will, in turn, result in different interior mappings. For example, the harmonic image shown in Fig. 3(b) has the starting vertex indicated by the black arrow, while in Fig. 5(b), the harmonic image of the same surface patch is different due to a different starting vertex.

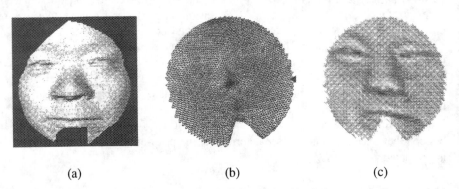

(a)   (b)   (c)

Fig. 5. The harmonic image(b) and Harmonic Shape Image(c) of the same surface patch(a) shown in Fig. 3(a) with a different a starting vertex for boundary mapping.

In fact, the harmonic images with different starting vertices are different by a planar rotation. The reason for this is that neither the angles θ_i(in (7)) nor the angles ψ_i (in (8)) will change with respect to different starting vertices. Nor will the radius r_j in (9) change. Therefore, the starting vertex can be selected randomly. The rotation difference will be found later by the matching process that is discussed in [23].

3.2.3 The Generation of Harmonic Shape Images

In Section 3.2.2, we have shown that, given a surface patch D, its harmonic image $HS(D)$ can be created using harmonic maps. There is one-to-one correspondence between the vertices in D and the vertices in $HS(D)$. Harmonic shape images, $HSI(D)$, are generated by associating a shape attribute at each vertex of $HS(D)$. In our current implementation, an approximation of the curvature at each vertex is used to generate Harmonic Shape Images. For details about the curvature approximation, please refer to [6].

Fig. 3(c) and Fig. 5(c) are examples of Harmonic Shape Images in which high intensity values represent high curvature values and low intensity values represent low curvature values.

Similar to Harmonic Shape Images, more images can be generated by associating other properties, e.g., color and texture, to each vertex in harmonic images. This shows that harmonic maps provide a general framework for storing surface-related information.

4. Properties of Harmonic Shape Images

Due to the way that Harmonic Shape Images are generated and the application of harmonic maps in the generation, Harmonic Shape Images have many good properties. In this section, we will discuss these properties and the advantages that benefit surface matching.

Local Representation. Harmonic Shape Images are created for surface patches on a given surface. In fact, Harmonic Shape Images can be generated for every vertex on the surface as long as there exists a valid patch at that vertex. For an arbitrary surface patch $D(v, R)$, the radius R controls the size of the patch. When R is small, the Harmonic Shape Image of $D(v, R)$ represents only the shape of the neighborhood around v. This shows that Harmonic Shape Images are local representations which do not depend on the overall topology of the underlying object.

Parameterization from D to P. At the first step of generating Harmonic Shape Images, a mapping needs to be created between the surface patch D and the 2D domain P. This mapping actually constructs a parameterization of D in P. Because the mapping is one-to-one and onto, correspondences between D and P can be established. This means that once two Harmonic Shape Images match, the correspondences between the two surface patches can be obtained immediately. In addition to having the properties of one-to-one and onto, this parameterization allows us to resample the original surface patch, if necessary. In practice, resampling such as raster scanning makes it easy to compare two Harmonic Shape Images.

Intrinsic to the Shape of the Underlying Surface. Different mappings between the surface patch D and the 2D domain P can be established. By using harmonic maps, the mapping constructed in this paper is intrinsic to the shape of the underlying surface patch D. In other words, this mapping is invariant to the position of D in 3D space. Further more, the approximation of harmonic maps used in our current implementation does not depend on a specific sampling strategy, e.g., uniform sampling. It can be applied to any triangular meshes. As long as the sampling rate is high enough, the mapping between D and P can be approximated with good accuracy. Harmonic Shape Images are obtained by associating curvature attributes at each vertex of the harmonic maps. Therefore, Harmonic Shape Images are invariant to rotation and translation, and are determined only by the shapes of the underlying surfaces.

Uniqueness. Following the above property, it can be seen that different surfaces of different shapes have different Harmonic Shape Images. No two surfaces with different shapes have the same Harmonic Shape Image. This means that Harmonic Shape Images are unique. This is an important property when applied to object recognition.

Existence. The existence of harmonic maps from a surface with disc topology to a planar domain ensures the existence of Harmonic Shape Image for a given surface patch D.

Robustness. One important issue for surface representations is their robustness with respect to occlusion. In this section, we first explain why Harmonic Shape Images are robust to occlusion and then, using real data, demonstrate the robustness under occlusion.

The reason for the robustness of Harmonic Shape Images with respect to occlusion lies in the way in which the boundary mapping is constructed. Recall that in Section 3.2.2 the boundary vertices are classified into radius boundary vertices and occluded boundary vertices. It is the radius boundary vertices that determine the angles a_i, which then determine the overall boundary mapping. The effect of occlusion is limited within the occlusion range; therefore, it does not propagate much outside of the occlusion range. This means that, as long as there are enough radius boundary vertices present in the surface patch, the overall harmonic image will remain approximately the same in spite of occlusion.

The following experiment was conducted to demonstrate the robustness of Harmonic Shape Images under occlusion. Let us use the surface patches D_1 and D_2 in Fig. 6(1) and (2) as an example to illustrate the experiment. D_2 is the same as D_1 except for the occluded region. Their Harmonic Shape Images are shown in Fig. 6(5) and (6), respectively. It can be seen that the Harmonic Shape Images of D_1 and D_2 are similar in spite of the occlusion on D_2. This means that the occlusion part of D_2 does not affect much of the shape representation for the non-occluded regions, thus making it possible to match D_2 to D_1 by matching the Harmonic Shape Images of their non-occluded regions. The normalized correlation coefficient of $HSI(D_1)$ and $HSI(D_2)$ is 0.9878, which verifies that D_2 can still be matched correctly to D_1 with occlusion.

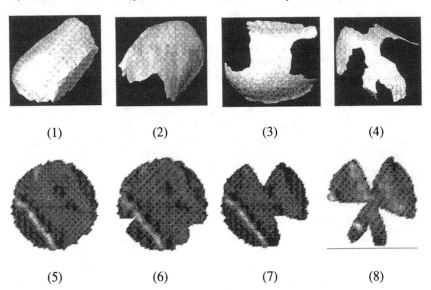

(1) (2) (3) (4)

(5) (6) (7) (8)

Fig. 6. A surface patch and its variations with different parts cut off to simulate occlusions.

A sequence of surface patches with different occlusion parts on D_1 is matched to D_1 to further test the robustness of Harmonic Shape Images under occlusion. Fig. 6(3) and (4) are two examples in the sequence. Their Harmonic Shape Images are shown in Fig. 6(7) and (8), respectively. Fig. 7 shows the matching result.

It can be seen from Fig. 7 that as the percentage of occlusion boundary increases, the occluded mesh area also increases. However, the normalized correlation coefficient remains stable in the range of [0.8, 1.0]. Fig. 6(8) shows the Harmonic Shape Image of the surface patch in Fig. 6(4). In spite of severe occlusion, the

Harmonic Shape Image for the non-occluded regions is still similar to that in Fig. 6(5).

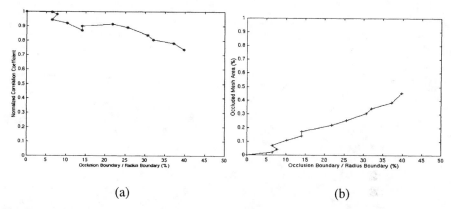

(a)   (b)

Fig. 7. Matching results under occlusion. The original surface is shown in Fig. 6(1). Some of the surfaces with occlusion are shown in Fig. 6(2)-(4). (a) Normalized correlation coefficient; (b) Percentage of occluded area.

5 Surface Registration Using Harmonic Maps

Given two 3D surfaces S_1, S_2 and no prior about the their relative position, the problem of surface matching is to find the rigid transformation between S_1 and S_2 and establish point correspondences between S_1 and S_2. In this section, the surface-matching problem is solved using Harmonic Shape Images. Experimental results using real data are presented. The algorithm for matching Harmonic Shape Images is discussed in detail in [23].

Two surfaces S_1 and S_2 shown in Fig. 8(a), (b) are to be registered. The surface patch in wireframe overlaid on S_1 is a selected surface patch $D_i(v_i, R)$ on S_1. After searching through all the valid surface patches on S_2 by comparing Harmonic Shape Images, the surface patch that best matches $D_i(v_i, R)$ is found. Fig. 8(c) shows the correspondences between the two matched surface patches. The transformation between S_1 and S_2 can be computed using those correspondences. The registered surfaces are shown in Fig. 8(d).

6 Conclusion and Future Work

In this paper, a shape representation called Harmonic Shape Images is proposed for 3D free-form objects. Owing to the application of harmonic maps, Harmonic Shape Images have many good properties, e.g., they preserve both the shape and continuity of the underlying surfaces. Preliminary results have shown that Harmonic Shape

Images are robust to occlusion. The experiment on surface registration has demonstrated the application of Harmonic Shape Images in surface matching.

There are two directions for future work. The first one is to develop a new similarity criterion for estimating Harmonic Shape Images. The second one is to apply the proposed Harmonic Shape Images to more applications such as object recognition and classification.

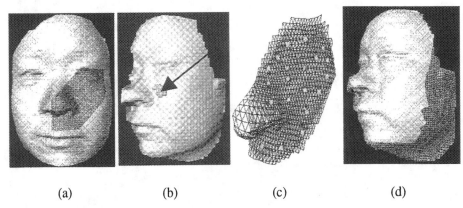

(a)   (b)   (c)   (d)

Fig. 8. Surface registration using Harmonic Shape Images. (a), (b) surfaces to be registered; (c) Some of the correspondences between the best matched surface patches; (d) The registered surfaces. The overlaid surface patch in (a) is randomly selected on S_1. For the surface patches on S_2 that are good matches of the overlaid surface patch in (a), their central vertices are marked on S_2. The best match is indicated by the black arrow.

References

1. Y. Xin: Geometry of Harmonic Maps. Birkhauser, 1996
2. J. Eells and L.H, Sampson: Harmonic Mappings of Riemannian Manifords. Amer. J. Math., 86:109-160, 1964
3. B. O'Neill: Elementary differential geometry. Academic Press, Inc., 1996
4. M. Eck, T. DeRose, T. Duchamp, H. Hoppe, M. Lounsbery, and W. Stuetzle,: Multi-resolution Analysis of Arbitrary Meshes. Proc. SIGGRAPH'96, 325-334
5. B.K.P. Horn: Extended Gaussian Image. *Proc.* IEEE, vol. 72, pp. 1671-1686, 1984
6. M. Hebert, K. Ikeuchi and H. Delingette,: A Spherical Representation for Recognition of Free-form Surfaces. IEEE Transactions on Pattern Analysis and Machine Intelligence, vol. 17, No.7, pp. 681-689, 1995
7. K. Higuchi, M. Hebert and K. Ikeuchi: Building 3D Models from Unregistered Range Images. CVGIP-Image Understanding, vol. 57, No.4, 1995
8. D. Zhang, M. Hebert: Multi-scale Classification of 3D Objects. Proc. CVPR'97, pp. 864-869
9. D. Zhang, M. Hebert, A. Johnson and Y. Liu: On Generating Multi-resolution Representations of Polygonal Meshes. Proc. ICCV'98 Workshop on Model-based 3-D Image Analysis, Jan. 3, 1998, Bombay, India
10. O.D. Faugeras and M. Hebert: The Representation, Recognition And Locating of 3-D Objects. Int'l J. of Robotics Research, vol. 5, No. 3, pp. 27-52, 1986.

11. C. Dorai, A. Jain: COSMOS - A Representation Scheme for 3D Free-form Objects. IEEE Transaction Pattern on Pattern Analysis and Machine Intelligence, vol. 19, No. 10: pp.1115-1130, 1997.

12. P.J. Besl: Triangles as A Primary Representation. Object Representation in Computer Vision, M. Hebert, J. Ponce, T. Boult and A. Gross, eds., Berlin, Springer-Verlag, pp.191-206, 1995.

13. T. Joshi, J. Ponce, B. Vijayakumar and D.J. Kriegman: HOT Curves for Modeling and Recognition of Smooth Curved 3D Objects. Proc. CVPR'94, pp.876-880, 1994

14. F. Stein and G. Medioni: Structural Indexing: Efficient 3-D Object Recognition, IEEE Transactions Pattern on Pattern Analysis and Machine Intelligence, vol. 14, No. 2: pp.125-145, 1992.

15. D. Keren, K. Cooper and J. Subrahmonia: Describing Complicated Objects by Implicit Polynomials. IEEE Transactions on Pattern Analysis and Machine Intelligence, vol. 16, No.1, pp. 38-53, 1994

16. A. Johnson and M. Hebert: Efficient Multiple Model Recognition in Cluttered 3-D Scenes. CVPR '98, pp. 671-677

17. C.S. Chua and R. Jarvis: 3D Free-form Surface Registration and Object Recognition. Int'l J. of Computer Vision, vol. 17, pp.77-99, 1996

18. P.J. Besl: The Free-form Surface Matching Problem. Machine Vision for Three-dimensional Scenes. H. Freeman, ed., Academic Press, pp.25-71, 1990

19. P.J. Besl and N.D. Mckay: A Method for Registration of 3-D Shapes. IEEE Transactions on Pattern Analysis and Machine Intelligence, vol. 14, No. 2, pp.239-256, 1992

20. Y. Chen and G. Medioni: Object Modeling by Registration of Multiple Range Images. Image Vision Computing, vol. 10, No. 3, pp.145-155, 1992

21. R. Bergevin, D. Laurendeau and D. Poussart: Estimating The 3D Rigid Transformation between Two Range Views of A Complex Object. Proc. 11th IAPR, Int'l Conf. Patt. Recog., The Hague, The Netherlands, pp. 478-482, Aug. 30 - Sep. 3, 1992

22. J. Deovre: Probability and statistics for engineering and science. Brooks/Cole, Belmont, CA, 1987

23. D. Zhang and M. Hebert: Harmonic Maps and Their Applications in Surface Matching. To appear in CVPR'99

Deformation Energy for Size Functions

Pietro Donatini[1], Patrizio Frosini[1], and Claudia Landi[2]

[1] Dipartimento di Matematica, Università di Bologna, P.zza P.ta S. Donato 5,
I-40126 Bologna, Italy, Tel.: ++39 51 354478, Fax: ++39 51 354490
donatini@dm.unibo.it, http://www.dm.unibo.it/~donatini
frosini@dm.unibo.it, http://www.dm.unibo.it/~frosini
[2] Dipartimento di Matematica, Università di Pisa, Via Buonarroti 2,
I-56127 Pisa, Italy, Tel.: ++39 50 844229, Fax: ++39 50 844224
landi@mail.dm.unipi.it

Abstract. *Size functions* are functions from the real plane to the natural numbers useful for describing shapes of objects. They allow to translate the problem of comparing shapes to the problem of comparing functions, that is a much simpler task. In order to perform the comparison between size functions we present a method to measure the energy necessary to deform size functions into each other. Minimizing such an energy allows for a measure of the similarity between shapes. Some experimental results concerning the comparison of free hand-drawn sketches are shown.

1 Introduction

Size functions are a mathematical tool originally conceived for representing shapes but applicable for studying all data that can be seen as topological spaces. Size functions have already been successfully used to deal with various recognition problems (see, e.g., [1], [2], [3], [4], [13], [14], [15], [16], [17], [18]). Basically, a size function is the output of a transform whose input is a topological space together with a real function defined on it. The topological space is the signal (e.g. B/W image, colour image, sound wave) under study, and the real function defined on it is the criterion used to study the signal. Each size function is a natural valued function of two real variables.

The purpose of this paper is to provide a mathematical method for measuring the energy necessary to deform size functions into each other. This gives a measure of the similarity of the corresponding topological spaces. In order to do so, size functions are preliminarily transformed into simpler objects, precisely into particular formal series. This representation contains almost the same amount of information about the space under study as the original size function does but it is much easier to handle. Then we show how to assign an energy to any deformation of a formal series associated with a size function. This way we define a deformation energy for size functions. Finally, we measure the extent to which two spaces resemble each other by minimizing the energy necessary to deform the corresponding size functions into each other.

This framework for the comparison of size functions is treated mainly from a theoretical point of view, nevertheless some experiments on B/W images are presented.

We shall confine ourselves to the study of a method for measuring the similarity between shapes, leaving out the problem of classification.

2 Size functions

Let $\mathcal{M}$ be a subset of some Euclidean space. We shall call any continuous function $\varphi : \mathcal{M} \to \mathbb{R}$ a *measuring function*. $\mathcal{M}$ is the signal we want to study and φ is a function chosen according to the properties of the signal we are interested in.

The *size function* of the pair $(\mathcal{M}, \varphi)$ is a function $\ell_{(\mathcal{M}, \varphi)} : \mathbb{R}^2 \to \mathbb{N} \cup \{+\infty\}$. For every pair $(x, y) \in \mathbb{R}^2$ consider the set of points in $\mathcal{M}$ at which φ takes value smaller or equal to x. Two such points are considered equivalent if they either coincide, or can be connected in $\mathcal{M}$ by a path at whose points φ takes value smaller or equal to y. Then $\ell_{(\mathcal{M}, \varphi)}(x, y)$ is defined to be equal to the number of equivalence classes so obtained.

For an example of size function see Figure 1. In this example the set $\mathcal{M}$ is a curve in the plane. By choosing the function ordinate of the point as measuring function, one obtains the size function shown on the right. The number displayed in each region of the domain of the size function denotes the value taken by the size function in that region.

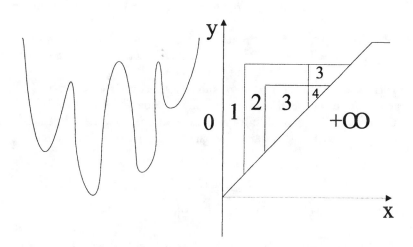

Fig. 1. A plane curve and the associated size function with respect to the measuring function ordinate of the point: $\varphi(x, y) = y$. The number displayed in each region of the domain of the size function denotes the value taken by the size function in that region. Continuous lines represent lines made of discontinuity points for the size function.

We remark that computing a size function is equivalent to counting the number of connected components of a graph: this gives the complexity of our task (see [6], [7]).

From the definition of size function one has immediately that if for $y < x$ there exists an infinite number of points $P \in M$ such that $y < \varphi(P) \leq x$ then $\ell_{(M,\varphi)}(x,y) = +\infty$. On the other hand, if M is a finite union of compact locally arcwise connected components then for $x < y$ we have $\ell_{(M,\varphi)}(x,y) < +\infty$ for every measuring function. Since this assumption is not very restrictive, in what follows we shall stick to it. Therefore in general we shall be interested only in the behaviour of size functions in the set $S_0 = \{(x,y) \in \mathbb{R}^2 : x < y\}$.

Another easy consequence of the definition of size functions is their monotonicity: size functions are non-decreasing in the first variable and non-increasing in the second one.

For more details and results about size functions we refer the reader to [5], [7], [8], [9], [12], [15] and [16]. Algorithms for the computation of size functions can be found in [6] and [17]. The resistance of size functions under noise and occlusions is treated in [11].

3 Size functions and formal series

Independently of the pair (M, φ), discontinuities of size functions satisfy some general properties. The most remarkable, but not immediate, one is that if $\ell_{(M,\varphi)}$ is discontinuous at (x,y) with $x < y$, then either x is a discontinuity point for $\ell_{(M,\varphi)}(\cdot, y)$ or y is a discontinuity point for $\ell_{(M,\varphi)}(x, \cdot)$ or both the statements hold.

Moreover, for $x < y$, discontinuities in the variable x propagate downward to the diagonal $\Delta = \{(x,y) \in \mathbb{R}^2 | x = y\}$ and discontinuities in the variable y propagate toward the right up to the diagonal.

For every size function it is then possible to select special points in S_0, called *cornerpoints*, from which discontinuity points propagate horizontally and vertically toward the diagonal. More precisely, cornerpoints are defined as those points $p = (x,y)$, with $x < y$, satisfying the following property: if we denote by $\mu_{\alpha,\beta}(p)$ the number

$$\ell_{(M,\varphi)}(x + \alpha, y - \beta) - \ell_{(M,\varphi)}(x + \alpha, y + \beta) - $$
$$\ell_{(M,\varphi)}(x - \alpha, y - \beta) + \ell_{(M,\varphi)}(x - \alpha, y + \beta),$$

it must hold $\mu(p) \overset{\text{def}}{=} \min\{\mu_{\alpha,\beta}(p) : \alpha > 0, \beta > 0, x + \alpha < y - \beta\} > 0$. We shall call $\mu(p)$ the *multiplicity of p*. It can be shown that $\mu(p) \geq 0$ for every $p \in S_0$.

Under our assumptions on M a size function has at most a countable set of cornerpoints in S_0. Actually, for any fixed $\rho > 0$, the region $S_\rho = \{(x,y) \in \mathbb{R}^2 : x < y - \rho\}$ contains only finitely many cornerpoints.

In Figure 2 we show that the only cornerpoints for the size function depicted on the left are the points a, b, c. All these cornerpoints have multiplicities equal

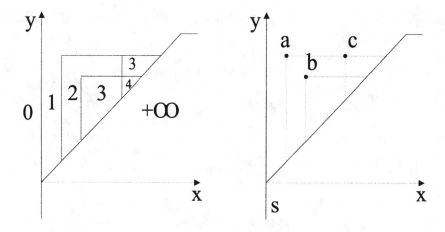

Fig. 2. A size function on the left and the corresponding cornerpoints (a, b, c) and cornerlines (s) on the right. The resulting formal series is $\alpha_0(\ell) = s + a + b + c$.

to 1 but it is easy to construct examples with points having multiplicity greater than 1.

Analogously one can select special vertical lines whose points above the diagonal are discontinuity points for the size function. These lines are called *cornerlines* and are defined as those lines l, with equation $x = k$, $k \in \mathbb{R}$, for which it holds

$$\mu(l) \overset{\text{def}}{=} \min_{\alpha > 0, k + \alpha < y} \ell_{(\mathcal{M}, \varphi)}(k + \alpha, y) - \ell_{(\mathcal{M}, \varphi)}(k - \alpha, y) > 0.$$

We shall call $\mu(l)$ the *multiplicity of* l. From the monotonicity of size functions it follows that $\mu(l) \geq 0$ for every vertical line l. Our assumptions on $\mathcal{M}$ imply that a size function can have at most a finite number of cornerlines.

The only cornerline for the size function of Figure 2 is the line s with multiplicity equal to 1. In general, it can be shown that every arcwise connected component of $\mathcal{M}$ generates a cornerline with multiplicity 1. Since different components can give rise to the same cornerline, one can obtain cornerlines with multiplicity greater than 1.

The importance of cornerpoints and cornerlines of a size function lies in the fact that for almost every point $p = (\bar{x}, \bar{y})$ lying above the diagonal, the value of the size function at p is equal to the sum of the multiplicities of cornerpoints (x, y) with $x < \bar{x}$ and $y > \bar{y}$ and of cornerlines $x = k$ with $k < \bar{x}$. That is to say that cornerpoints and cornerlines determine the size function almost everywhere.

Let us now denote by $\mathcal{R}$ the set of all lines with equation $x = k$, with $k \in \mathbb{R}$. For every $\rho \geq 0$, we shall call *formal series in* $\mathcal{S}_\rho \cup \mathcal{R}$ any object of the form $\sigma = \sum m(X)X$ with $m(X) \in \mathbb{Z}$, X varying in $\mathcal{S}_\rho \cup \mathcal{R}$ and $m(X) \neq 0$ only for a countable set of X's. The set of X's such that $m(X) \neq 0$ is called the *support* of the formal series.

It is then natural to define a map α_ρ from the set of all size functions into the set of formal series in $\mathcal{S}_\rho \cup \mathcal{R}$. For $\rho > 0$ it suffices to set α_ρ equal to the map which takes a size function $\ell_{(\mathcal{M}, \varphi)}$ into the formal series $\sum \mu(X) X$, where X varies in the set of all cornerpoints belonging to $\mathcal{S}_\rho$ and of all cornerlines for $\ell_{(\mathcal{M}, \varphi)}$, while $\mu(X)$ is its multiplicity (cf. Figure 2). For $\rho = 0$ one can take α_0 to be the map "extending" all the maps α_ρ for every $\rho > 0$. For every $\rho \geq 0$, α_ρ induces a well defined map $\tilde{\alpha}_\rho$ from the set $\mathcal{L}_\rho$ of all size functions quotiented by the relation of coincidence almost everywhere in $\mathcal{S}_\rho$ into the set of formal series in $\mathcal{S}_\rho \cup \mathcal{R}$.

Theorem 1. *For every real number $\rho \geq 0$ the map $\tilde{\alpha}_\rho : \mathcal{L}_\rho \longrightarrow \mathcal{S}_\rho \cup \mathcal{R}$ is injective.*

As a consequence it remains proven that size functions can be represented as particular formal series of points and lines of the plane.

Furthermore, for every $\rho \geq 0$, it is possible to characterize a particular subset of the set of formal series in $\mathcal{S}_\rho \cup \mathcal{R}$ on which $\tilde{\alpha}_\rho$ turns out to be also surjective.

For the proofs of the results stated in this section we refer to [10].

4 Deformation energy

It follows from Theorem 1 that we can reduce the problem of assigning an energy to the deformation of a size function to the problem of giving an energy to the deformation of the associated formal series. This is clearly an easier task since formal series allow a concise and algebraic representation of size functions.

The first step in order to assign an energy to the deformation of a size function is therefore that of defining what a deformation of the associated formal series is. A very natural way to do this is illustrated by the example in Figure 3. In such an example we consider two formal series $\sigma_1 = s + a + b + c$ and $\sigma_2 = s' + a' + b'$ obtained as images of two size functions by the map α_0 of Section 3. The discontinuity points of the two size functions are represented by dashed and continuous lines respectively. We deform σ_1 into σ_2 by transforming the line s into the line s', the points a and b into a' and b' respectively, and by sending c onto the diagonal (this last action has the same meaning as "destroying" the point c).

This example can be generalized as follows.

Let us fix a positive real number ρ. Then every formal series obtained as image of a size function by the map α_ρ has finite support in $\mathcal{S}_\rho \cup \mathcal{R}$. For the sake of simplicity in what follows the considered formal series will always have finite support in $\mathcal{S}_\rho \cup \mathcal{R}$. Nevertheless our approach can be extended to the general case when the support is not finite. Moreover, we shall consider each point p having multiplicity $m(p)$ as $m(p)$ distinct points.

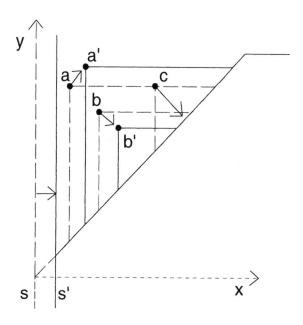

Fig. 3. A possible way to deform a size function, represented by dashed lines, into another one, represented by continuous lines.

Definition 1. We shall call *elementary deformation of a formal series σ* any of the following transformations:
(a) Identity transformation leaving σ unchanged.
(b) Transformation taking σ to $\sigma - X + X'$ with $X \neq X'$ and either both X and X' belonging to $\mathcal{S}_\rho$ or both belonging to $\mathcal{R}$.
(c) Transformation taking σ to $\sigma + X$ with $X \in \mathcal{S}_\rho \cup \mathcal{R}$.
(d) Transformation taking σ to $\sigma - X$ with $X \in \mathcal{S}_\rho \cup \mathcal{R}$.

Remark 1. We recall that coefficients of formal series are allowed to be negative. Applying an elementary deformation of type *(b)* means transforming either a point p into a different point p' or a vertical line l into a different line l'. An elementary deformation of type *(c)* has the effect of creating either a new point or a new line. Analogously, an elementary deformation of type *(d)* has the effect of eliminating either a point p or a line l.

Definition 2. We shall call *deformation of a formal series* any finite ordered sequence of elementary deformations of the considered formal series. We shall denote by $T(\sigma)$ the result of the deformation T applied to the formal series σ. Given two deformations T and S, the juxtaposition ST means applying first T and then S.

Remark 2. The effect of each deformation of formal series is independent of the order in which elementary deformations are carried out.

Remark 3. The inverse of an elementary deformation of type *(b)* is again an elementary transformation of the same type. The inverse of an elementary deformation of type *(c)* is an elementary deformation of type *(d)* and viceversa.

In general, the inverse of a deformation T will be denoted by the symbol T^{-1}.

Remark 4. Given two formal series (with finite support) there exists an infinite number of deformations transforming such formal series into each other. In fact, if T and S are deformations of a formal series σ, then $T(\sigma) = TSS^{-1}(\sigma)$ but $T \neq TSS^{-1}$.

The following step is that of defining what the energy of a deformation is.

Definition 3. For every deformation T of a formal series, the associated *energy* $\mathcal{E}(T)$ is a real number greater than or equal to 0 satisfying the following properties:
i) if T is the identity transformation, then $\mathcal{E}(T) = 0$;
ii) if $T = T_1 T_2 \cdots T_n$ then $\mathcal{E}(T) = \sum_{i=1}^{n} \mathcal{E}(T_i)$;
iii) $\mathcal{E}(T) = \mathcal{E}(T^{-1})$.
We allow the possibility that $\mathcal{E}(T)$ is equal to $+\infty$.

As an example of deformation energy, one can consider the following one: if T is an elementary deformation of type *(b)* sending a point p to a point p' (resp. a line l to a line l') then $\mathcal{E}(T)$ is equal to the Euclidean distance between p and p' (resp. l and l'); if T is an elementary deformation of type *(c)* or *(d)* creating or destroying a point p then $\mathcal{E}(T)$ is equal to the Euclidean distance between p and the diagonal Δ; if T is an elementary deformation of type *(c)* or *(d)* creating or destroying a line l then $\mathcal{E}(T)$ is equal to $+\infty$.

The deformation energy thus defined will be used for the experiments described in the following section.

The last step consists of deducing a measure of the similarity between two formal series from the measure of the energy necessary to deform them into each other.

Definition 4. For every pair of formal series σ_1 and σ_2 the *similarity between* σ_1 *and* σ_2 is the number $\Sigma(\sigma_1, \sigma_2) = \inf\{\mathcal{E}(T)\}$ where T runs in the set of all the deformations taking σ_1 to σ_2.

Remark 5. We point out that if $\mathcal{E}$ is the deformation energy defined in the above example, then $\inf\{\mathcal{E}(T)\}$ is actually a minimum. Anyway, in general $\inf\{\mathcal{E}(T)\}$ needs not to be attained.

Let us finally observe that Σ is actually well suited for a comparison between formal series, and hence between size functions, because it satisfies the following properties, making of Σ a pseudo-distance.

Proposition 1. *The following properties hold:*
1) for every pair of formal series σ_1 and σ_2, $\Sigma(\sigma_1, \sigma_2) \geq 0$;

2) for every formal series σ, $\Sigma(\sigma,\sigma) = 0$;

3) for every pair of formal series σ_1 and σ_2, $\Sigma(\sigma_1,\sigma_2) = \Sigma(\sigma_2,\sigma_1)$;

4) for every triad of formal series σ_1, σ_2 and σ_3, $\Sigma(\sigma_1,\sigma_3) \leq \Sigma(\sigma_1,\sigma_2) + \Sigma(\sigma_2,\sigma_3)$.

Proof. Property 1) is trivial. Since $\Sigma(\sigma,\sigma) \geq 0$ for every formal series σ, property 2) is easily proven by taking T equal to the identity transformation and by applying i) of Definition 3.

Property 3) is a direct consequence of property iii) of Definition 3.

Finally, property 4) is proven by observing that for every sequence of positive real numbers $\{\epsilon_i\}_{i \in \mathbb{N}}$ with $\lim_{i \to +\infty} \epsilon_i = 0$, there exists a sequence of deformations $\{T'_i\}_{i \in \mathbb{N}}$ taking σ_1 to σ_2 and verifying the inequality $\Sigma(\sigma_1,\sigma_2) \geq \mathcal{E}(T'_i) - \epsilon_i$. Analogously, there exists a sequence of deformations $\{T''_i\}_{i \in \mathbb{N}}$ taking σ_2 to σ_3 and verifying the inequality $\Sigma(\sigma_2,\sigma_3) \geq \mathcal{E}(T''_i) - \epsilon_i$. Thus for every $i \in \mathbb{N}$ it holds that $\Sigma(\sigma_1,\sigma_3) \leq \mathcal{E}(T''_i T'_i) = \mathcal{E}(T'_i) + \mathcal{E}(T''_i) \leq \Sigma(\sigma_1,\sigma_2) + \Sigma(\sigma_2,\sigma_3) + 2\epsilon_i$. The desired inequality is obtained for $i \to +\infty$.

5 Experimental results

We have tested the method here described in the task of recognizing B/W images. More precisely, we have considered 15 drawings belonging to three classes of objects: scissors, open end wrenches, screwdrivers. For each drawing we have computed the size function of its outer edge with respect to the measuring function distance from the center of mass. Then size functions have been transformed into formal series. These two operations have been carried out by means of the program SketchUp (for more details about this classifier of free-hand drawn sketches we refer the reader to [1]). Finally the similarity between images has been calculated by using the deformation energy suggested in Section 4.

The results are displayed in Table 1, showing that the deformation energy can be used to easily distinguish the considered shapes. Indeed, the average similarity between size functions corresponding to drawings of the same class, displayed in the gray rectangles, is smaller than that between drawings belonging to different classes.

6 Conclusions

In this paper we have shown how an energy minimization paradigm can be successfully used to compare size functions. This approach transforms the problem of shape comparison into a minimization problem with respect to deformations of finite sets of points and lines of the plane, that is a much simpler task. A concrete application of this technique has been given.

7 Acknowledgements

Research accomplished under support of CNR–GNSAGA, of ASI, and of the University of Bologna, funds for selected research topics.

SCISSORS — **OPEN END WRENCHES** — **SCREWDRIVERS**

	A1	A2	A3	A4	A5	B1	B2	B3	B4	B5	C1	C2	C3	C4	C5
A1	0	55,65	54,94	59,17	46,43	71,07	93,88	76,96	74,35	61,27	96,47	87,14	100,67	86,98	87,98
A2	55,65	0	29,42	47,17	31,56	116,07	132,99	108,2	128,05	102,34	115,8	120,8	111,6	112,6	121,89
A3	54,94	29,42	0	59,1	35,93	114,45	139,23	116,46	125,34	105,53	134,48	128,64	139,24	126,39	126,03
A4	59,17	47,17	59,1	0	31,32	114,23	124,01	107,05	122,58	98,14	106,96	97,17	102,34	103,83	101,21
A5	46,43	31,56	35,93	31,32	0	100,78	121,21	102,35	108,33	90,04	114,68	98,91	107,33	109,14	113,19
B1	71,07	116,07	114,45	114,23	100,78	0	26,31	20,72	16,19	17,73	52,95	45,27	55,2	45,47	44,47
B2	93,88	132,99	139,23	124,01	121,21	26,31	0	38,96	22,01	39,11	42,74	36,69	43,29	34,68	34,28
B3	76,96	108,2	116,46	107,05	102,35	20,72	38,96	0	32,62	19,4	48,79	56,61	49,3	54,37	49,6
B4	74,35	128,05	125,34	122,58	108,33	16,19	22,01	32,62	0	28,7	50,15	39,21	50,32	41,9	41,69
B5	61,27	102,34	105,53	98,14	90,04	17,73	39,11	19,4	28,7	0	53,42	48,42	56,17	47,39	45,24
C1	96,47	115,8	134,48	106,96	114,68	52,95	42,74	48,79	50,15	53,42	0	18,59	13,37	11,48	8,51
C2	87,14	120,8	128,64	97,17	98,91	45,27	36,69	56,61	39,21	48,42	18,59	0	17,89	13,7	11,48
C3	100,67	111,6	139,24	102,34	107,33	55,2	43,29	49,3	50,32	56,17	13,37	17,89	0	19,56	14,63
C4	86,98	112,6	126,39	103,83	109,14	45,47	34,68	54,37	41,9	47,39	11,48	13,7	19,56	0	5
C5	87,98	121,89	126,03	101,21	113,19	44,47	34,28	49,6	41,69	45,24	8,51	11,48	14,63	5	0

Table 1

References

1. Collina, C., Ferri, M., Frosini, P., Porcellini, E.: SketchUp: Towards Qualitative Shape Data Management. In: Chin, R., Pong, T. (eds.): Proc. ACCV'98, Lecture Notes in Computer Science 1351, Vol. I, Springer-Verlag, Berlin Heidelberg New York (1998) 338–345
2. Donatini, P., Frosini, P., Lovato, A.: Size Functions for Signature Recognition. In: Proc. SPIE Workshop "Vision Geometry VII" 3454, (1998) 178–183
3. Ferri, M., Frosini, P., Lovato, A., Zambelli, C.: Point Selection: A New Comparison Scheme for Size Functions (With an Application to Monogram Recognition). In: Chin, R., Pong, T. (eds.): Proc. ACCV'98, Lecture Notes in Computer Science 1351, Vol. I, Springer-Verlag, Berlin Heidelberg New York (1998) 329–337
4. Ferri, M., Lombardini S., Pallotti, C.: Leukocyte Classification by Size Functions. In: Proceedings of the second IEEE Workshop on Applications of Computer Vision, IEEE Computer Society Press, Los Alamitos, CA (1994) 223–229
5. Frosini, P.: Measuring Shapes by Size Functions. In: Proc. of SPIE, Intelligent Robots and Computer Vision X: Algorithms and Techniques, Boston, MA 1607 (1991), 122-133
6. Frosini, P.: Discrete Computation of Size Functions. J. Combin. Inform. System Sci. 17(3-4), (1992) 232–250
7. Frosini, P.: Connections Between Size Functions and Critical Points. Math. Meth. Appl. Sci., 19 (1996) 555–596
8. Frosini, P., Landi, C.: Size Functions and Morphological Transformations. Acta Appl. Math., 49 (1997) 85–104
9. Frosini, P., Ferri, M.: Range Size Functions. In: Proceedings of the SPIE's Workshop "Vision Geometry III", 2356 (1994), 243–251
10. Frosini, P., Landi, C.: Size Functions and Formal Series. Submitted for publication.
11. Frosini, P., Landi, C.: Size Theory as a Topological Tool for Computer Vision. To appear in Pattern Recognition and Image Analysis.
12. Frosini, P., Pittore, M.: New Methods for Reducing Size Graphs. Intern. J. Computer Math., 70 (1999) 505–517,
13. Uras, C., Verri, A.: Studying Shape through Size Functions. In: Y. O, A. Toet, D. Foster, H. Heijmans, P. Meer (editors) Shape in Picture, NATO ASI Series F 126, Springer-Verlag, Berlin Heidelberg (1994), 81–90
14. Uras, C., Verri, A.: On the Recognition of the Alphabet of the Sign Language through Size Functions. In: Proc. XII IAPR International Conference on Pattern Recognition, Jerusalem (Israel) II, IEEE Computer Society Press, Los Alamitos, CA (1994), 334–338
15. Verri, A., Uras, C., Frosini, P., Ferri, M.: On the Use of Size Functions for Shape Analysis. Biol. Cybern., 70 (1993) 99–107
16. Verri, A., Uras, C.: Metric-Topological Approach to Shape Representation and Recognition. Image Vision Comput., 14 (1996) 189–207
17. Verri, A., Uras, C.: Computing Size Functions from Edge Maps. Internat. J. Comput. Vision, 23 (1997) 1–15
18. Verri, A., Uras, C.: Invariant Size Functions. In: J. L. Mundy, A. Zisserman, D. Forsyth (editors) Applications of Invariance in Computer Vision, Lecture Notes in Computer Science 825, Springer-Verlag, Berlin Heidelberg (1994), 215–234

On Fitting Mixture Models

Mário A. T. Figueiredo[1], José M. N. Leitão[1], and Anil K. Jain[2]

[1] Instituto de Telecomunicações, and
Departamento de Engenharia Electrotécnica e de Computadores.
Instituto Superior Técnico, 1049-001 Lisboa
PORTUGAL
E-mail: mtf@lx.it.pt and jleitao@red.lx.it.pt

[2] Department of Computer Science and Engineering
Michigan State University, East Lansing, MI 48824
U.S.A.
E-mail: jain@cse.msu.edu

Abstract. Consider the problem of fitting a finite Gaussian mixture, with an unknown number of components, to observed data. This paper proposes a new minimum description length (MDL) type criterion, termed MMDL (for *mixture* MDL), to select the number of components of the model. MMDL is based on the identification of an "equivalent sample size", for each component, which does not coincide with the full sample size. We also introduce an algorithm based on the standard *expectation-maximization* (EM) approach together with a new agglomerative step, called *agglomerative* EM (AEM). The experiments here reported have shown that MMDL outperforms existing criteria of comparable computational cost. The good behavior of AEM, namely its good robustness with respect to initialization, is also illustrated experimentally.

1 Introduction

Finite mixtures are a flexible and powerful probabilistic modeling tool. In statistical pattern recognition, mixtures allow a formal (model-based) approach to (unsupervised) clustering [7], [15]; in fact, mixtures adequately describe situations where each observation is modeled as having been produced by one of a set of alternative mechanisms [31]. However, strict adherence to this interpretation is not required. Mixtures can simply be seen as models able to represent arbitrarily complex probability density functions (pdf's); this makes them an ideal tool for representing complex class-conditional pdf's in supervised learning scenarios (see [6], [22] and references therein).

This paper is devoted to the problem of fitting Gaussian mixtures with unknown number of components to multivariate observations. The two fundamental issues to be dealt with are: **(a)** how to estimate the number of components, for which several techniques (reviewed below) have been proposed; and **(b)** how to estimate the parameters defining the mixture model. For this second question, the standard answer is the *expectation-maximization* (EM) algorithm, but several authors have also advocated the (much more computationally demanding) *Markov chain Monte-Carlo* (MCMC) method.

We propose a new criterion to estimate the number of components which is shown experimentally to outperform existing methods of comparable computational cost. Our criterion is a modified version of the *minimum description length* (MDL) principle, based on what can be called the "equivalent sample size". We also introduce an (EM-based) algorithm aimed at mitigating the initialization dependence that makes EM difficult to use in practice. From a clustering perspective, our algorithm can be seen as an agglomerative hierarchical-type scheme, thus we term it *agglomerative EM* (AEM): we start with a large number of components (clusters) and evolve towards a small number of components. From a density estimation perspective, our algorithm has a multi-scale flavor: we go from a fine-scale representation with a large number of components, thus potentially irregular, to a smoother/coarser one with fewer components.

We review relevant previous work on mixture model fitting in Section 2, which also serves to introduce notation and the EM algorithm. Section 3 presents the MMDL criterion, while Section 4 is devoted to AEM. Section 5 reports experimental results, and Section 6 presents our conclusions.

2 Fitting Mixture Models

2.1 Introduction

Let $\mathbf{Y} = [Y_1, ..., Y_d]^T$ be a d-dimensional random variable, with $\mathbf{y} = [y_1, ..., y_d]^T$ representing one particular outcome of $\mathbf{Y}$. It is said that $\mathbf{Y}$ has a finite mixture distribution if its probability density function can be written as

$$f_\mathbf{Y}(\mathbf{y}|\Theta_{(k)}) = \sum_{m=1}^{k} \alpha_m f_\mathbf{Y}(\mathbf{y}|\theta_m), \qquad (1)$$

where k is the number of components, each $f_\mathbf{Y}(\mathbf{y}|\theta_m)$ is called a component density function, and the α_m ($\sum_{m=1}^{k} \alpha_m = 1$) are the mixing probabilities. Assuming that all the components have the same functional form (*e.g.*, they are all d-variate Gaussian), each one is fully characterized by the parameter vector θ_i. Let $\Theta_{(k)} = \{\theta_1, ..., \theta_k, \alpha_1, ..., \alpha_{k-1}\}$ be the parameter set defining a given mixture (notice that $\alpha_k = 1 - \sum_{m=1}^{k-1} \alpha_m$), and $\mathcal{M}_{(k)}$ be the space of all possible k-component mixtures built from a certain class of pdf's. This paper focuses on mixtures of Gaussian components, denoted as $f_\mathbf{Y}(\mathbf{y}|\theta_m) = \mathcal{N}(\mathbf{y}|\mu_m, \mathbf{C}_m)$, where $\theta_m = (\mu_m, \mathbf{C}_m)$, if arbitrary covariance $\mathbf{C}_m$ and mean μ_m are assumed; if a common covariance $\mathbf{C}$ is adopted, we simply write $\theta_m = \mu_m$.

The maximum likelihood (ML) estimate of $\Theta_{(k)}$, based on a set of n independent observations $\mathbf{y}_{\text{obs}} = \{\mathbf{y}^{(1)}, ..., \mathbf{y}^{(n)}\}$, is

$$\widehat{\Theta}_{(k)} = \arg\max_{\Theta_{(k)}} L\left(\Theta_{(k)}, \mathbf{y}_{\text{obs}}\right), \qquad (2)$$

where $L\left(\Theta_{(k)}, \mathbf{y}_{\text{obs}}\right)$ is the log-likelihood function

$$L\left(\Theta_{(k)}, \mathbf{y}_{\text{obs}}\right) = \log \prod_{i=1}^{n} f_{\mathbf{Y}}(\mathbf{y}^{(i)}|\Theta_{(k)}) = \sum_{i=1}^{n} \log \sum_{m=1}^{k} \alpha_m f_{\mathbf{Y}}(\mathbf{y}^{(i)}|\theta_m). \qquad (3)$$

In general, Eq. (2) has no closed-form solution but it can be approached quite easily via the *expectation-maximization* (EM) algorithm [16], [31].

2.2   The EM Algorithm for Gaussian Mixtures

Behind the EM algorithm is the interpretation of the set of observations $\mathbf{y}_{\text{obs}} = \{\mathbf{y}^{(1)}, ..., \mathbf{y}^{(n)}\}$ as incomplete data, with the missing information being a corresponding set of labels $\mathbf{z}_{\text{miss}} = \{\mathbf{z}^{(1)}, ..., \mathbf{z}^{(n)}\}$ [16], [31]. Each of these labels has the form $\mathbf{z}^{(i)} = [z_1^{(i)}, ..., z_k^{(i)}]^T$, with $z_m^{(i)} = 1$ and $z_p^{(i)} = 0$, for $p \neq m$, if and only if $\mathbf{y}^{(i)}$ was produced by the m-th component of the mixture. This complete data setup agrees with the interpretation of a mixture density as a model of a two-step data generation process: first, randomly choose one of the k available "data generators" with probabilities $\{\alpha_1, ..., \alpha_k\}$; then, produce a sample from the chosen "generator".

The loglikelihood function based on the complete data $\{\mathbf{y}_{\text{obs}}, \mathbf{z}_{\text{miss}}\}$, denoted $L_c(\Theta_{(k)}, \mathbf{y}_{\text{obs}}, \mathbf{z}_{\text{miss}})$, is easily found to be (for details see [31])

$$L_c\left(\Theta_{(k)}, \mathbf{y}_{\text{obs}}, \mathbf{z}_{\text{miss}}\right) = \sum_{j=1}^{n} \sum_{m=1}^{k} z_m^{(j)} \log\left[\alpha_m f_{\mathbf{Y}}(\mathbf{y}^{(j)}|\theta_m)\right]. \qquad (4)$$

In its general form, the EM algorithm proceeds by successively applying two steps to produce a sequence of parameter estimates $\{\widehat{\Theta}_{(k)}^{(1)}, \widehat{\Theta}_{(k)}^{(2)}, ..., \widehat{\Theta}_{(k)}^{(t)}, ...\}$:

E-step: Compute the expected value of the complete loglikelihood, conditioned on the observed data and on the current parameter estimate $\widehat{\Theta}_{(k)}^{(t)}$,

$$Q\left(\Theta_{(k)}, \widehat{\Theta}_{(k)}^{(t)}\right) = \int L_c\left(\Theta_{(k)}, \mathbf{y}_{\text{obs}}, \mathbf{z}_{\text{miss}}\right) f_{\mathbf{z}_{\text{miss}}}(\mathbf{z}_{\text{miss}}|\widehat{\Theta}_{(k)}^{(t)}, \mathbf{y}_{\text{obs}}) \, d\mathbf{z}_{\text{miss}}.$$

M-step: Update the parameter estimates according to

$$\widehat{\Theta}_{(k)}^{(t+1)} = \arg\max_{\Theta_{(k)}} Q\left(\Theta_{(k)}, \widehat{\Theta}_{(k)}^{(t)}\right). \qquad (5)$$

Under mild conditions [16], EM converges to a (local) maximum of $L\left(\Theta_{(k)}, \mathbf{y}_{\text{obs}}\right)$.

The key to the efficient implementation of this algorithm is the choice of an observed/missing data structure, *i.e.*, the function $L_c\left(\Theta_{(k)}, \mathbf{y}_{\text{obs}}, \mathbf{z}_{\text{miss}}\right)$, such that the E and M steps have simple closed-form expressions. This is the case in

Eq. (4), which is linear in the missing variables, thus reducing the E-step to the computation of the conditional expectation of the $z_m^{(i)}$ variables [16], [31],

$$w_m^{(i,t)} \equiv E\left[z_m^{(i)}|\widehat{\Theta}_{(k)}^{(t)}, \mathbf{y}_{\text{obs}}\right] = \frac{\widehat{\alpha}_m^{(t)} f_{\mathbf{Y}}(\mathbf{y}^{(i)}|\widehat{\boldsymbol{\theta}}_m^{(t)})}{\sum\limits_{j=1}^{k} \widehat{\alpha}_j^{(t)} f_{\mathbf{Y}}(\mathbf{y}^{(i)}|\widehat{\boldsymbol{\theta}}_m^{(t)})}. \tag{6}$$

The M-step also has a simple closed form solution (recall that $\boldsymbol{\theta}_m = \{\boldsymbol{\mu}_m, \mathbf{C}_m\}$):

$$\widehat{\alpha}_m^{(t+1)} = \frac{1}{n}\sum_{i=1}^{n} w_m^{(i,t)} \tag{7}$$

$$\widehat{\boldsymbol{\mu}}_m^{(t+1)} = \left(\sum_{i=1}^{n} w_m^{(i,t)}\right)^{-1} \sum_{i=1}^{n} \mathbf{y}^{(i)} \, w_m^{(i,t)} \tag{8}$$

$$\widehat{\mathbf{C}}_m^{(t+1)} = \left(\sum_{i=1}^{n} w_m^{(i,t)}\right)^{-1} \sum_{i=1}^{n} \left(\mathbf{y}^{(i)} - \widehat{\boldsymbol{\mu}}_m^{(t+1)}\right)\left(\mathbf{y}^{(i)} - \widehat{\boldsymbol{\mu}}_m^{(t+1)}\right)^T w_m^{(i,t)}. \tag{9}$$

The main difficulties in using EM for mixture model fitting are: its critical dependence on initialization; the possibility of convergence to a point on the boundary of the parameter space with unbounded likelihood (*i.e.*, one of the α_m parameters approaching zero with the corresponding covariance becoming arbitrarily close to singular).

2.3 Estimating the Number of Components

It is well known that the ML criterion can not be used to estimate the number of mixture components because $\mathcal{M}_{(k)} \subseteq \mathcal{M}_{(k+1)}$; for example, $\Theta_{(k)} = \{\boldsymbol{\theta}_1, ..., \boldsymbol{\theta}_k, \alpha_1, ..., \alpha_{k-1}\}$ and $\Theta'_{(k+1)} = \{\boldsymbol{\theta}'_1, ..., \boldsymbol{\theta}'_k, \boldsymbol{\theta}'_{k+1}, \alpha'_1, ..., \alpha'_{k-1}, \alpha'_k\}$, such that $\boldsymbol{\theta}_k = \boldsymbol{\theta}'_k = \boldsymbol{\theta}'_{k+1}$ and $\alpha_k = \alpha'_{k+1} + \alpha'_k$ (where, of course, $\alpha_k = 1 - \sum_{j=1}^{k-1}\alpha_j$, and $\alpha'_{k+1} = 1 - \sum_{j=1}^{k}\alpha'_j$) represent intrinsically indistinguishable mixture densities. Consequently, the maximized likelihood $L(\widehat{\Theta}_{(k)}, \mathbf{y}_{\text{obs}})$ is a non-decreasing function of k, thus useless as a model selection criterion. This is a particular instance of the *identifiability* problem (see, *e.g.*, [31]). As also pointed out in [31], classical (χ^2 based) hypothesis testing is not useful here because the necessary regularity conditions are not met.

Several approaches are available to estimate the number of components of a mixture; from an algorithmic standpoint, they can be divided into two main classes: EM-based techniques and stochastic techniques.

EM-based approaches use the (fixed k) EM algorithm to obtain a sequence of parameter estimates for a range of values of k, $\{\widehat{\Theta}_{(k)}, \; k = k_{\min}, ..., k_{\max}\}$, with the estimate of k being defined as the minimizer of some cost function,

$$\widehat{k} = \arg\min_{k} \left\{ \mathcal{C}\left(\widehat{\Theta}_{(k)}, k\right), \; k = k_{\min}, ..., k_{\max} \right\}. \tag{10}$$

Most often, this cost function includes the maximized log-likelihood function plus an additional term whose role is to penalize large values of k.

Under this general formulation, we find the MDL criterion [23] in which the cost function is

$$C_{\text{MDL}}\left(\widehat{\Theta}_{(k)}, k\right) = -L\left(\widehat{\Theta}_{(k)}, \mathbf{y}_{\text{obs}}\right) + \frac{N(k)}{2}\log n, \tag{11}$$

where $N(k)$ is the number of parameters needed to specify a k-component mixture. For arbitrary means and covariances, $N(k) = (k-1) + k(d + d(d+1)/2)$ (recall that d is the dimension of $\mathbf{Y}$); if a common covariance is assumed, then $N(k) = (k-1) + kd + d(d+1)/2$.

Several EM-based approaches also use approximate versions of the *Bayes factor* (the correct Bayesian model selection criterion [9]), such as the evidence-based Bayesian (EBB) criterion [25], the *approximate weight of evidence* (AWE) [1], and Schwarz's *Bayesian inference criterion* (BIC) [5]. Although derived in a different framework, BIC formally coincides with MDL and is also given by Eq. (11). The *minimum message length* (MML) criterion [20], Akaike's *information criterion* (AIC) [35], and Bezdek's *partition coefficient* (PC) [3] are other approaches in this class. As pointed out in [25], EBB, MDL/BIC, and MML perform comparably and outperform all other methods against which they were tested. Concerning AWE, it is argued in [5] that MDL/BIC provides a better approximation to the true Bayes factor. The AIC and PC criteria were shown in [20] (based on tests on 20 different mixtures) to be outperformed by MML and MDL/BIC. Accordingly, any new method in this class need only be compared against EBB, MDL/BIC, or MML. Finally, drawbacks of MML and EBB are: MML can not be used for certain values of d (for example $d = 9$ and $d > 24$) [25]; both EBB and MML depend on arbitrarily chosen parameters which can critically influence its results.

Resampling-based schemes [14] (which have also been used in a clustering framework [8]) and cross-validation approaches [30] are (computationally) much closer to stochastic techniques (see below) than to the methods in the previous paragraph and will not be further considered here.

Stochastic approaches involve Markov chain Monte Carlo (MCMC) sampling and are far more computationally intensive than EM. MCMC is used in two different ways: to implement model selection criteria to actually estimate k (*e.g.*, [2], [18], [26]); and, in a more "fully Bayesian" way, to sample from the full *a posteriori* distribution with k considered unknown [19], [21]. Despite their formal appeal, we think that MCMC-based techniques are still far too computationally demanding to be useful in pattern recognition applications. For example, tests reported in [21], using small samples ($n = 245, 155, 82$) of univariate data, require 100000 MCMC sweeps following a so-called *burn-in* period of another 100000 sweeps; this is a huge amount of computation for such small problems.

2.4 Initialization of EM

The EM algorithm requires an initial parameter setting $\widehat{\Theta}_{(k)}^{(1)}$ or an initial association of each observation to one of the components (*i.e.*, an initial setting of $w_m^{(i,1)}$) [16], [31]. This is a critical issue because EM converges to a local maximum of the likelihood function: the final estimate depends on the initialization. There are several different approaches to deal with this difficulty. Running EM several times, from random initializations, and then choosing the final estimate that leads to the highest local maximum of the likelihood is a commonly used technique (*e.g.*, [17] and [25]). Another common procedure is to use some clustering method to provide an initial partition of the data [17]. Finally, we mention the deterministic annealing (DA) EM algorithm (DAEM); DA is a fast surrogate of the (stochastic) simulated annealing approach to global optimization, which has been successfully applied in several problems [27]. In particular, for mixture estimation, DAEM avoids some of the initialization dependence of EM [10], [32], [36]. All these choices pay a high price in terms of computational efficiency.

3 The MMDL Criterion

It was shown in [25] that MDL/BIC (although simpler) performs comparably with EBB and MML, although it sometimes slightly underestimates the true k. A similar conclusion can be obtained from the many (20) tests described in [20]. It was also reported in [11] and [29] that MDL/BIC tends to slightly underestimate the true order. In order to overcome this problem, let us look again at the MDL criterion in Eq. (11). The meaning of the MDL cost function is the total code length of a two-part code for the observed data $\mathbf{y}_{\mathrm{obs}}$ and the parameter estimate $\widehat{\Theta}_{(k)}$ (see [23], for details and motivation): first encode the data, given $\widehat{\Theta}_{(k)}$; then, encode $\widehat{\Theta}_{(k)}$. Formally, Eq. (11) is of the form

$$
\begin{aligned}
\mathcal{C}_{\mathrm{MDL}}\left(\mathbf{y}_{\mathrm{obs}},\widehat{\Theta}_{(k)}\right) &= \mathcal{L}\left(\mathbf{y}_{\mathrm{obs}},\widehat{\Theta}_{(k)}\right) \\
&= \mathcal{L}\left(\mathbf{y}_{\mathrm{obs}}|\widehat{\Theta}_{(k)}\right) + \mathcal{L}\left(\widehat{\Theta}_{(k)}\right),
\end{aligned}
\tag{12}
$$

where $\mathcal{L}(\mathbf{y}_{\mathrm{obs}}|\widehat{\Theta}_{(k)}) = -L(\widehat{\Theta}_{(k)},\mathbf{y}_{\mathrm{obs}})$ is the well-known Shannon's optimal code length[1]. The second code-length, $\mathcal{L}(\widehat{\Theta}_{(k)})$, results from the following reasoning. To obtain finite-length codewords for $\widehat{\Theta}_{(k)}$, its (real-valued) elements are truncated to some finite precision. With a coarse precision, $\mathcal{L}(\widehat{\Theta}_{(k)})$ is small but the encoded parameters may be far from the optimal ones and so the first part of the code may become longer. With a finer resolution, the encoded parameters will be close to the optimal ones, but longer codewords are required. As shown

---

[1] As is usually done, we are ignoring the integer constraint on code-lengths and disregarding that we are dealing with densities, not probability masses. Discretization would lead to probability masses and a common (thus irrelevant) additional code length term [23].

in [23], the optimal code-length for each real parameter, asymptotically for large n, is $(1/2)\log n$; this leads to Eq. (11).

In most problems where the MDL/BIC criterion is used, all data points have equal importance in estimating each component of the parameter vector. This is not the case in mixtures, where each data point has its own weight in estimating different parameters, as is clear from Eqs. (8) and (9). This fact is revealed if we compute the Fisher information of a parameter of the m-th mode of the mixture (denoted generically as θ_m) which leads to (see [31])

$$I(\theta_m) = n\,\alpha_m\,I_1(\theta_m), \tag{13}$$

where $I_1(\theta_m)$ denotes the Fisher information associated with a single observation known to have been produced by the m-th component density, $i.e.$,

$$I_1(\theta_m) = -E\left[\frac{\partial^2}{\partial\theta_m^2}\log f_Y\left(y|\theta_m\right)\right].$$

What Eq. (13) shows is that a parameter θ_m "sees" an $equivalent\ sample\ size$ equal to $n\alpha_m$, rather than n. This is intuitively acceptable because θ_m will basically be estimated from the data that "was generated" by the m-th component of the mixture; the expected amount of this data is precisely $n\,\alpha_m$. Applying this fact, while keeping the classical MDL code-length for the mixing probabilities (because these are estimated from all the data), we finally obtain the MMDL cost function

$$\mathcal{C}_{\mathrm{MMDL}}\left(\widehat{\Theta}_{(k)}, k\right) = -L\left(\widehat{\Theta}_{(k)}, \mathbf{y}_{\mathrm{obs}}\right) + \frac{k-1}{2}\log n + \frac{N(1)}{2}\sum_{m=1}^{k}\log\left(n\alpha_m\right)$$

$$= -L\left(\widehat{\Theta}_{(k)}, \mathbf{y}_{\mathrm{obs}}\right) + \frac{N(k)}{2}\log n + \underbrace{\frac{N(1)}{2}\sum_{m=1}^{k}\log\alpha_m}_{<\,0} \tag{14}$$

where $N(1)$ is the number of real parameters defining each component (see the paragraph after Eq. (11)). The MMDL cost function can also be interpreted from a BIC-type perspective as the inclusion of some of the $o(1)$ terms that are dropped to obtain the classical form.

In summary, the MMDL criterion introduces a lower penalty than MDL/BIC; notice that the new term that appears in Eq. (14) when compared with Eq. (11) is necessarily negative. This is a result of the identification of the amount of data which is effectively used in estimating the parameters of each component of the mixture.

4   The Algorithm

To implement the MMDL criterion we propose a new (EM-based) algorithm. Let $k_{\max}$ be some number known to be considerably larger than the true/optimal k

(say, k_{true}) and $k_{\min}$ be another number such that, for sure, $k_{\min} < k_{\text{true}}$. The basic structure of the algorithm is as follows:

Initialization:

Set $k \leftarrow k_{\max}$.

Let $\widehat{\Theta}_{(k)}^{(1)}$ be some initial k-component mixture estimate.

Main Loop:

While $k \geq k_{\min}$, repeat:

- Run EM, using $\widehat{\Theta}_{(k)}^{(1)}$ as initialization, until a stopping condition is met. Store the resulting mixture parameter estimate $\widehat{\Theta}_{(k)}$.
- Compute and store $\mathcal{C}_{\text{MMDL}}\left(\widehat{\Theta}_{(k)}, k\right)$.
- Obtain a $(k-1)$-component mixture, "close" (in a sense to be specified below) to the k-component one specified by $\widehat{\Theta}_{(k)}$.

  Let $\widehat{\Theta}_{(k-1)}^{(1)}$ represent this $(k-1)$-component mixture.
- Set $k \leftarrow k - 1$

Choosing the optimal k:

Find the minimum of the stored MMDL cost function values:

$$\widehat{k}_{\text{MMDL}} = \arg\min_{k}\left\{\mathcal{C}_{\text{MMDL}}\left(\widehat{\Theta}_{(k)}, k\right), \; k = k_{\min}, k_{\min} + 1, ..., k_{\max}\right\}.$$

The final mixture parameter estimate is $\widehat{\Theta}_{(\widehat{k}_{\text{MMDL}})}$.

The crucial aspect of the algorithm is the use of a $(k-1)$-component mixture, "close" to the current k-component one, to initialize the next run of EM. This is done by looking for the pair of components that are closer to each other and less probable and merge them into a single new component (see details below). For this reason, our algorithm shares some of the spirit of agglomerative hierarchical clustering schemes [7], thus we call it *agglomerative* EM (AEM). The first run of EM, due to the excessive number of components, is somewhat insensitive to initialization. Of course we are not claiming that AEM is guaranteed to find the globally optimal mixture estimate; it is known that even MCMC may have difficulties escaping from local maxima of the likelihood function [24].

AEM can be used with any criterion other than MMDL, or even when k_{true} is known: in this case, simply set $k_{\min} = k_{\text{true}}$ and skip the phase where the optimal k is chosen. Naturally, AEM can also be based on modified versions of EM [16]. Finally, observe that the computational requirements of AEM are the minimum possible for any EM-based method doing unknown order mixture fitting. EM only has to be applied once for each value of k, instead of the common approach of using a set of random initializations for each k.

4.1 Initialization

For low dimensions ($d = 1, 2$), the initial mixture is composed of $k_{\max}$ components uniformly spread over the region occupied by the observed data (defined by the minimum and maximum observed values of each coordinate). For higher dimensions, a better initialization is obtained by clustering the data into $k_{\max}$ groups using successive binary splitting and K-means optimization at each stage [7]. As long as $k_{\max}$ is large enough, AEM is quite insensitive to initialization.

4.2 Stopping Conditions for EM

Each run of EM is stopped if at least one of the following two conditions is true:

$$
\text{Condition 1:} \quad
\begin{cases}
\max\left\{ \dfrac{\| \widehat{\boldsymbol{\mu}}_m^{(t)} - \widehat{\boldsymbol{\mu}}_m^{(t-1)} \|}{\| \widehat{\boldsymbol{\mu}}_m^{(t)} \|}, \; m = 1, 2, ..., k \right\} < \delta_\mu \\
\text{and} \\
\max\left\{ \dfrac{\| \widehat{\mathbf{C}}_m^{(t)} - \widehat{\mathbf{C}}_m^{(t-1)} \|}{\| \widehat{\mathbf{C}}_m^{(t)} \|}, \; m = 1, 2, ..., k \right\} < \delta_C
\end{cases}
\tag{15}
$$

$$
\text{Condition 2:} \quad \min\left\{ \widehat{\alpha}_m^{(t)}, \; m = 1, 2, ..., k \right\} < \alpha_{\min}.
\tag{16}
$$

Condition 1 checks if consecutive parameter estimates do not differ significantly; in all the examples below, we set $\delta_\mu = \delta_C = 0.001$ and use infinity norms $\| \cdot \|_\infty$. Condition 2 looks for a component whose probability is becoming too small; we typically use $\alpha_{\min} = 5\,d/n$. Condition 2 avoids one of the known problems of EM mentioned earlier (convergence to the boundary of the parameter space).

4.3 Obtaining the $(k-1)$-Component Mixture

The $(k-1)$-component mixture is obtained by merging two components of the k-component one. We start by locating the pair of mixture components, say m_1 and m_2, that are closer to each other and, simultaneously, less probable. Specifically, we choose m_1 and m_2 as

$$
(m_1, m_2) = \arg\min_{(i,j)} \left\{ (\widehat{\alpha}_i + \widehat{\alpha}_j)\, \mathcal{D}_s \left[f_{\mathbf{Y}}(\mathbf{y}|\widehat{\boldsymbol{\theta}}_i), f_{\mathbf{Y}}(\mathbf{y}|\widehat{\boldsymbol{\theta}}_j) \right], \; i \neq j \right\},
\tag{17}
$$

where $\mathcal{D}_s[f_{\mathbf{Y}}(\mathbf{y}|\widehat{\boldsymbol{\theta}}_i), f_{\mathbf{Y}}(\mathbf{y}|\widehat{\boldsymbol{\theta}}_j)]$ is the *symmetric Kullback-Leibler* (KL) *divergence* [12], the standard dissimilarity measure between probability densities [12]. The Jensen-Shannon divergence (see [13]) would be a natural candidate, because it allows weighting differently the two probability functions being compared; however, it does not have a closed form expression for Gaussian densities and so we settled for the KL divergence. In the Gaussian case, the symmetric KL divergence is [12]:

$$
\mathcal{D}_s \left[\mathcal{N}(\mathbf{y}|\boldsymbol{\mu}_i, \mathbf{C}_i), \mathcal{N}(\mathbf{y}|\boldsymbol{\mu}_j, \mathbf{C}_j) \right] = \frac{1}{2}\,\mathrm{tr}\left[(\mathbf{C}_i - \mathbf{C}_j)\left(\mathbf{C}_j^{-1} - \mathbf{C}_i^{-1}\right) \right]
$$
$$
+ \frac{1}{2}\,(\boldsymbol{\mu}_i - \boldsymbol{\mu}_j)^T \left[\mathbf{C}_i^{-1} + \mathbf{C}_j^{-1}\right]^{-1} (\boldsymbol{\mu}_i - \boldsymbol{\mu}_j).
$$

If EM was stopped by Condition 2 (Eq. (16)), we force m_1 to be the component responsible for making it true. We then choose m_2 by Eq. (17), fixing $i = m_1$.

Consider now the sub-mixture $\alpha'_{m_1} f_{\mathbf{Y}}(\mathbf{y}|\boldsymbol{\theta}_{m_1}) + \alpha'_{m_2} f_{\mathbf{Y}}(\mathbf{y}|\boldsymbol{\theta}_{m_2})$, where $\alpha'_{m_1} = \alpha_{m_1}/(\alpha_{m_1} + \alpha_{m_2})$ and $\alpha'_{m_2} = 1 - \alpha'_{m_1}$. Merging the two components of this submixture is equivalent to finding the parameters $\boldsymbol{\mu}^*$ and $\mathbf{C}^*$ of the "closest" Gaussian density. If "closeness" is taken in the KL sense, then

$$(\boldsymbol{\mu}^*, \mathbf{C}^*) = \arg\min_{\boldsymbol{\mu}, \mathbf{C}} \mathcal{D}\left[\alpha'_{m_1} \mathcal{N}(\mathbf{y}|\boldsymbol{\mu}_{m_1}, \mathbf{C}_{m_1}) + \alpha'_{m_2} \mathcal{N}(\mathbf{y}|\boldsymbol{\mu}_{m_1}, \mathbf{C}_{m_1}), \mathcal{N}(\mathbf{y}|\boldsymbol{\mu}, \mathbf{C})\right],$$

which has a simple solution (see [34], Chapp. 12): $\boldsymbol{\mu}^*$ and $\mathbf{C}^*$ are the global mean and covariance of the given two-component mixture, i.e.,

$$\boldsymbol{\mu}^* = \alpha'_{m_1} \boldsymbol{\mu}_{m_1} + \alpha'_{m_2} \boldsymbol{\mu}_{m_2} \tag{18}$$

$$\mathbf{C}^* = \alpha'_{m_1}(\mathbf{C}_{m_1} + \boldsymbol{\mu}_{m_1}\boldsymbol{\mu}_{m_1}^T) + \alpha'_{m_2}(\mathbf{C}_{m_2} + \boldsymbol{\mu}_{m_2}\boldsymbol{\mu}_{m_2}^T) - \boldsymbol{\mu}^*\boldsymbol{\mu}^{*T}. \tag{19}$$

This means that when merging components m_1 and m_2 of the mixture, the resulting component must retain the combined probability, mean, and covariance. Assume, without loss of generality, that $m_2 = k$, which can always be achieved by resorting the components. Merging component $m_1 < k$ and $m_2 = k$ of the k-component mixture given by $\{\alpha_m, \boldsymbol{\mu}_m, \mathbf{C}_m, m = 1, ..., k\}$ then yields a $(k-1)$-component mixture defined by $\{\alpha'_m, \boldsymbol{\mu}'_m, \mathbf{C}'_m, m = 1, ..., k-1\}$, where

$$\alpha'_m = \begin{cases} \alpha_m, & m \neq m_1 \\ \alpha_{m_1} + \alpha_{m_2}, & m = m_1, \end{cases}$$

$$\boldsymbol{\mu}'_m = \begin{cases} \boldsymbol{\mu}_m, & m \neq m_1 \\ \dfrac{\alpha_{m_1}\boldsymbol{\mu}_{m_1} + \alpha_{m_2}\boldsymbol{\mu}_{m_2}}{\alpha_{m_1} + \alpha_{m_2}}, & m = m_1, \end{cases}$$

$$\mathbf{C}'_m = \begin{cases} \mathbf{C}_m, & m \neq m_1 \\ \dfrac{\alpha_{m_1}(\mathbf{C}_{m_1} + \boldsymbol{\mu}_{m_1}\boldsymbol{\mu}_{m_1}^T) + \alpha_{m_2}(\mathbf{C}_{m_2} + \boldsymbol{\mu}_{m_2}\boldsymbol{\mu}_{m_2}^T)}{\alpha_{m_1} + \alpha_{m_2}} - \boldsymbol{\mu}'_{m_1}\boldsymbol{\mu}'^T_{m_1}, & m = m_1. \end{cases}$$

5 Experimental Results

This section is divided into two parts: the first one basically illustrates the working of the AEM algorithm showing how it evolves from a redundant mixture to successively lower order ones, and how this avoids the need for careful initialization. The second part focuses on MMDL by presenting examples (with synthetic and real data) where it overcomes the under-fitting tendency of MDL/BIC.

5.1 The AEM Algorithm

The first example uses 1000 samples from a mixture of 3 univariate Gaussians with means $\mu_1 = \mu_2 = 0$, and $\mu_3 = 6$, and standard deviations $\sigma_1 = 1$, $\sigma_2 = \sqrt{6}$, and $\sigma_3 = 1$; mixing probabilities are $\alpha_1 = 0.3$, $\alpha_2 = 0.4$, and $\alpha_3 = 0.3$. Fig. 1 shows AEM evolving from an 8-component mixture (after starting at $k_{\max} = 12$) to just two components. Observe the above mentioned multi-scale flavor of the method in the evolution from more erratic density estimates to smoother ones.

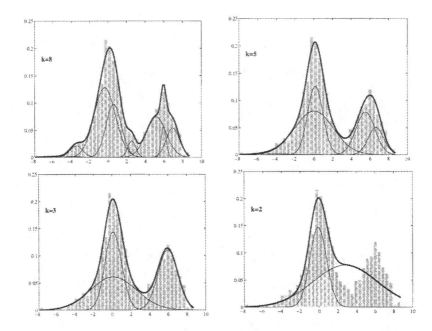

Fig. 1. Mixture estimates for $k = 8, 5, 3$ (the true value), and 2, obtained by AEM. Thin lines show the component densities multiplied by the corresponding probabilities, while the thick line plots the resulting mixture. The gray bars represent a (normalized) histogram of the observations.

The MMDL estimates are $\widehat{k} = 3$, $\widehat{\mu}_1 = 0.09$, $\widehat{\mu}_2 = 0.11$, $\widehat{\mu}_3 = 5.97$, $\widehat{\sigma}_1 = \sqrt{0.87}$, $\widehat{\sigma}_2 = \sqrt{6.12}$, $\widehat{\sigma}_3 = \sqrt{1.11}$, $\widehat{\alpha}_1 = 0.32$, $\widehat{\alpha}_2 = 0.38$, and $\widehat{\alpha}_3 = 0.30$.

For the next example, 1500 samples were drawn from a mixture of 3 bivariate Gaussians with $\alpha_1 = \alpha_3 = 0.3$, $\alpha_2 = 0.4$, $\mu_1 = \mu_2 = [-4, -4]^T$, $\mu_3 = [3, 3]^T$,

$$\mathbf{C}_1 = \begin{bmatrix} 1 & 0.5 \\ 0.5 & 1 \end{bmatrix}, \quad \mathbf{C}_2 = \begin{bmatrix} 6 & -2 \\ -2 & 6 \end{bmatrix}, \text{ and } \mathbf{C}_3 = \begin{bmatrix} 2 & -1 \\ -1 & 2 \end{bmatrix}.$$

Fig. 2 shows the algorithm evolving from its initialization (a set of $k_{\max} = 9$ similar and uniformly spread Gaussians) to the correct 3-component mixture. Notice how different the initial mixture is from the true one and how AEM was able to overcome this poor initialization. The final parameter estimates are $\widehat{\mu}_1 = [-4.03, -4.12]^T$, $\widehat{\mu}_2 = [-4.01, -3.90]^T$, $\widehat{\mu}_3 = [3.08, 2.91]^T$,

$$\widehat{\mathbf{C}}_1 = \begin{bmatrix} 1.07 & 0.56 \\ 0.56 & 0.88 \end{bmatrix}, \quad \widehat{\mathbf{C}}_2 = \begin{bmatrix} 5.4 & -1.89 \\ -1.89 & 6.12 \end{bmatrix}, \text{ and } \widehat{\mathbf{C}}_3 = \begin{bmatrix} 2.10 & -1.14 \\ -1.14 & 2.17 \end{bmatrix}.$$

Finally, we study the well known IRIS data set[2] that consists of 50 (4-dimensional) samples of each of the three classes present: *Versicolor*, *Virginica*, and *Setosa*. Starting with $k_{\max} = 8$, both MMDL and MDL/BIC correctly selected $\widehat{k} = 3$. Using the corresponding parameter estimates to build a *maximum*

---

[2] Available, *e.g.*, at http://www.ics.uci.edu/pub/machine-learning-databases/

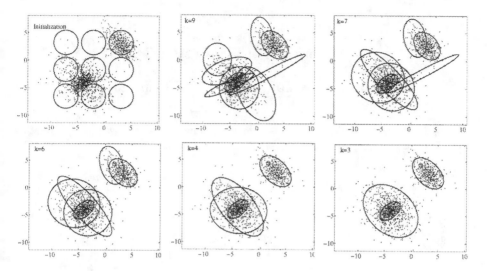

Fig. 2. Initialization and sequence of mixture estimates for $k = 9, 7, 6, 4$, and 3 (the ellipses are isodensity curves of each component).

a posteriori classifier according to

$$\widehat{m}(\mathbf{y}^{(i)}) = \arg\max_m \left\{ \widehat{\alpha}_m \, f_{\mathbf{Y}}(\mathbf{y}^{(i)}|\widehat{\boldsymbol{\theta}}_m) \right\},$$

we find that only two samples get misclassified (one *Versicolor* is classified as *Virginica* and one *Virginica* as *Versicolor*). This is even a little better than the three errors reported in [25]; more importantly, it is obtained without multiple random starts of EM.

5.2 Comparing MMDL versus MDL/BIC

Univariate Data. We start by considering two real univariate data sets for which MMDL and MDL/BIC yield different estimates of the number of Gaussian components: the Old Faithful geyser eruption durations (well known in the density estimation literature [28]), and the enzyme activity data from [21]. Table 1 reports the values of $\mathcal{C}_{\text{MMDL}}(\cdot)$ and $\mathcal{C}_{\text{MDL/BIC}}(\cdot)$ for several values of k for these two data sets. Fig. 3 shows the resulting mixture density estimates. For the Old Faithful data, MMDL allows an extra component ($\widehat{k}_{\text{MMDL}} = 4$) with which the resulting mixture adjusts better to the skewness of the right portion of the histogram. For the enzyme data, the additional component in the mixture selected by MMDL yields a clearly better fit to the observed histogram. Of course, in this real data cases, there is no underlying true mixture, and so there is no way to tell what is the correct number of components and we must rely on visual evaluation. An alternative would be to perform a (leave-one-out type) cross validation study comparing the MDL/BIC and MMDL criteria.

	k	1	2	3	4	5
Old Faithful	MMDL	429.8	288.8	283.6	**282.2**	286.7
	MDL/BIC	429.8	293.2	**289.4**	291.1	287.8
	k	1	2	3	4	5
Enzyme	MMDL	236.3	66.9	**65.8**	67.4	73.9
	MDL/BIC	236.3	**71.2**	72.5	77.4	87.5

Table 1. MMDL and MDL/BIC cost function values for several values of k for the Old Faithful and enzyme data sets.

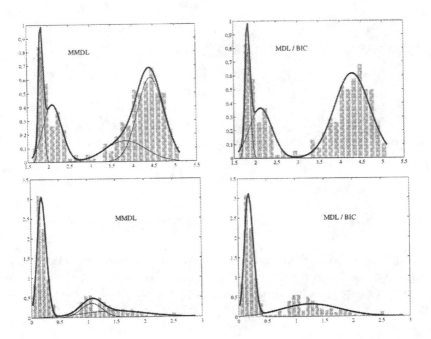

Fig. 3. Mixture estimates produced by the MMDL and MDL/BIC criteria for the Old Faithful (top row) and the enzyme data (bottom row).

Multivariate Data. To test the MMDL criterion on multivariate data, we have considered a mixture with 8 components on a 3D sample space. The component means are located at the vertices of a cube of side Δ,

$$\mu_1 = \begin{bmatrix} 0 \\ 0 \\ 0 \end{bmatrix}, \ \mu_2 = \begin{bmatrix} \Delta \\ 0 \\ 0 \end{bmatrix}, \ \mu_3 = \begin{bmatrix} 0 \\ \Delta \\ 0 \end{bmatrix}, \dots, \mu_7 = \begin{bmatrix} 0 \\ \Delta \\ \Delta \end{bmatrix}, \ \mu_8 = \begin{bmatrix} \Delta \\ \Delta \\ \Delta \end{bmatrix}$$

and all have unit covariance matrix $\mathbf{C}_i = \text{diag}\{1, 1, 1\}$, for $i = 1, 2, ..., 8$. We obtained 50 sets of 1200 samples each, for three different separations among the mixture components: $\Delta = 3$, 3.5, and 4. Figure 4 shows, for these three values of Δ, the number of times that each value of k was chosen by MMDL and MDL/BIC. Notice how the performance of MDL/BIC degrades faster than that of MMDL. For this test, since the goal is to study the behavior of the MMDL and MDL/BIC criteria, not of the AEM algorithm, we have used $k_{max} = 8$ and the true parameters as initialization.

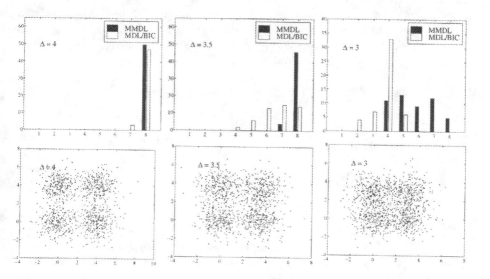

Fig. 4. Top row: histograms of the estimates of k (true value is 8) for $\Delta = 4$, 3.5, 3. Bottom row: examples of the first two components of the sample sets.

The MMDL criterion was also used in [33], for a Bayesian image classification problem. The class-conditional densities are represented by Gaussian mixtures, learned via a vector-quantization (VQ) approach, with the MMDL criterion controlling the size of each VQ. Given the very high dimensionality of the feature space (> 100), $N(1)$ is very high and MDL/BIC always yielded uselessly small estimates of k. With the estimates provided by MMDL, the resulting Bayesian classifier exhibited very good performance.

6 Conclusions and Further Work

We have proposed a new criterion to select the number of components in Gaussian mixtures and a new algorithm specially suited for mixture model estimation with an unknown number of components. The new criterion, called *mixture MDL* (MMDL), is a simple modification of the standard MDL/BIC, resulting from the identification of what can be called the *equivalent sample size* for each component. The proposed algorithm is based on EM together with an agglomerative step, thus it is called *agglomerative EM* (AEM). We have presented examples illustrating the behavior of AEM and its robustness with respect to initialization (although a more complete set of tests is still required). To compare MMDL versus MDL/BIC, we have performed experiments on real and synthetic data. All the experiments confirm that MMDL allows a better fit to the observed data.

Finally, we mention the parameterization of the covariance matrices (based on eigen-decomposition), introduced in [1] (see also [4]). That parameterization allows taking selected characteristics of the components to be common (for example, same shape, arbitrary orientation). MMDL can also be used to perform model selection among the options provided by that approach. The goal is to simultaneously choose the number of components and decide which characteristics (if any) should be assumed common.

References

1. J. Banfield and A. Raftery. Model-based Gaussian and non-Gaussian clustering. *Biometrics*, 49:803–821, 1993.
2. H. Bensmail, G. Celeux, A. Raftery, and C. Robert. Inference in model-based cluster analysis. *Statistics and Computing*, 7:1–10, 1997.
3. J. Bezdek. *Pattern Recognition with Fuzzy Objective Function Algorithms*. Plenum Press, New York, 1981.
4. G. Celeux and G. Govaert. Gaussian parsimonious clustering models. *Pattern Recognition*, 28(5):781–793, 1995.
5. C. Fraley and A. Raftery. How many clusters? Which clustering method? Answers via model-based cluster analysis. Technical Report 329, Department of Statistics, University of Washington, Seattle, WA, 1998.
6. T. Hastie and R. Tibshirani. Discriminant analysis by Gaussian mixtures. *Journal of the Royal Statistical Society (B)*, 58:155–176, 1996.
7. A. Jain and R. Dubes. *Algorithms for Clustering Data*. Prentice Hall, Englewood Cliffs, N. J., 1988.
8. A. Jain and J. Moreau. Bootstrap techniques in cluster analysis. *Pattern Recognition*, 20(5):547–568, 1987.
9. R. Kass and A. Raftery. Bayes factors. *Journal of the American Statistical Association*, 90:733–795, 1995.
10. M. Kloppenburg and P. Tavan. Deterministic annealing for density estimation by multivariate normal mixtures. *Physical Review E*, 55:R2089–R2092, 1997.
11. P. Kontkanen, P. Myllymäki, and H. Tirri. Comparing bayesian model class selection criteria in discrete finite mixtures. In *Proceedings of Information, Statistics, and Induction in Science – ISIS'96*, pp. 364-374, Singapore, 1996. World Scientific.
12. S. Kullback. *Information Theory and Statistics*. J. Wiley & Sons, N. York, 1959.
13. J. Lin. Divergence measures based on the Shannon entropy. *IEEE Trans. Information Theory*, 37:145–151, 1991.
14. G. McLachlan. On bootstrapping the likelihood ratio test statistic for the number of components in a normal mixture. *Jour. Roy. Stat. Soc. (C)*, 36:318–324, 1987.
15. G. McLachlan and K. Basford. *Mixture Models: Inference and Application to Clustering*. Marcel Dekker, New York, 1988.
16. G. McLachlan and T. Krishnan. *The EM Algorithm and Extensions*. John Wiley & Sons, New York, 1997.
17. G. McLachlan and D. Peel. MIXFIT: an algorithm for the automatic fitting and testing of normal mixture models. In *Proceedings of the 14th IAPR International Conference on Pattern Recognition*, volume II, pages 553–557, 1998.
18. K. Mengersen and C. Robert. Testing for mixtures: a Bayesian entropic approach. In J. Bernardo, J. Berger, A Dawid, and F. Smith, editors, *Bayesian Statistsics 5: Proceedings of the Fifth Valencia International Meeting*, pages 255–276. Oxford University Press, 1996.

19. R. Neal. Bayesian mixture modeling. In *Proceedings of the 11th International Workshop on Maximum Entropy and Bayesian Methods of Statistical Analysis*, pages 197–211. Kluwer, Dordrecht, The Netherlands, 1992.

20. J. Oliver, R. Baxter, and C. Wallace. Unsupervised learning using MML. In *Proceedings of the Thirtheenth International Conference on Machine Learning*, pages 364–372. Morgan Kaufmann, San Francisco, CA, 1996.

21. S. Richardson and P. Green. On Bayesian analysis of mixtures with unknown number of components. *Jour. of the Royal Statist. Soc. B*, 59:731–792, 1997.

22. B. Ripley. *Pattern Recognition and Neural Networks*. Cambridge University Press, Cambridge, U.K., 1996.

23. J. Rissanen. *Stochastic Complexity in Stastistical Inquiry*. World Scientific, 1989.

24. C. Robert. Mixtures of distributions: Inference and estimation. In W. Gilks, S. Richardson, and D. Spiegelhalter, editors, *Markov Chain Monte Carlo in Practice*, London, 1996. Chapman & Hall.

25. S. Roberts, D. Husmeier, I. Rezek, and W. Penny. Bayesian approaches to gaussian mixture modelling. *IEEE Transactions on Pattern Analysis and Machine Intelligence*, 20(11), November 1998.

26. K. Roeder and L. Wasserman. Practical Bayesian density estimation using mixtures of normals. *Journal of the American Statistical Association*, 92:894–902, 1997.

27. K. Rose. Deterministic annealing for clustering, compression, classification, regression, and related optimization problems. *Proc. of IEEE*, 86:2210–2239, 1998.

28. B. Silverman. *Density Estimation for Statistics and Data Analysis*. Chapman & Hall, London, 1986.

29. P. Smyth. Clustering using Monte-Carlo cross-validation. In *Proceedings of the Second International Conference on Knowledge Discovery and Data Mining*, pages 126–133. AAAI Press, Menlo Park, CA, 1996.

30. P. Smyth. Model selection for probabilistic clustering using cross-validated likelihood. Technical Report UCI-ICS 98-09, Information and Computer Science, University of California, Irvine, CA, 1998.

31. D. Titterington, A. Smith, and U. Makov. *Statistical Analysis of Finite Mixture Distributions*. John Wiley & Sons, Chichester (U.K.), 1985.

32. N. Ueda and R. Nakano. Deterministic annealing EM algorithm. *Neural Networks*, 11:271–282, 1998.

33. A. Vailaya, M. Figueiredo, A. K. Jain, and H. Jiang Zhang. A bayesian framework for semantic classification of outdoor vacation images. In *Proceedings of the 1999 SPIE Conference on Storage and Retrieval for Image and Video Databases VII*, pages 415–426. San Jose, CA, 1999.

34. M. West and J Harrison. *Bayesian Forecasting and Dynamic Models*. Springer-Verlag, New York, 1989.

35. M. Whindham and A. Cutler. Information ratios for validating mixture analysis. *Journal of the American Satistical Association*, 87:1188–1192, 1992.

36. A. Yuille, P. Stolorz, and J. Utans. Statistical physics, mixtures of distributions, and the EM algorithm. *Neural Computation*, 6:332–338, 1994.

Bayesian Models for Finding and Grouping Junctions

M.A. Cazorla, F. Escolano, D. Gallardo, and R. Rizo

Grupo Vgia: Visión, Gráficos e Inteligencia Artificial
Departamento de Ciencia de la Computación e Inteligencia Artificial
Universidad de Alicante
E-03690, San Vicente, Spain
Phone +34 96 590 39 00, Fax +34 96590 39 02
e-mail: miguel@dccia.ua.es

Abstract. In this paper, we propose two Bayesian methods for detecting and grouping junctions. Our junction detection method evolves from the Kona approach, and it is based on a competitive greedy procedure inspired in the region competition method. Then, junction grouping is accomplished by finding connecting paths between pairs of junctions. Path searching is performed by applying a Bayesian A^* algorithm that has been recently proposed. Both methods are efficient and robust, and they are tested with synthetic and real images.

1   Introduction

As is well known, junctions are the atoms of more complex processes or tasks – depth estimation, matching, segmentation, and so on – because these features provide useful local information about geometric properties and occlusions. Hence, methods for extracting these low-level features from real-world images must be efficient and reliable. Moreover the relation between junctions and specific tasks must be investigated. In this context, mid-level representations, that encode spatial relations between junctions, may be useful to reduce the complexity of these tasks. In this paper we propose two Bayesian methods to detect and group junctions along connecting edges.

Previous works on junction extraction can be classified as: edge-grouping methods, template-matching methods, and mixed strategies. Grouping algorithms use gradient information to build junctions (e.g. [6] and [13]), whereas template-based methods (e.g. [5] and [22]) are based on local filters. Mixed approaches are based on a local filter followed by template fitting. Two examples of these methods are [20] and [17]. In the latter approach, whose implementation is called Kona, junctions are modeled as piecewise constant regions – wedges – emanating from a central point. Junction detection is performed in two steps: center extraction – based on a local operator – and radial partition detection – based on a template deformation framework that uses the minimum description length (MDL) principle [19]. However, the proposed strategy for finding wedges – dynamic programming – may be too slow for real-time purposes, and also the robustness of the method may be improved. In the first part of this paper we propose a junction detector that evolves from Kona, and therefore it also pays attention to MDL principle. As in Kona, we first perform corner detection by using a robust filter. Then, we find the optimal number of wedges and also their image properties and location. In this case our strategy is based on the local competition of wedges and search can be done with a simple greedy procedure. This strategy is inspired on the region competition method [26] recently developed for image segmentation. Our method is fast and reliable. Robustness is provided by the use of sound statistics.

The use of junctions in segmentation, matching, and recognition, is the subject of several recent works. In [11] junctions are used as breaking points to locate and classify edges as straight or curved. Junctions are used as stereo cues in [15]. In [16] junctions are used as fixation cues. In this work, fixation is driven by a grouping strategy which forms groups of connected junctions separated from the background at depth discontinuities. The role of corners in recognition appears in [12], where a mixed bottom-up/top-down strategy is used to combine information derived from corners and the results of contour segmentation. Finally, junctions are used in [9] to constrain the grey level of image regions in segmentation. In the second part of this paper we propose a method to connect junctions along edges. This method is based on recent results on edge tracking using non-linear filters under a statistical framework [7], [24], [2], and [25].

The rest of the paper is organized as follows: In section 2, we present our junction detector and some experimental results with synthetic and real images. The analysis of these results motivates our junction-connecting approach presented in section 3. Grouping results are presented at the end of this section. Finally we present our conclusions and future work.

2 Junction Detection and Classification

2.1 Junction Model: Parameters and Regions

The relation between the real configurations of junctions and their appearance is well documented in the literature [23], [14]. A generic junction model can be encoded by a parametric template $\Theta = (x, y, r, M, \{\theta_i\}, \{T_i\})$, where: (x, y) is the center, r is the radius, M is the number of wedges, $\{\theta_i\}$, with $i = 1, 2, \ldots, M$, are the limits of the angular sections, and $\{T_i\}$ are the intensity distributions associated to these angular sections (see Fig 1).

We assume that potential junction centers (x, y) can be localized by a local filter. Examples of this operators are: the Plessey detector for corners [8] and the filters proposed in [10], and [1]. Here, we use SUSAN, a robust and fast non-linear filter that has been recently proposed [21]. SUSAN estimates the intensity homogeneity inside a circular domain (SUSAN measure). Corners have low homogeneity. In consequence they can be detected, provided that we use a good threshold. This principle can also be applied to find edges in the image.

In order to avoid distortions near the junction center, we also discard a small circular domain centered at (x, y) with radius R_{min}, as suggested by Parida et al. Then, $r = R_{max} - R_{min}$, where R_{max} is the scope of the junction. Moreover, although Kona provides a method for estimating the optimal value of r around a given center, its cost is prohibitive for real-time purposes. Then we assume that r can be estimated by the user.

We also consider that a junction is defined by several regions of homogeneous intensity around the circular domain defined by (x, y) and r. Then, the problem of finding M, $\{\theta_i\}$ and $\{T_i\}$ can be solved by analyzing the piecewise constant function associated to the junction. We compute a one-dimensional intensity profile by estimating, for each angle $\theta \in [0, 2\pi]$, the averaged accumulated intensity $\tilde{\mathcal{I}}_\theta$ along the radius r. An example of circular domain and its associated profile is showed in Fig 1.

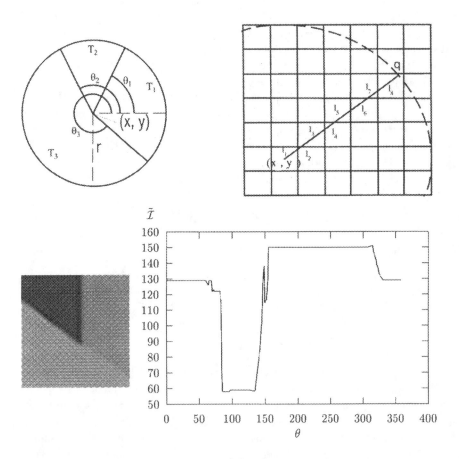

Fig. 1. Top left: Parametric model. Top right: Discrete accumulation of intensity along a direction. Bottom left: Example of a junction without noise. Bottom right: Intensity profile of the junction.

Angular sections are mapped to intervals $S_i = \{\theta | \theta \in [\theta_{S_i}, \theta_{S_{i+1}}]\}$ in the profile. Then, we can formulate the problem of junction detection as the *segmentation of the profile* into homogeneous intervals. An interval S_i is considered homogeneous when its intensity values are consistent with a given probability distribution T_i. Here, we assume a Gaussian model, so that $T_i = P(\tilde{\mathcal{I}}_\theta | \mu_i, \sigma_i)$, where μ_i is the mean, and σ_i^2 is the variance. This model allows us to cope with noise in real images.

Applying the Region Competition framework [26] to the intensity profile, the segmentation task consists of minimizing the following energy function:

$$\mathcal{E}_J(\Theta, \{\mu_i, \sigma_i\}) = \sum_{i=1}^{M} \left\{ -\log P(\{\tilde{\mathcal{I}}_\theta : \theta \in S_i\} | \mu_i, \sigma_i) \right\} \tag{1}$$

where: M is the number of angular sections, $P(\{\tilde{\mathcal{I}}_\theta : \theta \in S_i\} | \mu_i, \sigma_i)$ is the sum of the cost of coding each value $\tilde{\mathcal{I}}_\theta$ within the interval S_i according to a distribution $P(\tilde{\mathcal{I}}_\theta | \mu_i, \sigma_i)$. Assuming independent probability models for each interval, we have:

$$\log P(\{\tilde{\mathcal{I}}_\theta : \theta \in S_i\} | \mu_i, \sigma_i) = \int_{S_i} \log P(\tilde{\mathcal{I}}_\theta | \mu_i, \sigma_i) d\theta \tag{2}$$

and the global energy function is (if $i = M$ then $S_{i+1} = S_1$):

$$\mathcal{E}_J(\Theta, \{\mu_i, \sigma_i\}) = \sum_{i=1}^{M} - \int_{\theta_{S_i}}^{\theta_{S_{i+1}}} \log P(\tilde{\mathcal{I}}_\theta | \mu_i, \sigma_i) d\theta \tag{3}$$

where θ_{S_i} are the limits of the interval. As this function depends of M, this criterion pays attention to the MDL principle.

2.2 Wedge Identification with Competitive Descent

Wedge identification is performed by a greedy algorithm, that minimizes $\mathcal{E}_J(\Theta, \{\mu_i, \sigma_i\})$. Previously, we compute initial guesses for each interval. These guesses are given by the SUSAN edge detector applied to the intensity profile. In consequence, the initial number of wedges is always greater than or equal to the optimal one. Then, we will eventually need to merge regions.

Then, we calculate the μ_i, σ_i values of each interval, and perform gradient descent with respect to the limits θ_{S_i}. For each limit θ_{S_i} we have:

$$\frac{d\theta_{S_i}}{dt} = -\frac{\partial \mathcal{E}(\Theta, \{\mu_i, \sigma_i\})}{\partial \theta_{S_i}} = \frac{\partial}{\partial \theta_{S_i}} \left\{ \int_{\theta_{S_{i-1}}}^{\theta_{S_i}} \log P(\tilde{\mathcal{I}}_\theta | \mu_{i-1}, \sigma_{i-1}) d\theta \right.$$

$$\left. + \int_{\theta_{S_i}}^{\theta_{S_{i+1}}} \log P(\tilde{\mathcal{I}}_\theta | \mu_i, \sigma_i) d\theta \right\} = \frac{\partial}{\partial \theta_{S_i}} \left\{ [F(\theta)]_{\theta_{S_{i-1}}}^{\theta_{S_i}} + [F(\theta)]_{\theta_{S_i}}^{\theta_{S_{i+1}}} \right\}$$

$$= \log P(\tilde{\mathcal{I}}_{\theta_{S_i}} | \mu_{i-1}, \sigma_{i-1}) - \log P(\tilde{\mathcal{I}}_{\theta_{S_i}} | \mu_i, \sigma_i) = \log \left\{ \frac{P(\tilde{\mathcal{I}}_{\theta_{S_i}} | \mu_{i-1}, \sigma_{i-1})}{P(\tilde{\mathcal{I}}_{\theta_{S_i}} | \mu_i, \sigma_i)} \right\} \tag{4}$$

where $F()$ is the primitive function of $\log P(\mathcal{I}|\mu, \sigma)$. This ratio determines the change of θ_{S_i} and it is equivalent to the classification rule for two categories [4]. If $P(\tilde{\mathcal{I}}_{\theta_{S_i}}|\mu_i, \sigma_i) > P(\tilde{\mathcal{I}}_{\theta_{S_i}}|\mu_{i-1}, \sigma_{i-1})$, i.e., if $\tilde{\mathcal{I}}_{\theta_{S_i}}$ fits better to the distribution of the angular section S_i than to the distribution of S_{i-1}, then the value of the limit θ_{S_i} is decreased. Otherwise, it is increased.

Considering our Gaussian assumption we have:

$$P(\tilde{\mathcal{I}}_\theta|\mu, \sigma) = \frac{1}{\sqrt{2\pi}\sigma} \exp\left\{ -\frac{(\tilde{\mathcal{I}}_\theta - \mu)^2}{2\sigma^2} \right\} \tag{5}$$

and replacing 5 in 4:

$$\frac{d\theta_{S_i}}{dt} = \log\left(\frac{\sigma_i}{\sigma_{i-1}}\right) + \frac{1}{2}\left(\frac{(\tilde{\mathcal{I}}_{\theta_{S_i}} - \mu_i)^2}{\sigma_i^2} - \frac{(\tilde{\mathcal{I}}_{\theta_{S_i}} - \mu_{i-1})^2}{\sigma_{i-1}^2}\right) \tag{6}$$

In order to provide robutness we use the median value and the trimmed variance, instead of considering the mean and the variance.

Interval merging is based on an statistical test. Instead of applying the Fisher distance as it is done in the Region Competition method, our algorithm merges two adjacent intervals when the difference between their medians is below a given threshold D (typically $D = 10 - 15$ units). Additional merging occurs when the length of the interval is below $\pi/9$. Moreover, false junctions with $M = 2$ are removed when their angle is close to π.

2.3 Junctions: Detection Results

We have selected the following values for our experiments: $R_{min} = 2$, and $R_{max} = 5$, and $D = 10$. In these conditions we obtain a average error of 9 degrees. The processing time was 0.5 seconds in a Pentium II 233MHz under Linux.

Our algorithm obtains good results with synthetic images (see Fig 2(top)). However, experiments with real images show several problems that deserve more attention: (a) Our corner detection may generate a imprecise localization of the center, or it may not be detected. (b) Bad choices of r may generate several false limits (see Fig 2(bottom)): when r is high we can invade another region, and generate distortions in the intensity profile. (c) Several false junctions may not have geometric meaning. These problems can be observed in Fig 2(bottom), and also in Fig 3.

These problems motivate the extension of our work to perform junction grouping along connecting edges. As we will see in the second part of this paper, junction grouping allows us to remove false positives due to locality, to detect curved junctions, to localize undetected corners, and to correct poor localization.

Fig. 2. Top: Results with a synthetic image. Bottom: Example of a image from a corridor with curved junctions (left), and several zoomed areas from the same image (right).

76

Fig. 3. More results with real images. Most of X-junctions have false wedges.

3 Connecting and Filtering Junctions

3.1 Path Modeling and Edge Tracking

Now we are interested in finding *connecting paths*, i.e. paths that connect pairs of junctions along the edges between them, provided that these edges exist. More precisely a connecting path P of length N, rooted on a junction center (x, y) and oriented with the angle θ between two angular sections, is defined by a collection of connected segments $p_1, p_2, \ldots, p_M$ with fixed or variable length. We assume that the curvature of these paths must be smooth, so we also define orientation variables $\alpha_1, \alpha_2, \ldots, \alpha_{M-1}$, where α_j is the angle between two consecutive segments p_j and p_{j+1}. Following the bayesian approach of Yuille and Coughlan [25], the optimal path P^* maximizes

$$\mathcal{E}_P(\{p_j, \alpha_j\}) = \sum_{j=1}^{N} \log\{\frac{P_{on}(f_{(p_j)})}{P_{off}(f_{(p_j)})}\} + \sum_{j=1}^{N-1} \log\{\frac{P_{\Delta G}(\alpha_{j+1} - \alpha_j)}{U(\alpha_{j+1} - \alpha_j)}\} \qquad (7)$$

The first term of this function is the *intensity reward*, and it depends on the edge strength along each segment p_j. Edge strength is modeled by a probability distribution $P(f_{(p_j)})$ of the responses of a non-linear filter $f_{(p_j)} = \phi(|\nabla I(p_j)|)$, where $|\nabla I(p_j)|$ is the magnitude of the gradient in the neighbourhood of the segment. As the distribution of these responses depends on the relative position between a segment p_j and the edge, $P(f_{(p_j)})$ is defined in the following terms:

$$P(f_{(p_j)}) = \begin{cases} P_{on}(f_{(p_j)}) & p_j \in P^* \\ P_{off}(f_{(p_j)}) & \text{otherwise.} \end{cases} \qquad (8)$$

where $P_{on}(f_{(p_j)})$, and $P_{off}(f_{(p_j)})$ are the probability distributions of the responses of segments lying on and off the path. These distributions are obtained by gathering statistics of the responses of the filter when a segment is placed on and off the edges given by the gradient operator.

The second term is the *geometric reward*: $P_G(\alpha_{j+1} \mid \alpha_j) = P_{\Delta G}(\alpha_{j+1} - \alpha_j)$ models a first order Markov chain on orientation variables α_j. Curvature smoothing is provided by a negative exponential density function

$$P_{\Delta G}(\Delta \alpha_j) \propto \exp\{-\frac{C}{2A}|\Delta \alpha_j|\} \qquad (9)$$

where: $\Delta \alpha_j = \alpha_{j+1} - \alpha_j$, A is the maximum angle between two consecutive segments, and C modulates the rigidity of the path. Additionally, $U(\alpha_{j+1} - \alpha_j)$ is the uniform distribution of the angular variation, and it is included to keep both the geometric and the intensity terms in the same range.

The Yuille and Coughlan approach evolves from the work of Geman and Jedinak [7] on road tracking, and introduces the analysis of the gradient operator and the consideration of the geometric term. For the design of our intensity reward we have used the original filter of Geman and Jedinak, and we have applied it to the gradient obtained with the SUSAN edge detector. Our geometric reward is designed as suggested by Yuille and Coughlan. These distributions are showed in Fig 4.

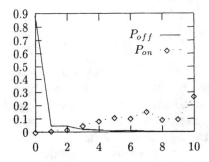 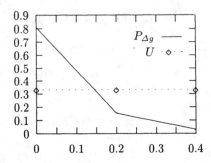

Fig. 4. Left: Intensity distributions P_{on} and P_{off}, for the SUSAN gradient operator, obtained with 200 samples. Righ: Geometric distributions $P_{\triangle G}$, for $C = 5.0$, and $A = 0.2 = \pi/15$ radians; and $U_{\triangle G}$.

3.2 Path Searching and Junction Grouping

Finding straight or curved connecting paths in cluttered scenes may be a difficult task, and it must be done in a short time, specially when real-time constraints are imposed. Coughlan and Yuille [2] have recently proposed a method, called bayesian A^*, that exploits the statistical knowledge associated to the intensity and geometric rewards. This method is rooted on a previous theoretical analysis [24] about the connection between the Twenty Question algorithm of Geman and Jedinak and the classical A^* algorithm [18].

Given an initial junction center (x_0, y_0) and an orientation θ_0, the algorithm explores a tree in which each segment p_j can expand Q succesors, so there are Q^N possible paths. The bayesian A^* reduces the conservative breadth-first behaviour of the classical A^* by exploiting the fact that we want to detect a target path against clutter, instead of finding the best choice from a population of paths. In consequence there is one true path and a lot of false paths. Then it is possible to reduce the complexity of the search by pruning partial path with low rewards. The algorithm evaluates the averaged intensity and geometric rewards of the last N_0 segments of a path (the *segment block*) and discards them when one of these averaged rewards is below a threshold, i.e. when

$$\frac{1}{N_0} \sum_{j=zN_0}^{(z+1)N_0-1} \log\{\frac{P_{on}(p_j)}{P_{off}(p_j)}\} < T, \ or \ \frac{1}{N_0} \sum_{j=zN_0}^{(z+1)N_0-1} \log\{\frac{P_{\triangle G}(\triangle\alpha_j)}{U(\triangle\alpha_j)}\} < \hat{T} \quad (10)$$

where T and $\hat{T}$ are the intensity and geometric thresholds that modulate the pruning behaviour of the algorithm. These parameters establish the minimum averaged reward that a path needs to survive, and in consequence they are closely related to the probability distributions used to design both the intensity and the geometric rewards. They must satisfy the following conditions:

$$-D(P_{off}||P_{on}) < T < D(P_{on}||P_{off}), -D(U_{\triangle G}||P_{\triangle G}) < \hat{T} < D(P_{\triangle G}||P_{\triangle G})$$
$$(11)$$

where D is the Kullback-Leibler divergence. The algorithm finds the best path that survives the pruning, and the expected convergence rate is $O(N)$. Typically the values of T and $\hat{T}$ are set close to their higher bounds. Additionally, if P_{on} diverges from P_{off}, the pruning rule will be very restrictive. Conversely, if these distributions are similar, the algorithm will be very conservative. The same reasoning follows for $P_{\triangle G}$ and $U_{\triangle G}$.

The application of this algorithm in the context of junction grouping motivates the extension of the basic pruning rule. We also consider the *stability of long paths* against shorter paths. Long paths are more probable to be close to the target that shorter ones, because they have survived to more reward prunes. Then, if N_{best} is the length of the best partial path, we will also prune paths with lengths N_j when

$$N_{best} - N_j > Z N_0 \tag{12}$$

where $Z > 0$ sets the minimum allowed difference between the best path and the rest of the paths. Low values of Z introduce more pruning, and the risk of loosing the true path is higher. When Z is large, shorter paths can survive.

The algorithm selects for expansion the best partial path that survives to the extended pruning rule. These paths are stored in a sorted queue. We consider that we have reached the end of a connecting path, when the center (x_f, y_f) of a junction is found in a small neighbourhood around the end of the selected path. In order to perform this test, we use a *range tree*, a representation from Computational Geometry [3] that is suitable to search efficiently within a range. The cost of generating the tree is $O(J \log J)$, where J is the number of detected junctions. Using this representation, a range query can be performed with cost $O(\log J)$ in the worst case.

Once a new junction is reached, the last segment of the path must lie on the limit θ_f between two wedges. Then, we use this condition to label the closest limit as *visited*. If the last segment falls between two limits and the angle between them is below a given threshold B, then both limits are labeled as visited. As the search of a new path can be started only along a non-visited limit, this mechanism avoids tracking the same edge in the opposite direction.

However, the search may finish without finding a junction. This event is indicated by an empty queue. In this case, if the length of the last path expanded by the algorithm is below the block size N_0, we consider that this path emanates from a false limit, and this limit is cancelled. Otherwise, the search has reached a termination point and its coordinates must be stored. If we find another termination point in a given neighbourhood, both paths are connected. This connection is associated to a potential undetected junction when the angle between the last segments of the paths is greater than $\pi/9$, the minimum angle to declare a junction.

Our *local-to-global* grouping algorithm starts from a given junction and performs path searching for each non-visited limit. When a new junction is reached, its corresponding limit is labeled as visited. Once all paths emanating from a junction are tested, the algorithm selects a new junction. Connected junctions are grouped. Labeling avoids path duplicity. Robustness is provided by the fact that an edge can be tracked in a given direction, if the search from the opposite direction fails. As we have seen previously, it is possible to join partial paths at termination points. False limits are filtered, and it is possible to discover new junctions, and also to correct the existing ones. False junctions are removed when all their paths fail. This method generates a mid-level representation. We can use the connectivity and the information contained in the paths for segmentation, tracking and recognition tasks.

3.3 Grouping: Connecting Results

We have tested our grouping algorithm both with the synthetic and the real images processed by our junction detector. Some results are showed in Fig 5. We have used the following parameters: branching factor $Q = 3$, with $Q_0 = 5$ at the first step of the algorithm; block size $N_0 = 3$ segments; maximum angle $A = 0.2 = \pi/15$ radians; rigidity $C = 5.0$; divergences and thresholds: $-D(P_{off}\|P_{off}) = -5.980$, $D(P_{on}\|P_{off}) = 3.198$, $T = 0.0$, $-D(U_{\triangle G}\|P_{\triangle G}) = -0.752$, $D(P_{\triangle G}\|P_{\triangle G}) = 0.535$, $\hat{T} = 0.40$, $B = \pi/6$; and extended pruning parameter $Z = 1.0$. The approximate processing time is $t = 3.0$ secs. in a Pentium II 233Mhz under Linux.

4 Conclusions and Future Work

There are two main contributions in this paper: the junction detection method and the grouping approach. Both methods are inspired in Bayesian techniques and they are tested with synthetic and real images. Our junction detector is fast – it is based on greedy search – and reliable – because we use sound statistics. Our junction grouping approach finds connecting paths between pairs of junctions and it is also efficient and robust. This method allows us to filter false junctions and to discover undetected ones. Then, a mid-level representation is obtained. Future work includes the refinement of this structure and its use in segmentation and reconstruction tasks, specially in the context of robot navigation.

References

1. P.R. Beaudet. "Rotational invariant image operators" In Proc. of the 4th. International Conference on Pattern Recognition (1978)
2. J. Coughlan and A. Yuille "A Probabilistic Formulation of A^*: O(N) Expected Convergence Rates for Visual Search" Submitted to Artificial Intelligence (1998)
3. M. de Berg, M. van Kreveld, M. Overmars and O. Schwarkopf "Computational Geometry: Algorithms and Applications" Springer-Verlag. (1997)
4. R. O. Duda and P. E. Hart "Pattern Classification and Scene Analisys" John Wiley and Sons. (1973)
5. W. Frstner. "A Framework for Low Level Feature Extraction." Proc. of the European Conference on Computer Vision (1994)
6. W. Freeman and E. Adelson. "Junctions Detection and Classification" Proc. of ARVO (1991)

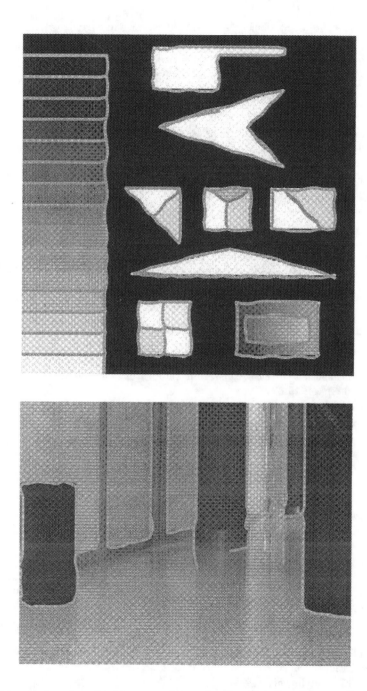

Fig. 5. Experimental results for connecting paths. Top: Results with a synthetic image. Bottom: Finding connecting paths in the corridor image.

7. D. Geman and B. Jedynak "An Active Testing Model for Tracking Roads in Satellite Images." IEEE Transaction on Pattern Analysis and Machine Intelligence.(1996) vol. 18 n. 1 pp. 1-14.

8. C. G. Harris and M. Stephens. "A combined corner and edge detection" 4th Alvey Vision Conference (1988) pp. 147-151

9. H. Ishikawa and D. Geiger "Segmentation by Grouping Junctions". In Proc. of the IEEE Computer Society Conference on Computer Vision and Pattern Recognition (1998)

10. L. Kitchen and A. Rosenfeld. "Gray Level Corner Detection" Pattern Recognition Letters. (1982) vol. 1 n. 2 pp. 95-102

11. T. Lindeberg and M-X. Li "Segmentation and Classification of Edges Using Minimum Description Length Aproximation and Complementary Cues". Technical Report ISRN KTH/NA/P-96/01-SE, Royal Institute of Technology. Jan.(1996)

12. T-L Liu and D. Geiger "Visual Deconstruction: Recognizing Articulated Objects". Proc. International Workshop on Energy Minimization Methods in Computer Vision and Pattern Recognition. Eds. M. Pelillo and E. Hancock. Venice, Italy. Springer-Verlag. May (1997)

13. J. Matas and J. Kittler "Contextual Junction Finder" In J.L. Crowley and H. Christensen (Eds) Vision as Process (1995) pp. 133-141

14. J. Malik "Interpreting Line Drawings of Curved Objects". International Journal of Computer Vision. (1996) vol. 1 n.1 pp. 73-104.

15. J. Malik "On Binocularly Viewed Occlusion Junctions". Proc. of the European Conference of Computer Vision (1996)

16. K. Brunnstrm, J-0. Eklundh and T. Uhlin "Active Fixation and Scene Exploration". International Journal of Computer Vision.(1996) vol. 17 n. 2 pp. 137-162

17. L. Parida, D. Geiger, and R. Hummel "Junctions: Detection, Classification, and Reconstruction" IEEE Transaction on Pattern Analysis and Machine Intelligence.(1998) vol. 20 n. 7

18. J. Pearl "Heuristics" Addison-Wesley (1984)

19. J. Rissanen "A Universal Prior for Integers and Estimation by Minimum Description Length" Annals Statistics (1983) vol. 11 pp. 416-431

20. K. Rohr "Recognizing Corners by Fitting Parametric Models" International Journal of Computer Vision (1992) vol. 9 pp. 213-230

21. S. M. Smith and J. M. Brady "SUSAN= A New Approach to Low Level Image Processing" Int. J. Comp. Vision (1997) vol. 23 n. 1 pp. 45-78

22. B. Vasselle, G. Giraudon, and M. Berthod "Following Corners on Curves and Surfaces in the Scale Space" In Proc. of the European Conference on Computer Vision 94 (1994) vol. 800 pp. 109-114

23. D. Waltz "Understanding Line Drawings of Scenes with Shadows" The Psychology of Computer Vision. McGraw-Hill. (1972)

24. A. Yuille and J. Coughlan "Twenty Questions, Focus of Attention and A^*: A Theoretical Comparison of Optimization Strategies". Proc. International Workshop on Energy Minimization Methods in Computer Vision and Pattern Recognition. Eds. M. Pelillo and E. Hancock. Venice, Italy. Springer-Verlag. May (1997)

25. A. Yuille and J. Coughlan "Visual Search: Fundamental Bounds, Order Parameters, and Phase Transitions" Submitted to IEEE Transactions on Pattern Analysis and Machine Intelligence (1998)

26. S.C. Zhu and A. Yuille "Region Competition = Unifying Snakes, Region Growing, and Bayes/MDL for Multiband Image Segmentation" IEEE Transactions on Pattern Analysis and Machine Intelligence (1996) vol. 18 n. 9 pp. 884-900

Semi-iterative Inferences with Hierarchical Energy-Based Models for Image Analysis

Annabelle Chardin and Patrick Pérez

IRISA/INRIA
Campus de Beaulieu, F-35042 Rennes cedex, France.

tel: (+33) 2.99.84.72.73
fax: (+33) 2.99.84.71.71

{achardin,perez}@irisa.fr

Abstract. This paper deals with hierarchical Markov Random Field models. We propose to introduce new hierarchical models based on a hybrid structure which combines a spatial grid of a reduced size at the coarsest level with sub-trees appended below it, down to the finest level. These models circumvent the algorithmic drawbacks of grid-based models (computational load and/or great dependance on the initialization) and the modeling drawbacks of tree-based approaches (cumbersome and somehow artificial structure). The hybrid structure leads to algorithms that mix a non-iterative inference on sub-trees with an iterative deterministic inference at the top of the structure. Experiments on synthetic images demonstrate the gains provided in terms of both computational efficiency and quality of results. Then experiments on real satellite spot images illustrate the ability of hybrid models to perform efficiently the multispectral image analysis.

1 Background: Hierarchical Energy-based Models

Many inverse problems from image analysis can be managed by designing an *energy* function $U(x, y)$ which captures the interaction between a large number of unknown variables $x = (x_i)_i$ to be estimated, and the observed variables –the measurements or data–, $y = (y_j)_j$. The manipulation of this function is made tractable by its usual decomposition as a sum of *local* terms involving just a few variables at a time. This kind of problem is encountered in Markov random field-based approaches as well as in partial differential equation (PDE)-based approaches, where, in the first stage, the energy depends on a continuous function x. Within the framework of Markov random fields, x and y are random vectors and we have the following relation between the joint distribution and the energy function: $P(x, y) \propto \exp\{-U(x, y)\}$. As far as the inference of x is concerned, the Bayesian estimation theory provides two standard estimators: the Maximum A Posteriori (MAP) estimator which corresponds to the global minimizer of the energy fonction ($\hat{x} = \arg\max_x P(x|y) = \arg\min_x U(x, y)$) and the Modes of Posterior Marginals (MPM) estimator ($\forall\, i,\ \hat{x}_i = \arg\max_{x_i} P(x_i|y)$)

which requires the computation of marginals by summation over huge sets of configurations.

It turns out that for most energy-based models suitable for image analysis problems, one has to devise deterministic or stochastic iterative algorithms exploiting the locality of the model in order to conduct the MAP or MPM inference. While permitting tractable single-step computations, the locality results in a very slow propagation of information. As a consequence, these iterative procedures may converge very slowly. This motivates the search either for improved algorithms of generic use, or for specific models allowing non-iterative or more efficient inference.

So far, the more fruitful approaches in both cases have relied on some notion of *hierarchy*. Hierarchical models or algorithms allow the information to be integrated in a progressive and efficient way (especially in the case of multiresolution data, when images come into a hierarchy of scales) providing gains in terms of both computational efficiency and quality of results.

Algorithm-based hierarchical approaches are usually related to well-known *multigrid* resolution techniques from Numerical Analysis, where an increasing sequence of nested spaces is explored in a number of possible ways. The particular case of *coarse-to-fine* exploration has been successfully extended to discrete image models [3, 8]. Within this framework, reduced versions of an original (spatial) model can be deduced in a consistent way (the form of the energy and associated parameters are deduced at once). The "stack" of models thus obtained can then be used for inference purposes, the estimate *iteratively* obtained at a given level being used as an initialization for the processing at the next level.

On the other hand, model-based hierarchical approaches aim at defining a new global hierarchical model which has nothing to do with any original (spatial) model. It has to be manipulated as a whole, but according to procedures of reduced complexity. These models usually lie on the nodes of a quad-tree whose leaves fit the pixels of (maximum resolution) images [4, 9, 10]. In this case, the peculiar dependency structure, like in case of Markov chains, allows *non-iterative* inference procedures made of two sweeps: a bottom-up sweep propagating all information to the root, and a top-down one which in turn allows optimal estimate to be obtained at each node given *all the data*.

One of the drawbacks of these tree-based approaches lies in the structural constraints they impose: first of all they might appear artificial for certain types of problems or data; in any case the relevance of the inferred variables at coarsest levels is not obvious (especially at the root). Second, the complete tree-structure is cumbersome in case of large images.

To circumvent this, here we propose a hierarchical model based on a "hybrid" structure which combines a spatial grid of reduced size at a coarser level with "sub-trees" appended below it, down to the finest level (see Fig.1). This paper investigates the use of the MAP estimator and of the MPM estimator, known as more reliable than MAP, on the hybrid structure. It should be noticed that

MPM estimator relies on the computation of posterior marginals, which are a key ingredient for EM-type parameter estimation techniques.

The section 2 describes the hybrid structure and its associated energy function. In the section 3, we explain the procedures to achieve the MAP and MPM estimates by mixing a non-iterative inference on sub-trees with iterative deterministic inference of reduced cost at the top of the structure. The section 4 illustrates these procedures for image classification problems with synthetic and real images.

2   Hierarchical Model and energy function

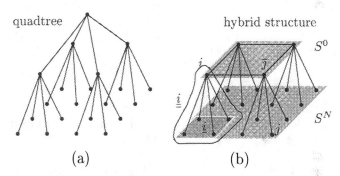

quadtree

hybrid structure

(a)   (b)

Fig. 1. Two hierarchical structures: (a) quadtree with three levels; (b) truncated tree with two levels.

The hierarchical model we use is based on a hybrid structure [6]. One example is shown in Fig.1(b) for a single level below the coarsest grid. To describe this graph, we shall introduce some notations.

First, we define the coarsest level S^0 as a rectangular grid with a 1st-order neighborhood. Then each site of S^0 initiates a quadtree, so that the grid S^n $(0 < n \leq N)$ made up by the nodes at the level n is $2^n \times 2^n$ times larger than S^0. Now each site i of S^n has four natural correspondents in S^{n+1} (provided that i does not belong to the finest level S^N), its children, forming site set $\underline{i}$, and one natural correspondent in S^{n-1} (provided that i does not belong to the coarsest level S^0), its parent, denoted as $\bar{i}$. Finally, the site set forming the tree rooted at i is denoted $\underline{i}$ (Fig.1). Vectors x and y are now indexed by the nodes of $S \triangleq \bigcup_{n=0}^{N} S^n$.

Given this graphical structure consider an energy function of the following form:

$$U(x,y) \triangleq \sum_{<i,j> \in S^0} v_{ij}(x_i, x_j) + \sum_{i \notin S^0} w_i(x_i, x_{\bar{i}}) + \sum_{i \in S} l_i(x_i, y_i) , \qquad (1)$$

where $< i,j >$ designates pairs of neighbors in S^0, v_{ij} and w_i are local functions capturing respectively the spatial prior and the hierarchical prior (they will usually encourage identity between neighbors and between parents and children, resp.), and l_i expresses the point-wise relation between the observed variable y_i [1] and the unknown one x_i. The MAP estimator for this model amounts to minimizing $U(x,y)$ in (1) with respect to x for a given y.

From a probabilistic point of view, the associated posterior distribution of (x,y) is:

$$P(x|y) \propto \prod_{<i,j>\in S^0} \underbrace{g_{ij}(x_i,x_j)}_{\triangleq \exp\{-v_{ij}(x_i,x_j)\}} \times \prod_{i\notin S^0} \underbrace{f_i(x_i,x_{\bar{\imath}})}_{\triangleq \exp\{-w_i(x_i,x_{\bar{\imath}})\}} \times \prod_{i\in S} \underbrace{h_i(x_i,y_i)}_{\triangleq \exp\{-l_i(x_i,y_i)\}} \quad . \tag{2}$$

The MPM estimator requires to deduce from this global posterior distribution, each of the local posterior distributions $P(x_i|y)$, for $i \in S$.

3 Semi-iterative inferences

3.1 MAP computation

The global minimization of $U(x,y)$ w.r.t. x can be written in a way that distinguishes variables on S^0 (interacting in a non-causal fashion) from the others (interacting in a tree-based causal fashion):

$$\min_x U(x,y) = \min_{x_{S^0}} \left\{ \sum_{<i,j>\in S^0} v_{ij}(x_i,x_j) + \sum_{i\in S^0} l_i(x_i,y_i) \right. $$
$$\left. + \sum_{i\in S^0} \min_{x_{\underline{i}\backslash\{i\}}} \sum_{j\in\underline{i}\backslash\{i\}} [w_j(x_j,x_{\bar{\jmath}}) + l_j(x_j,y_j)] \right\}. \tag{3}$$

Each tree-based minimization w.r.t. $x_{\underline{i}\backslash\{i\}}$, for $i \in S^0$ in the left-hand-side can be conducted exactly, with fixed complexity per node, using an extension of the chain-based Viterbi algorithm [7]. The first (upward) sweep of this algorithm computes recursively the optimal value $x_i^*(x_{\bar{\imath}})$ at a node i given the value of its parent, and the associated value $V_i(x_{\bar{\imath}})$ for the energy term that concerns $\underline{i}$:

$$V_i(x_{\bar{\imath}}) \triangleq \min_{x_{\underline{i}}} \sum_{j\in\underline{i}} [w_j(x_j,x_{\bar{\jmath}}) + l_j(x_j,y_j)] \tag{4}$$

$$= \min_{x_i} [w_i(x_i,x_{\bar{\imath}}) + l_i(x_i,y_i) + \sum_{j\in\underline{i}} V_j(x_i)] \tag{5}$$

---

[1] In the following we consider that measurements are available at each node $i \in S$, to take into account the case of multiresolution images. Derivations are readily adapted to the cases where data are only available on a subset of S. In practice data are, for instance, often defined at the finest level S^N only. In this paper our experimentations were led on monoresolution images which were either monospectral (synthetic cases) or multispectral (real satellite data).

$$\text{and } x_i^*(x_{\bar{i}}) \triangleq \arg\min_{x_i}[w_i(x_i, x_{\bar{i}}) + l_i(x_i, y_i) + \sum_{j\in\underline{i}} V_j(x_i)]. \tag{6}$$

Once this recursion is completed it remains to perform in (3) the minimization w.r.t. x_{S^0}, which amounts to

$$\min_{x_{S^0}} \sum_{<i,j>\in S^0} v_{ij}(x_i, x_j) + \sum_{i\in S^0}[l_i(x_i, y_i) + \sum_{j\in\underline{i}} V_j(x_i)].$$

Apart if S^0 is very small, an iterative ICM-type procedure [2] must be introduced here due to the non-causal term $\sum_{<i,j>\in S^0} v_{ij}(x_i, x_j)$. This provides estimates $\hat{x}_i$, for $i \in S^0$, from which all the other estimates are recursively recovered thanks to functions x_j^*: $\hat{x}_j = x_j^*(\hat{x}_{\bar{j}})$. The whole procedure goes as follows:

Semi-iterative energy minimization

▲ upward sweep

Leaves ($i \in S^N$)
$$\begin{cases} V_i(x_{\bar{i}}) \triangleq \min_{x_i}[w_i(x_i, x_{\bar{i}}) + l_i(y_i, x_i)] \\ x_i^*(x_{\bar{i}}) \triangleq \arg\min_{x_i}[w_i(x_i, x_{\bar{i}}) + l_i(y_i, x_i)] \end{cases}$$

Recursion (for $n = N-1\ldots1$, $i \in S^n$)
$$\begin{cases} V_i(x_{\bar{i}}) \triangleq \min_{x_i}[w_i(x_i, x_{\bar{i}}) + l_i(x_i, y_i) + \sum_{j\in\underline{i}} V_j(x_i)] \\ x_i^*(x_{\bar{i}}) \triangleq \arg\min_{x_i}[w_i(x_i, x_{\bar{i}}) + l_i(x_i, y_i) + \sum_{j\in\underline{i}} V_j(x_i)] \end{cases}$$

◀▶ coarse ICM: update a few times all sites of S^0 in turn, for energy
$$\sum_{<i,j>\in S^0} v_{ij}(x_i, x_j) + \sum_{i\in S^0}[l_i(x_i, y_i) + \sum_{j\in\underline{i}} V_j(x_i)]$$
$\Rightarrow \hat{x}_i, \forall i \in S^0$

▼ downward sweep
$$\hat{x}_i = x_i^*(\hat{x}_{\bar{i}})$$

If an exact minimization can be obtained at the coarsest level (which is especially the case for the complete tree where S^0 reduces to a single site), the final estimate $\hat{x}$ is exactly the global minimizer of U. Note that the functions $V_i(x_{\bar{i}})$, which appear in the upward sweep, progressively collect dependencies with respect to the data, even though this is not made explicit by abuse of notation: $V_i(x_{\bar{i}})$ (as well as $x^*(x_{\bar{i}})$) actually depends on $y_{\underline{i}}$. This means that ICM at the level 0 and downward sweep provide inferences based on all data.

When y is only attached to S^N, this procedure can be compared to the multigrid method in [8], where the non-hierarchical energy

$$U^N(x^N, y) = \sum_{<i,j>\in S^N} v_{ij}^N(x_i, x_j) + \sum_{i\in S^N} l_i(x_i, y_i) \tag{7}$$

is minimized within the set of configurations which are piece-wise constant over $2^{N-n} \times 2^{N-n}$ blocks, for $n = 0\ldots N$; equivalently

$$U^n(x^n, y) = \sum_{<i,j>\in S^n} 2^{N-n} v_{ij}^N(x_i^n, x_j^n) + \sum_{j\in\underline{i}\cap S^N} l_j(x_i^n, y_j) \tag{8}$$

is minimized, where x^n is defined on S^n. Let us note that U^0, thus defined for $n = 0$ in the multigrid approach, corresponds to the energy which is manipulated at the coarsest level of the semi-iterative approach for $w_i(x_i, x_{\bar{\imath}}) \triangleq \beta[1 - \delta(x_i, x_{\bar{\imath}})]$ with $\beta \to +\infty$ (in this case the optimal configuration x is constant over each tree $\underline{i}$, $i \in S^0$, and $\sum_{j' \in \underline{i}} V_{j'}(x_i) = \sum_{j \in \underline{i} \cap S^N} l_j(x_i^n, y_j)$, $\forall i \in S^0$). Moreover, if data are only available at the finest level, the initialization of the multigrid coarse ICM is given by the minimization of $\sum_{j \in \underline{i} \cap S^N} l_j(x_i^n, y_j)$ for each site $i \in S^0$ while the initialization of the hybrid coarse ICM is given by the minimization of $\sum_{j \in \underline{i}} V_j(x_i)$ for each site $i \in S^0$. To confirm this statement, we will just have to compare the coarsest initializations and estimates provided by the two methods, taking the same number of levels, the same functions v_{ij} and l_i, and $w_i(x_i, x_{\bar{\imath}}) \triangleq \beta[1 - \delta(x_i, x_{\bar{\imath}})]$ with β very large for the hybrid energy.

3.2 MPM computation

In the case of a complete tree where S^0 reduces to a single site, the exact MPM estimates can be computed on each node [9] using an extension of Baum-Welch algorithm on a chain [1]. In the general case, the downward recursion is now based on the following relation:

$$\forall i \notin S^0, \; \mathsf{P}(x_i|y) = \sum_{x_{\bar{\imath}}} \mathsf{P}(x_i|x_{\bar{\imath}}, y)\mathsf{P}(x_{\bar{\imath}}|y) \;,$$

where

$$\mathsf{P}(x_i|x_{\bar{\imath}}, y) = \mathsf{P}(x_i|x_{\bar{\imath}}, y_{\underline{i}})$$

due to separation property. The use of this recursion requires that a previous upward sweep provides $\mathsf{P}(x_i|x_{\bar{\imath}}, y_{\underline{i}})$ for $i \notin S^0$ and $\mathsf{P}(x_i|y)$ for $i \in S^0$. This is achieved by successivly summing out x_i's for all $i \notin S^0$. The resursion is based on:

$$\mathsf{P}(x_i|x_{\bar{\imath}}, y_{\underline{i}}) \propto f_i(x_i, x_{\bar{\imath}})h_i(x_i, y_i) \sum_{x_{\underline{i}\setminus\{i\}}} \prod_{j \in \underline{i}\setminus\{i\}} f_j(x_j, x_{\bar{\jmath}})h_j(x_j, y_j)$$

$$\propto f_i(x_i, x_{\bar{\imath}})h_i(x_i, y_i) \prod_{j \in \underline{i}} \sum_{x_{\underline{j}}} \prod_{k \in \underline{j}} f_k(x_k, x_{\bar{k}})h_k(x_k, y_k) \;, \qquad (9)$$

with

$$\mathbb{F}_j(x_i) \triangleq \sum_{x_{\underline{j}}} \prod_{k \in \underline{j}} f_k(x_k, x_{\bar{k}})h_k(x_k, y_k)$$

$$= \sum_{x_j} f_j(x_j, x_{\bar{\jmath}})h_j(x_j, y_j) \prod_{k \in \underline{j}} \mathbb{F}_k(x_j). \qquad (10)$$

This eventually provides the probability

$$\mathsf{P}(x_{S^0}|y) = \sum_{x_{S\setminus S^0}} \mathsf{P}(x|y).$$

Because of the non-causal structure on S^0, $P(x_i|y)$ for $i \in S^0$ have to be approximated with the help of a Gibbs sampling of the distribution $P(x_{S^0}|y)$. This procedure provides approximated local posterior marginals in a semi-iterative way. If a MPM estimator is employed, an approximation of it is obtained as a by-product.

Then the whole procedure goes as follows:

Semi-iterative local posterior marginal computation and MPM estimation

▲ upward sweep

Leaves $(i \in S^N)$

$\mathbb{F}_i(x_{\bar{i}}) = \sum_{x_i} f_i(x_i, x_{\bar{i}}) h_i(x_i, y_i)$

Recursion (for $n = N - 1 \ldots 1,\ i \in S^n$)

$\mathbb{F}_i(x_{\bar{i}}) = \sum_{x_i} f_i(x_i, x_{\bar{i}}) h_i(x_i, y_i) \prod_{j \in \underline{i}} \mathbb{F}_j(x_i)$

◄► coarse posterior marginal computation:

draw samples $x_{S^0}(1), \ldots, x_{S^0}(m)$ from:

$P(x_{S^0}|y) \propto \prod_{<i,j> \in S^0} g_{ij}(x_i, x_j) \times \prod_{i \in S^0} h_i(x_i, y_i) \times \prod_{i \in S^0} \prod_{j \in \underline{i}} \mathbb{F}_j(x_i)$

approximation of $P(x_i|y) \approx \frac{1}{m-k} \sum_{j=k+1}^{m} \delta[x_i(j), x_i]$

▼ downward sweep

Recursion (for $n = 1 \ldots N,\ i \in S^n$)

$P(x_i|y) = \sum_{x_{\bar{i}}} P(x_{\bar{i}}|y) \frac{f_i(x_i, x_{\bar{i}}) h_i(x_i, y_i)}{\mathbb{F}_i(x_{\bar{i}})} \prod_{j \in \underline{i}} \mathbb{F}_j(x_i)$

MPM at leaves

$\hat{x}_i \triangleq \arg\min_{x_i} P(x_i|y)$

Here, the same abuse of notation is made as in the MAP procedure: in fact the functions $\mathbb{F}_i(x_{\bar{i}})$ depend on $y_{\underline{i}}$, so that the progressive summations out x_i's for all $i \notin S^0$ are made with respect to data associated to $\underline{i}$ and the downward sweep provides local posterior marginal based on all data.

4 Supervised Classification Comparisons

To demonstrate the practicability and the relevance of the approach for discrete low-level image analysis, we first report comparative experiments for supervised classification. To this end, we considered a Potts-type prior with potentials

$$v_{ij}(x_i, x_j) \triangleq 2^N \alpha [1 - \delta(x_i, x_j)], \tag{11}$$

$$w_i(x_i, x_{\bar{i}}) \triangleq \beta [1 - \delta(x_i, x_{\bar{i}})], \tag{12}$$

along with Gaussian likelihoods

$$l_i(x_i = k, y_i) \triangleq \begin{cases} \frac{(y_i - \mu_k)^2}{2\sigma_k^2} + \log(\sigma_k) & \text{if } i \in S^N, \\ 0 & \text{otherwise.} \end{cases} \tag{13}$$

Comparative experiments were led for $N \in \{0, 2, 3, 8\}$ for the synthetic scene and for $N \in \{0, 3, 4, 8\}$ for the real spot scene. For $N = 0$ the inferences are based on standard non-hierarchical models using iterative ICM for the approximation of the MAP or an iterative Gibbs sampler for the approximation of the MPM. While $N = 8$, when the size of S^N is $2^8 \times 2^8$, corresponds to the complete tree ($|S^0| = 1$) allowing an exact non-iterative computations. $N = 2, 3, 4$ correspond to three-, four- and five-level hybrid structures on which semi-iterative MAP and MPM inference are performed.

4.1 Synthetic images

First, the experiments were carried out on a 256×256 synthetic image involving 5 classes (Fig. 2). We applied an additive Gaussian white noise with a different standard deviation for each class, thus the gray level means and variances $(\mu_k, \sigma_k^2)_{k=1}^5$ are known.

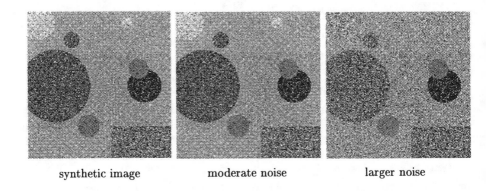

synthetic image   moderate noise   larger noise

Fig. 2. Synthetic data.

First we stress on the MAP procedure to support experimentally the statement given in the section 3, according to which the semi-iterative MAP inference with data at the finest level only tends to constrained non-hierarchical minimization when $\beta \to +\infty$. It is easy to see (Fig.3) that both the initializations and the results of the coarse ICM for the multigrid method and the semi-iterative one are the same as soon as β is large enough (greater than 80). Accordingly, the multigrid approach and the semi-iterative procedure with $\beta \to +\infty$ really behave in the same way.

Secondly, we worked on a moderately corrupted and a fairly corrupted images, and then paid attention to how our algorithms behave for a noisier image (see Fig.2). The obtained classification results are shown in Fig.4 and in Fig.5 with their respective percentages of misclassification and cpu times in seconds (on a 170 MHz Ultra Sparc I workstation) in tables 1 and 2.

As can be seen from the figures 4 and 5, the hierarchical models provide good results. As the noise level is low, the resulting classifications of all the methods

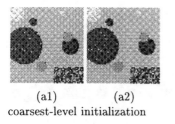

(a1)   (a2)

coarsest-level initialization

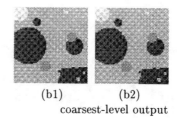

(b1)   (b2)

coarsest-level output

Fig. 3. comparative results at the coarsest level between the multigrid method (a1-b1) and the semi-iterative method (a2-b2) with $N = 2$ and $\beta \to +\infty$.

are of an almost equivalent quality. Whereas the degradation of the synthetic image increases, the hierarchical models and the multigrid approach seem more robust to noise than the plain non-hierarchical iterative procedures. Moreover, it can be noticed that the non- or semi-iterative algorithms take much fewer cpu time: for the MAP, they take three times less cpu time than the ICM, and for the MPM, more than forty times less cpu time than the Gibbs sampler. For the estimation of the posterior marginals in the sampling procedure, we fit the number of retained samples ($m - k$, see the description of the algorithm) to the size of the concerned grid (200 for the complete grid and 20 for the coarse grid of the hybrid structure).

To go further in the comparisons, we can focus on the results for the complete tree ($N = 8$) and for the two examples of the hybrid structure for each estimator. First, if we just look at the tables, we can say that the semi-iterative estimation provides slightly better classification than a non-iterative one for a comparable cpu time. Moreover, the MPM classifications are slightly better than the MAP ones, especially for $N = 2$, but they enquire more calculations because the MAP downward recursion is simplier (one has only to read look-up tables built in the upward recursion, whereas there are some calculations during the MPM downward step). Anyway, whatever estimator is used, the use of a coarse iterative procedure on top of the hybrid structures does not seem to imply an extra computational load, while improving the results.

As far as the visual aspect is concerned, the semi-iterative classifications still reveal a blocky aspect (as observed in all tree-based approaches). But these artifacts are less and less pronounced as the number of levels in the hybrid structure decreases.

On top of the fact that the MPM is doing better than the MAP, the hierarchical MPM provide a measure of relative confidence associated to the estimated value at each site, through the entropy

$$c_i \triangleq \sum_{x_i} P(x_i|y) \log P(x_i|y).$$

These measures (Fig.6) allow us to better appreciate the quality of the obtained estimates and to use them in consequence.

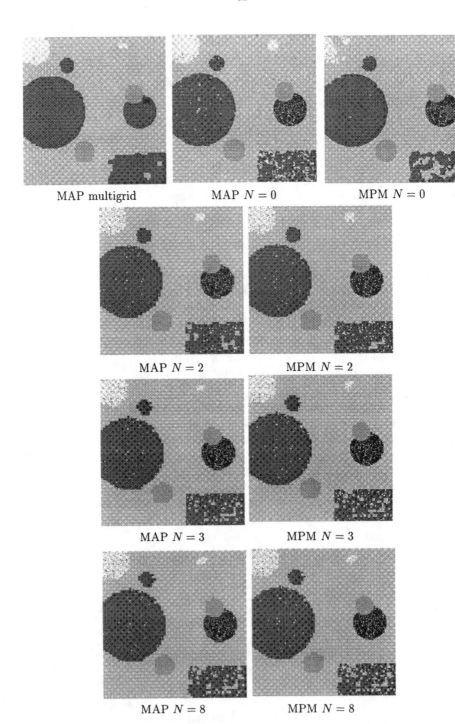

Fig. 4. Comparative results for the classification problem with a moderately corrupted image.

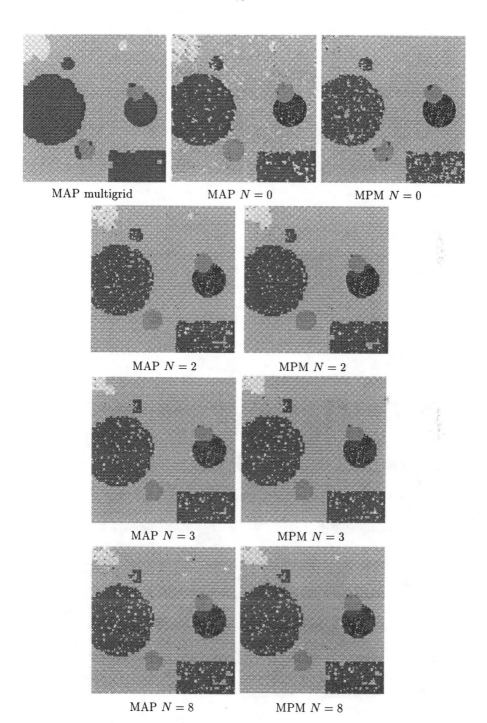

MAP multigrid   MAP $N = 0$ MPM $N = 0$

MAP $N = 2$ MPM $N = 2$

MAP $N = 3$ MPM $N = 3$

MAP $N = 8$ MPM $N = 8$

Fig. 5. Comparative results for the classification problem with a more corrupted image.

Model	MAP misclassification	cpu time	MPM misclassification	cpu time
iterative $N = 0$	5.3%	11s	7%	7min
semi-iterative $N = 2$	4.15%	3.6s	3.6%	10.5s
semi-iterative $N = 3$	4.2%	3.6s	4.5%	9.3s
non-iterative $N = 8$	4.8%	3.6s	4.7%	9s
multigrid	5.9%	11.6s		

Table 1. Performances of the different algorithms with the synthetic image in fig.4

Model	MAP misclassification	cpu time	MPM misclassification	cpu time
iterative $N = 0$	9.65%	11s	10%	7min
semi-iterative $N = 2$	6.6%	3.6s	6.2%	10.5s
semi-iterative $N = 3$	7.9%	3.6s	7.9%	10s
non-iterative $N = 8$	8%	3.6s	8%	9s
multigrid	7.3%	11.6s		

Table 2. Performances of the different algorithms with the synthetic image in fig.5

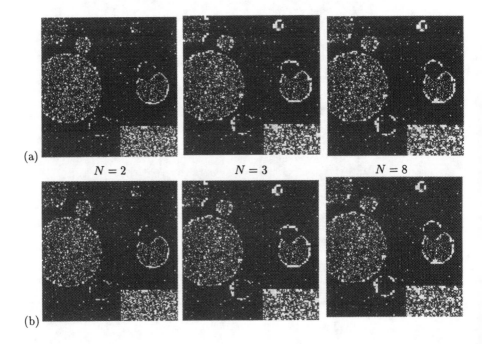

Fig. 6. Confidence maps associated to classification (a) of the moderately noisy image and (b) of the noisier image.

4.2 Real satellite images

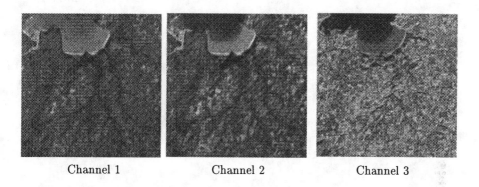

Channel 1   Channel 2   Channel 3

Fig. 7. 512×512 Spot images (courtesy of Costel, University of Rennes 2 and GSTB).

The previous algorithms were applied to SPOT satellite images provided by the Costel laboratory (University of Rennes 2) in the context of remote sensing researches. The scene (Fig.7) is composed of three 512×512 images with different wavelenghts in the visible spectrum and represents the Bay of Lannion in France in December 1996. The goal of this study was to determine the land cover of this area. So as to reach this aim, the geographers of the Costel laboratory built a list of eight classification categories: Sea and water, Sand and bare soil, Urban area, Forests and heath, Temporary meadows, Permanent meadows, Colza, Vegetables.

Thanks to both tests on the lands and photointerpretations, they were also able to provide samples of these eight categories on the three SPOT images of the scene. Some of them were used to learn the gray levels means and the variances of each category for each image, so that we could perform supervised classifications, and the left samples were kept to assess the accuracy of the classifications.

As for the model, we considered the three channels as independant. As a consequence, the form of the relation between the observed variables (here the multispectral scene) and the unknown variables (class label at each pixel) becomes:

$$l_i(x_i = k, y_i^c \; c \in \{1,2,3\}) \triangleq \sum_{c=1}^{c=3} \left(\frac{(y_i^c - \mu_{k_c})^2}{2\sigma_{k_c}^2} + \log(\sigma_{k_c}) \right) \quad \text{if } i \in S^N , \quad (14)$$

where $(\mu_{k_c}, \sigma_{k_c})$ are the gray level mean and variance of the class k within channel number $c \in \{1,2,3\}$.

The eight algorithms provide quite similar results of a good quality (see Fig.8 and 9, and Tab.3). About 92% of the pixels of the samples are well classified. The three hierarchical MAP inferences are achieved in almost twice less cpu time than the iterative ICM. In comparison with the iterative MPM which takes more than one hour to be performed, the three hierarchical MPM infernces need reasonable cpu time. This is encouraging for the EM-type parameter estimation algorithms for which the form of one recursion step is close to the MPM algorithm form.

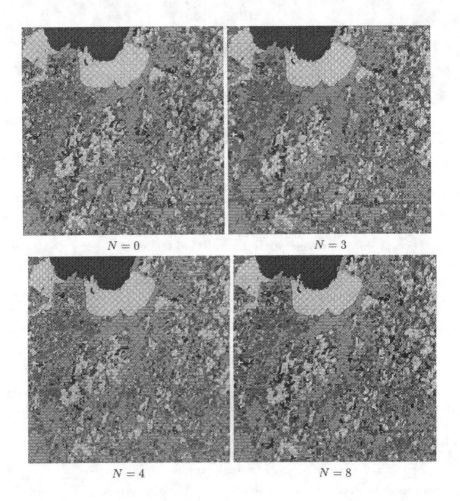

Fig. 8. Comparative results for the classification problem with multispectral spot images and the MAP estimator (see Fig.10 for the legend).

5   Conclusion and extensions

In this paper, we presented a hybrid hierarchical structure which is an interesting compromise between standard spatial models and hierarchical models based on a complete quadtree. We introduce two inference algorithms devoted to this new

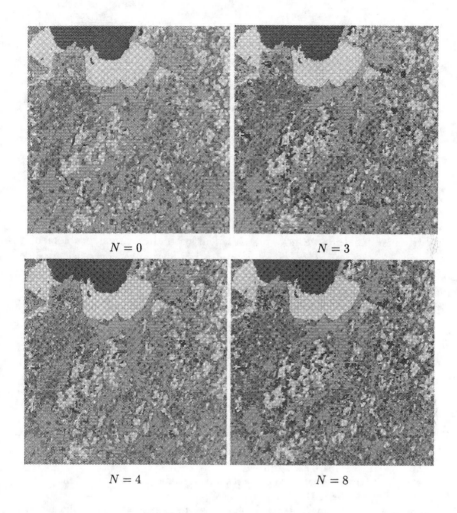

$N = 0$ $N = 3$

$N = 4$ $N = 8$

Fig. 9. Comparative results for the classification problem with multispectral spot images and the MPM estimator (see Fig.10 for the legend).

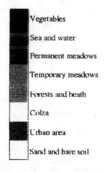

Vegetables

Sea and water

Permanent meadows

Temporary meadows

Forests and heath

Colza

Urban area

Sand and bare soil

Fig. 10. Legend for the classifications shown in Fig.8 and 9

Model	MAP		MPM	
	misclassification	cpu time	misclassification	cpu time
iterative $N = 0$	8%	150s	8%	1h10min
semi-iterative $N = 3$	7.5%	85s	7.5%	165s
semi-iterative $N = 4$	7.5%	85s	7.5%	165s
non-iterative $N = 8$	8%	65s	7.8%	160s

Table 3. Performances of the different algorithms with the real multispectral images in fig.7.

structure: the first one computes the MAP estimate and the second one computes the local posterior marginals and the MPM estimate.

With this structure we now plan to deal with the critical problem of parameter estimation: using marginal computation introduced here, we should be able to design specific EM-like algorithms on the hybrid structure, as already done with trees [4, 9] and with spatial grids [5]. In the classification problem, this will concern both the data parameters (number of classes, gray level means and variances of each class) whose automatic estimation will allow unsupervised classification, and the parameters of the prior model (α, β). We also plan to address the issue of automatically estimating the optimal number of levels in the struture.

References

1. L.E. Baum, T. Petrie, G. Soules, and N. Weiss. A maximization technique occuring in the statistical analysis of probabilistic functions of Markov chains. *Ann. Math. Stat.*, Vol 41: pp. 164–171, 1970.
2. J. Besag. Spatial interaction and the statistical analysis of lattice systems. *J. Royal Statist. Soc.*, 36, Série B:192–236, 1974.
3. C. Bouman and B. Liu. Multiple resolution segmentation of textured images. *IEEE Trans. Pattern Anal. Machine Intell.*, Vol. 13, No. 2 : pages 99–113, Février 1991.
4. C. Bouman and M. Shapiro. A multiscale random field model for Bayesian image segmentation. *IEEE Trans. Image Processing*, 3, No. 2 :162–177, March 1994.
5. B. Chalmond. An iterative Gibbsian technique for reconstruction of m-ary images. *Pattern Recognition*, Vol. 22, No 6: pages 747–761, 1989.
6. A. Chardin and P. Pérez. Semi-iterative inference with hierarchical models. In *Int. Conf. on Image Processing*, pages 630–634, Chicago, USA, october 1998.
7. G.D. Forney. The Viterbi algorithm. *Proc. IEEE*, Vol. 13: pages 268–278, March 1973.
8. F. Heitz, P. Pérez, and P. Bouthemy. Multiscale minimization of global energy functions in some visual recovery problems. *CVGIP : Image Understanding*, Vol. 59, No 1: pages 125–134, Jan. 1994.
9. J-M Laferté, P. Pérez, and F. Heitz. Discrete markov image modeling and inference on the quad-tree. *IEEE Trans. Image Processing*, Accepted for publication, 1999.
10. M. Luettgen, W. Karl, and A. Willsky. Efficient multiscale regularization with applications to the computation of optical flow. *IEEE Trans. Image Processing*, 3(1):41–64, 1994.

Metropolis vs Kawasaki Dynamic for Image Segmentation Based on Gibbs Models

Xavier Descombes[†], Eugène Pechersky[‡] [*]

† INRIA, 2004, route des Lucioles BP 93, 06902 Sophia Antipolis Cedex, France,
tel. (33) 92 38 76 63, Fax (33) 92 38 76 43.
‡ IPIT, 19, Bolchoj Karatnyj, 101447 Moscow GSP-4, Russia, tel. (7/095) 313 3870.
web : `www.inria.fr/ariana`
mail : `Xavier.Descombes@sophia.inria.fr, pech@iitp.ru`

Abstract. In this paper we investigate several dynamics to optimize a posterior distribution defined to solve segmentation problems. We first consider the Metropolis and the Kawasaki dynamics. We also compare the associated Bayesian cost functions. The Kawasaki dynamic appears to provide better results but requires the exact values of the class ratios. Therefore, we define alternative dynamics which conserve the properties of the Kawasaki dynamic and require only an estimation of the class ratios. We show on synthetic data that these new dynamics can improve the segmentation results by incorporating some information on the class ratios. Results are compared using a Potts model as prior distribution.

1 Introduction

Since one decade, Gibbs fields are widely used in image processing especially for image segmentation [1]. The segmented image is given by a ground state of the model composed of a prior model (the Potts model for instance) and a likelihood term depending on the data. This latter term can be interpreted as an inhomogeneous external field. The ground state is obtained by sampling the model at zero temperature using a decreasing temperature parameter in a simulated annealing scheme [2]. The Metropolis dynamic or the Glauber dynamic are usually used in the simulated annealing algorithm. These dynamics consist in sampling the model in the canonical ensemble, i.e. in the whole configuration space.

In case of severe noise, the external field (given by the data) is less reliable. Therefore, the prior model (the interactions) have a greater influence in the results. Artifacts due to this prior can then appear in the segmented image. Without external field the Potts model presents a phase transition at low temperature [3]. In a typical sample, one label prevails and at low temperature, a uniform configuration is obtained. With a uniform external field this phenomenum does not exist. However, for inhomogeneous external fields, phase

[*] This work was partially supported by the French-Russian Institute A.M. Liapunov (Grant 98-02)

transitions may appear, as shown for the Ising model with a chess-board external field in [4]. In an image segmentation framework, this leads to over-regularization if the signal to noise ratio is too low. Fine objects are erased during the segmentation scheme and big objects tend to spread in their environment in case of severe noise.

The main idea of this paper is to avoid this effect by sampling the model in a micro-canonical ensemble, i.e. a subspace of the configuration space which exclude undesired configurations. In this paper, we first consider that the proportions between the different labels are fixed. The Kawasaki dynamic allows us to derive such a sampling [5]. With this dynamic we can expect to preserve objects because an unique label cannot fill the whole volume. However, in practice the ratio between the different classes is unknown. Therefore, we first have to estimate the class ratios. To take into account the bias of the estimates, we have to modify the Kawasaki dynamic.

Herein we study different dynamics for image segmentation. We first compare Metropolis and Kawasaki dynamics. Embedded in a Bayesian framework, these dynamics correspond to two different cost functions. To tackle the problem of the error in the class ratios estimation, we define a new Bayesian risk function leading to optimize the model in a "soft" micro-canonical ensemble. This cost function allows configurations having class ratios slightly different from the estimated one. We then study a mixture of Metropolis and Kawasaki dynamics. Finally, we compare the results obtained with the different dynamics on a synthetic image corrupted by a severe Gaussian noise.

2 Metropolis and Kawasaki dynamics

2.1 Background and notations

Herein we consider the problem of image segmentation using a Potts model. Denote $X = \{x_s\}$ the data on the lattice $S = \{s\}$ and $Y = \{y_s\}$ the segmented image.

The segmentation problem consists in optimizing the a posteriori probability $P(Y|X)$ using the Bayesian rule which gives:

$$P(Y|X) \propto P(X|Y)P(Y). \tag{1}$$

$P(X|Y)$ refers to the likelihood and $P(Y)$ to the prior.

We consider the pixels to be conditionally independent, i.e.:

$$P(X|Y) = \prod_{s \in S} p(x_s|y_s). \tag{2}$$

The different classes are supposed to be Gaussian, i.e.:

$$p(x_s|y_s) = \sum_{l \in \Lambda} \frac{1}{\sqrt{2\pi\sigma_l^2}} \exp -\frac{(x_s - \mu_l)^2}{2\sigma_l^2} \Delta(y_s, l), \tag{3}$$

where l indexes the different classes and (μ_l, σ_l^2) are the mean and variance of the corresponding Gaussian distribution. $\Delta(a, b)$ is equal to 1 if $a = b$ and to 0 otherwise.

We consider a prior defined by a 8-connected Potts model [6–8], written as follows:

$$P(Y) = \frac{1}{Z} \exp - \sum_{<s,s'>} \beta(1 - \Delta(y_s, y_{s'})), \tag{4}$$

where $< s, s' >$ defines a clique, i.e. is an element of the set:

$$\mathcal{C} = \{< s, s' >\in S \times S : d(s, s') \in \{1, \sqrt{2}\}\}. \tag{5}$$

Therefore the posterior distribution can be written as a Gibbs distribution:

$$P(Y|X) \propto \exp - [U(X|Y) + U(Y)], \tag{6}$$

and the local conditional probabilities are written as follows:

$$p(y_s|y_t, t \neq s, x_s) = p(y_s|y_t, t \in \mathcal{N}_s, x_s) \propto \exp -U(y_s|\mathcal{N}_s), \tag{7}$$

where $\mathcal{N}_s$ is the neighborhood of s (pixels in interactions with s) and the local conditional energy $U(y_s|\mathcal{N}_s)$ is written:

$$U(y_s|\mathcal{N}_s) = \sum_{t \in \mathcal{N}_s} \beta(1 - \Delta(y_s, y_t)) - \sum_{l \in \Lambda} \left(\frac{(x_s - \mu_l)^2}{2\sigma_l^2} + \frac{1}{2}\log(2\pi\sigma_l^2) \right) \Delta(y_s, l). \tag{8}$$

To optimize the a posteriori distribution $P(Y|X)$ we consider the Bayesian framework. The optimum is the configuration which minimizes the Bayes risk:

$$\hat{Y} = arg \min_{Y'} \int_{\Omega = \Lambda^S} R(Y, Y')P(Y)dY, \tag{9}$$

where $R(Y, Y')$ is the cost function.

2.2 Algorithms and associated Bayesian cost functions

The most used Bayesian criterion for image segmentation purpose is the Maximum A Posteriori (MAP) criterion which maximizes the a posteriori probability:

$$\hat{Y}_{MAP} = arg \max_{Y'} P(Y'), \tag{10}$$

which corresponds to the following cost function:

$$R_1(Y, Y') = 1 - \Delta(Y, Y'), \tag{11}$$

i.e. $R_1(Y, Y')$ equal to 0 if $Y = Y'$ and to 1 otherwise.

To obtain the MAP criterion, we run a simulated annealing scheme using a Metropolis dynamic [2]:

ALGORITHM 1

1 Initialize a random configuration $Y = (y_s)$, set $T = T_0$
2 For each site s:
  2.a Choose a random value new different from the current value cur
  2.b Compute the local energies $U(y_s = cur|\mathcal{N}_s)$ and $U(y_s = new|\mathcal{N}_s)$
  2.c If $U(y_s = new|\mathcal{N}_s) < U(y_s = cur|\mathcal{N}_s)$ set y_s to new, otherwise set y_s to
    new with probability $\exp -[\frac{\Delta U}{T}]$ where
    $\Delta U = U(y_s = new|\mathcal{N}_s) - U(y_s = cur|\mathcal{N}_s)$.
3 If the stopping criterion is not reached decrease T and go to 2

To adhere to the theoretical properties of convergence we should consider a logarithmic decrease of the temperature. However, to obtain a faster algorithm we have considered a linear decrease of rate 0.95. For the kind of energy functions we use, this decrease rate is slow enough to achieve a good solution.

The aim of the prior model is to regularize the solution, i.e. to avoid noise in the segmented image. In case of severe noise or when two classes are strongly mixed, the prior becomes preponderating the likelihood that leads to unreliable results. This may introduce some artifact due to the prior in the segmented image. Among them is the loss of small objects [9, 10]. To overcome this problem we propose to add an information which consists in keeping the proportionality ratios of the different classes. In that way, the volume (number of pixels) of each class will be constant during the optimization. In this section, we suppose that these ratios are known.

We consider the criterion defined by the following cost function:

$$R_2(Y, Y') = 1 - \Delta(Y, Y')\Delta(\gamma', \gamma_0), \tag{12}$$

where γ' is a vector containing the proportionality cœfficients of the different classes in the configuration Y' ($\gamma'(i) = \frac{number\ of\ pixels\ such\ that\ y'_s = i}{total\ number\ of\ pixels}$), and γ_0 represents the class ratios of the expected solution which are supposed to be known. Notice that the defined function is symmetric as $(Y = Y') \Rightarrow (\gamma = \gamma')$.

To obtain the associated Bayesian criterion we run a simulated annealing algorithm using the Kawasaki dynamic. The Metropolis dynamic is based on a spin-flip procedure which consists in changing the spin value of a site (the label of a pixel). The Kawasaki dynamic is based on a spin-exchange procedure which consists in exchanging the spin values of two sites. The induced algorithm is written as follows:

ALGORITHM 2

1 Initialize a random configuration $Y = (y_s)$ with $\gamma = \gamma_0$, set $T = T_0$
2 During N iterations:
  2.a Choose randomly two sites s and s', denote cur_s (resp. $cur_{s'}$) the current value of y_s (resp. $y_{s'}$)
  2.b Compute the local energies $U(y_s = cur_s|\mathcal{N}_s)$, $U(y_{s'} = cur_{s'}|\mathcal{N}_{s'})$, $U(y_s = cur_{s'}|\mathcal{N}_s)$ and $U(y_{s'} = cur_s|\mathcal{N}_{s'})$

2.c If $U(y_s = cur_{s'}|\mathcal{N}_s) + U(y_{s'} = cur_s|\mathcal{N}_{s'}) < U(y_s = cur_s|\mathcal{N}_s) + U(y_{s'} = cur_{s'}|\mathcal{N}_{s'})$ set y_s to $cur_{s'}$ and $y_{s'}$ to cur_s, otherwise set y_s to $cur_{s'}$ and $y_{s'}$ to cur_s with probability $\exp -[\frac{\Delta U}{T}]$ where
$\Delta U = U(y_s = cur_{s'}|\mathcal{N}_s) + U(y_{s'} = cur_s|\mathcal{N}_{s'})$
$-U(y_s = cur_s|\mathcal{N}_s) + U(y_{s'} = cur_{s'}|\mathcal{N}_{s'})$.

3 If the stopping criterion is not reached decrease T and go to 2

2.3 Experiments

We consider the synthetic image shown on figure 1.a. A Gaussian noise is added to this image. We consider severe noise corresponding to a Signal to Noise Ratio (SNR) equal to $-7.5dB$ (figure 1.b) and $-11dB$ (figure 1.c). A low value of the prior distribution parameter β leads to noisy results for both the Metropolis (figures 2.a and 5.a) and the Kawasaki dynamic (figures 2.d and 5.d). However, the results obtained with the Kawasaki dynamic are closer to the expected results. When increasing the value of β the regularization is increased. For high values of β the segmentation tends to "forget" the data and the prior distribution induces artifacts, especially for low SNR (figures 5.c and 5.f). A good compromise must be found. For low SNR, such a compromise cannot be found with the Metropolis dynamic. Imposing the class ratios avoids the mixing between objects. We can notice that, in the Kawasaki dynamic experiments, some isolated pixels are still misclassified. In fact, to have a fair comparison we have set the same number of iterations for all experiments. The convergence with the Kawasaki dynamic is slower than with the Metropolis dynamic.

3   A new Bayesian cost function

3.1   Algorithm

From the previous section, it appears that fixing the ratios between the classes improves the segmentation especially in case of low signal to noise ratio. However, in practice these parameters are not known. Therefore we propose to estimate them in a first step. Then, we cannot use the Kawasaki dynamic as it stands because we only have some estimates of the class ratios. Some deviations of the class ratios from these estimates must be allowed. Therefore, we define a new Bayesian cost function which penalize configurations with class ratios different from the estimates. However, this penalty is not "hard" in the sense that small deviations from the initial estimates are not highly penalized.

The proposed criterion is defined by the following cost function:

$$R_3(Y,Y') = 1 - \Delta(Y,Y')f(||\gamma' - \hat{\gamma_0}||)f(||\gamma - \hat{\gamma_0}||), \tag{13}$$

where $\hat{\gamma_0}$ represents the estimated class ratios and $f(.)$ is a decreasing function on $[0,\infty)$ taking its values in $[0,1]$ with maximum at 0. The Kawasaki dynamic is obtained for $f(||\gamma' - \hat{\gamma_0}||) = \Delta(\gamma', \gamma_0)$.

To optimize the posterior distribution with respect to this criterion we have to compute:

$$\hat{Y}_3' = arg\min_{Y'} \int_\Omega R_3(Y',Y)P(Y|X)dY, \tag{14}$$

$$= arg\min_{Y'} \int_\Omega [1 - \Delta(Y,Y')f(||\gamma' - \hat{\gamma_0}||)f(||\gamma - \hat{\gamma_0}||)]\, P(Y|X)dY,$$

$$= arg\max_{Y'} f^2(||\gamma' - \hat{\gamma_0}||)P(Y'|X),$$

$$= arg\max_{Y'} \frac{1}{Z} \exp\left[-U(X|Y') - U(Y') + \log f^2(||\gamma' - \hat{\gamma_0}||)\right],$$

$$= arg\max_{Y'} exp - [U(X|Y') + U(Y') - 2*\log f(||\gamma' - \hat{\gamma_0}||)].$$

Therefore, the defined criterion is equivalent to the MAP criterion using the prior:

$$P(Y') \propto \exp - [U(Y') - 2*\log f(||\gamma' - \hat{\gamma_0}||)]. \tag{15}$$

This new Bayesian cost function is equivalent to add a prior which favorizes the configurations having class ratios close to the estimated one. To reach this criterion, we can use a simulated annealing scheme with a Metropolis dynamic on this new prior:

ALGORITHM 3

1 *Estimate the class ratios* $\hat{\gamma_0}$
2 *Initialize a random configuration* $Y = (y_s)$, *set* $T = T_0$
3 *For each site* s:
    3.a *Choose a random value new different from the current value cur*
    3.b *Compute the local energies* $U(y_s = cur|\mathcal{N}_s)$ *and* $U(y_s = new|\mathcal{N}_s)$ *and the corresponding class ratios* γ_{cur} *and* γ_{new}
    3.c *If* $U(y_s = new|\mathcal{N}_s) - 2*\log f(||\gamma_{new} - \hat{\gamma_0}||) < U(y_s = cur|\mathcal{N}_s) - 2*\log f(||\gamma_{cur} - \hat{\gamma_0}||)$ *set* y_s *to new, otherwise set* y_s *to new with probability* $\exp - [\frac{\Delta U}{T}]$ *where*
    $\Delta U = U(y_s = new|\mathcal{N}_s) - 2*\log f(||\gamma_{new} - \hat{\gamma_0}||)$
    $-U(y_s = cur|\mathcal{N}_s) + 2*\log f(||\gamma_{cur} - \hat{\gamma_0}||).$
4 *If the stopping criterion is not reached decrease* T *and go to 3*

For the simulations, we have considered the following function:

$$f(||\gamma - \hat{\gamma_0}||) \propto \exp - \frac{||\gamma - \hat{\gamma_0}||^2}{2\sigma^2}. \tag{16}$$

In this paper, we have considered that the mean and variance of each class are known. The different classes are supposed to be Gaussian. Therefore the histogram of the data is written as follows:

$$h(x) = \sum_{l \in \Lambda} \gamma_0(l) \frac{1}{\sqrt{2\pi\sigma_l^2}} \exp{-\frac{(x - \mu_l)^2}{2\sigma_l^2}}. \tag{17}$$

Denote the vectors $H = (h(j))$, $\Gamma = (\gamma_0(l))$ and the matrix $P = (p_{lj})$ where $p_{lj} = \frac{1}{\sqrt{2\pi\sigma_l^2}} \exp{-\frac{(j-\mu_l)^2}{2\sigma_l^2}}$. Then we have:

$$H = \Gamma P. \tag{18}$$

The class ratios are then estimated by a singular value decomposition computing:

$$\Gamma = HP^{-1}. \tag{19}$$

Remark that, if we do not know the mean and variance of the classes, we can fit a sum of Gaussian to the histogram using the Levenberg-Marquardt method [11].

3.2 Experiments

For the first experiment (SNR $= -7.5dB$), the modified Bayesian cost function leads to results very similar to those obtained with the Kawasaki dynamic, for both real (figures 3.a and 3.b) and estimated (figures 3.d and 3.e) class ratios. We can remark that here the convergence is as fast as with the Metropolis dynamic and we do not get misclassified isolated pixels. With a SNR equal to $-11dB$, small deviations from the class ratios produce some perturbations on the results even with the real class ratios (figures 6.b and 6.c). However, the results are still much better than with the original Metropolis dynamic.

4 Mixing the Metropolis and the Kawasaki dynamics

4.1 Algorithm

Finally, we propose to mix the Metropolis and the Kawasaki dynamics. Using a simulated annealing scheme we compute alternatively long sequences of iterations using the Kawasaki dynamics and short sequences of the Metropolis dynamic. The Kawasaki dynamic sequences allows us to optimize the a posteriori probability at constant class ratios. The short Metropolis dynamic sequences provide some variations from the estimated class ratios. The algorithm is written as follows:

ALGORITHM 4

1 Estimate the class ratios $\hat{\gamma}_0$
2 Initialize a random configuration $Y = (y_s)$ with $\gamma = \hat{\gamma}_0$, set $T = T_0$
3 During αN iterations:
   3.a Choose randomly two sites s and s', denote cur_s (resp. $cur_{s'}$) the current value of y_s (resp. $y_{s'}$)

3.b Compute the local energies $U(y_s = cur_s|\mathcal{N}_s)$, $U(y_{s'} = cur_{s'}|\mathcal{N}_{s'})$, $U(y_s = cur_{s'}|\mathcal{N}_s)$ and $U(y_{s'} = cur_s|\mathcal{N}_{s'})$

3.c If $U(y_s = cur_{s'}|\mathcal{N}_s) + U(y_{s'} = cur_s|\mathcal{N}_{s'}) < U(y_s = cur_s|\mathcal{N}_s) + U(y_{s'} = cur_{s'}|\mathcal{N}_{s'})$ set y_s to $cur_{s'}$ and $y_{s'}$ to cur_s, otherwise set y_s to $cur_{s'}$ and $y_{s'}$ to cur_s with probability $\exp -[\frac{\Delta U}{T}]$ where
$\Delta U = U(y_s = cur_{s'}|\mathcal{N}_s) + U(y_{s'} = cur_s|\mathcal{N}_{s'})$
$-U(y_s = cur_s|\mathcal{N}_s) + U(y_{s'} = cur_{s'}|\mathcal{N}_{s'})$.

4 During $(1 - \alpha)N$ iterations:

4.a Choose randomly a site s and a random value new different from the current value cur

4.b Compute the local energies $U(y_s = cur|\mathcal{N}_s)$ and $U(y_s = new|\mathcal{N}_s)$

4.c If $U(y_s = new|\mathcal{N}_s) < U(y_s = cur|\mathcal{N}_s)$ set y_s to new, otherwise set y_s to new with probability $\exp -[\frac{\Delta U}{T}]$ where
$\Delta U = U(y_s = new|\mathcal{N}_s) - U(y_s = cur|\mathcal{N}_s)$.

5 If the stopping criterion is not reached decrease T and go to 3

4.2 Experiments

The mixing between the Metropolis and the Kawasaki dynamics leads to results similar to those obtained with the new Bayesian cost function defined in the previous section, especially for a SNR equal to $-7.5dB$ (figures 4.b, 4.c, 4.e and 4.f). For a SNR equal to $-11dB$, the results are better with the real class ratios (figures 7.b and 7.c) but they are less robust with respect to the bias in the class ratios estimation (figures 7.e and 7.f).

5   Conclusion

Markov Random Fields (Gibbs Fields) are widely used in image segmentation. Using a Bayesian framework allows us to include prior information on the expected result which turns to regularize the solution. However, in case of severe noise or low SNRs, these models show their limits. One field of investigation is to defined more efficient prior distributions [12, 9, 13]. In this paper, we have investigated another aspect of the problem. Once the model is defined, we have to define a Bayesian criterion and a dynamic to optimize the posterior distribution with respect to this criterion. The most used criteria are the MAP and the MPM which can be reached using a Metropolis dynamic. We have shown the limit of the Potts model within these dynamic and the improvement which can be expected by using a Kawasaki dynamic. This dynamic considers configurations with given class ratios. Because the class ratios are not known in a practical application, we have to use an estimation of these quantities. Therefore, we have proposed some adaptations of the Kawasaki dynamic to tackle the problem of the bias in the class ratios estimation. The defined Bayesian cost function appears to be a tool to incorporate some prior knowledge on the expected result.

The first approach consists in generalizing the Bayesian cost function associated with the Kawasaki dynamics. The derivation of an optimal cost function is an open problem. The second approach consists in mixing the Metropolis and the Kawasaki dynamics. Here also, the optimal mixing is an open problem. We have shown the relevance of these approaches on synthetic data. We are currently investigating some applications on real data such as SAR or sonar images which are characterized by a strong noise. Besides, this work motivates researchs on accurate estimators of the class ratios. We currently investigate the EM algorithm.

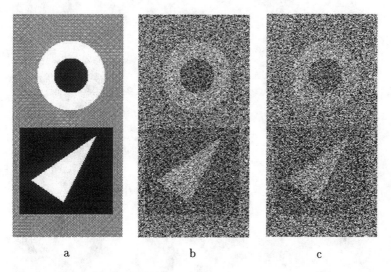

a   b   c

Fig. 1. Synthetical image (a), noisy version with SNR = -7.5dB (b) and -11dB (c)

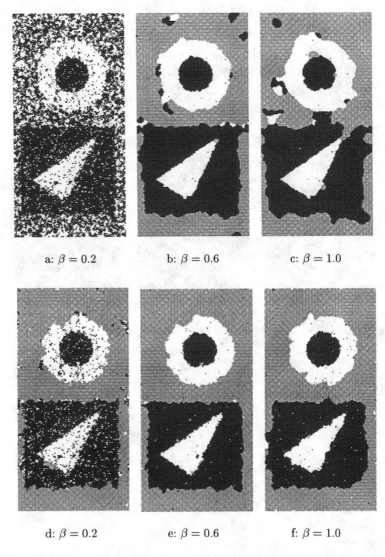

a: $\beta = 0.2$ b: $\beta = 0.6$ c: $\beta = 1.0$

d: $\beta = 0.2$ e: $\beta = 0.6$ f: $\beta = 1.0$

Fig. 2. Results with the Metropolis dynamic (a,b,c) and the Kawasaki dynamic (d,e,f) on figure 1.b (SNR = -7.5dB)

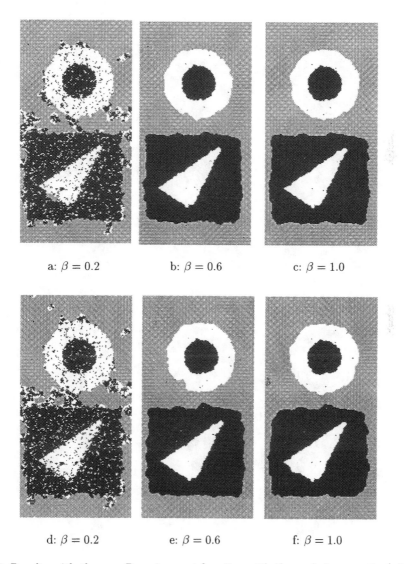

a: $\beta = 0.2$ b: $\beta = 0.6$ c: $\beta = 1.0$

d: $\beta = 0.2$ e: $\beta = 0.6$ f: $\beta = 1.0$

Fig. 3. Results with the new Bayesian cost function with the real classes ratio (a,b,c) and estimated class ratio (d,e,f) on figure 1.b (SNR = -7.5dB)

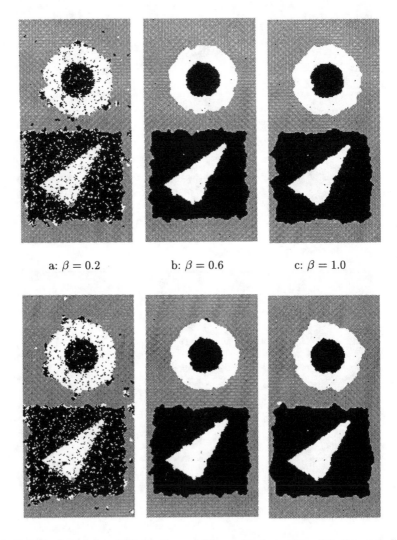

a: $\beta = 0.2$ b: $\beta = 0.6$ c: $\beta = 1.0$

Fig. 4. Results by mixing Metropolis and Kawasaki dynamics with the real classes ratio (a,b,c) and estimated class ratio (d,e,f) on figure 1.b (SNR = -7.5dB)

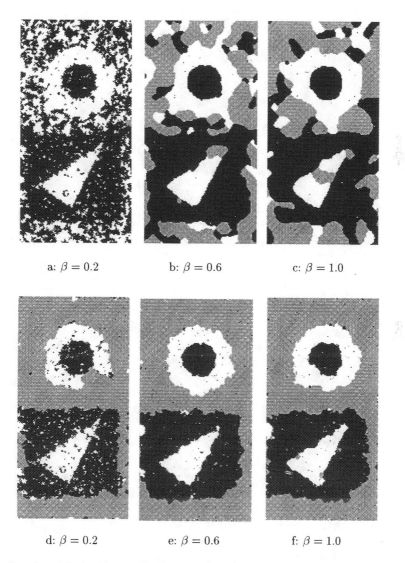

a: $\beta = 0.2$ b: $\beta = 0.6$ c: $\beta = 1.0$

d: $\beta = 0.2$ e: $\beta = 0.6$ f: $\beta = 1.0$

Fig. 5. Results with the Metropolis dynamic (a,b,c) and the Kawasaki dynamic (d,e,f) on figure 1.c (SNR = -11dB)

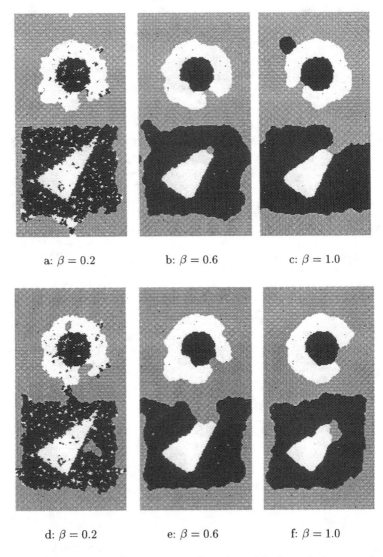

a: $\beta = 0.2$ b: $\beta = 0.6$ c: $\beta = 1.0$

d: $\beta = 0.2$ e: $\beta = 0.6$ f: $\beta = 1.0$

Fig. 6. Results with the new Bayesian cost function with the real classes ratio (a,b,c) and estimated class ratio (d,e,f) on figure 1.c (SNR = -11dB)

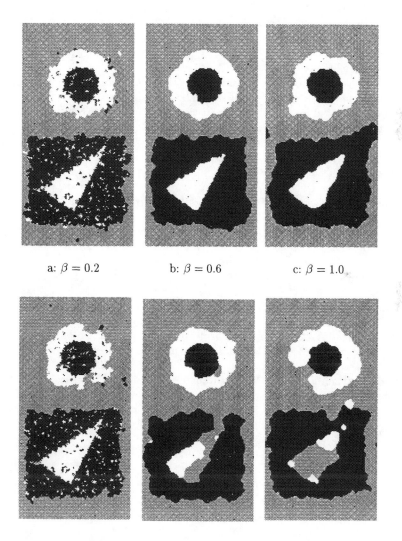

a: $\beta = 0.2$ b: $\beta = 0.6$ c: $\beta = 1.0$

Fig. 7. Results by mixing Metropolis and Kawasaki dynamics with the real classes ratio (a,b,c) and estimated class ratio (d,e,f) on figure 1.c (SNR = -11dB)

References

1. R. Chellappa, A. Jain. *Markov random fields : theory and application.* Academic Press, Inc., 1993. Collective work.
2. S. Geman, D. Geman. Stochastic relaxation, Gibbs distribution, and the Bayesian restoration of images. *IEEE trans. on Pattern Analysis and Machine Intelligence,* 6(6):721–741, 1984.
3. H.O. Georgii. *Gibbs Measures and Phase Transitions.* De Gruyter - Studies in Mathematics, 1988. Vol. 9.
4. A. Maruani, E. Pechersky, M. Sigelle. On Gibbs fields in image processing. *Markov Processes and Related Fields,* 1:419–442, 1995.
5. B. Prum. *Processus sur un réseau et mesures de Gibbs (in French),* chapter 9, Dynamic of Ising systems, pages 150–162. Masson, 1986.
6. J. Besag. Spatial interaction and statistical analysis of lattice systems. *Journal of the Royal Statistical Society Series B,* 36:721–741, 1974.
7. H. Derin, H. Elliott. Modelling and segmentation of noisy and textured images using Gibbs random fields. *IEEE trans. on Pattern Analysis and Machine Intelligence,* 9(1):39–55, January 1987.
8. S. Lakshmanan, H. Derin. Simultaneous parameter estimation and segmentation of Gibbs random fields using simulated annealing. *IEEE trans. on Pattern Analysis and Machine Intelligence,* 11(8):799–813, August 1989.
9. X. Descombes, J.F. Mangin, E. Pechersky, M. Sigelle. Fine structures preserving model for image processing. In *Proc. 9th SCIA 95 Uppsala, Sweden,* pages 349–356, 1995.
10. R.D. Morris, X. Descombes, J. Zerubia. The Ising/Potts model is not well suited to segmentation tasks. In *proc. IEEE Digital Signal Processing Worshop,* 1996. Sept. 1-4 1996 , Norway.
11. W. Press, S. Teukolski, W. Vetterling, B. Flannery. *Numerical Recipes in C: The Art of Scientific Computing.* Cambridge University Press, 2nd ed., 1992.
12. G. Wolberg, T. Pavlidis. Restoration of binary images using stochastic relaxation with annealing. *Pattern Recognition Letters,* 3:375–388, 1985.
13. H. Tjelmeland, J. Besag. Markov Random Fields with higher order interactions. submitted to JASA. Preprint.

Hyperparameter Estimation for Satellite Image Restoration by a MCMCML Method

André Jalobeanu, Laure Blanc-Féraud, Josiane Zerubia *

The authors of this paper are part of Ariana, a joint research group between CNRS, INRIA and UNSA, located at INRIA, BP 93, 06902 Sophia Antipolis Cedex, France. L. Blanc-Féraud is also with Laboratoire I3S, CNRS-UNSA.
web : `www.inria.fr/ariana`
mail : `Firstname.Lastname@inria.fr`

Abstract. Satellite images can be corrupted by an optical blur and electronic noise. Blurring is modeled by convolution, with a known linear operator H, and the noise is supposed to be additive, white and Gaussian, with a known variance. The recovery problem is ill-posed and therefore must be regularized. Herein, we use a regularization model which introduces a φ-function, avoiding noise amplification while preserving image discontinuities (i.e. edges) of the restored image. This model involves two hyperparameters. Our goal is to estimate the optimal parameters in order to reconstruct images automatically.

In this paper, we propose to use the Maximum Likelihood estimator, applied to the observed image. To evaluate the derivatives of this criterion, we must estimate expectations by sampling (samples are extracted from a Markov chain). These samples are images whose probability takes into account the convolution operator. Thus, it is very difficult to obtain them directly by using a standard sampler. We have developed a new algorithm for sampling, using an auxiliary variable based on Geman-Yang algorithm, and a cosine transform. We also present a new reconstruction method based on this sampling algorithm. We detail the Markov Chain Monte Carlo Maximum Likelihood (MCMCML) algorithm which ables to simultaneously estimate the parameters, and to reconstruct the corrupted image.

1 Introduction

The problem of reconstructing an image X from noisy and blurred data Y is ill-posed in the sense of Hadamard [18]. Knowing the degradation model is not sufficient to obtain satisfying results : it is necessary to regularize the solution by introducing *a priori* constraints [6,8]. Y is defined by $Y = HX + N$. When the blur operator H (corresponding to the Point Spread Function h) and the variance σ^2 of the Gaussian additive noise N are known, the Maximum Likelihood (ML) estimate of X consists in searching for the image which minimizes the energy $U_0(X) = ||Y - HX||^2/2\sigma^2$. The regularization constraint is expressed

---

* This work has been conducted in relation with the GdR ISIS (CNRS).

through a function $\Phi(X)$ added to U_0, which represents a roughness penalty on X. This function could be quadratic (as suggested by Tikhonov in [30]), assuming that images are globally smooth, but it yields oversmooth solutions. A more efficient image model assumes that only homogeneous regions are smooth, and edges must remain sharp. To get this *edge-preserving* regularization, we use a *non-quadratic* φ-function, as introduced in [5] and [12]. Properties of the φ-function have been studied in a variational approach in order to preserve the edges, avoiding noise amplification. We use a convex φ-function, ensuring the uniqueness of the solution, so that restoration is made by a deterministic minimization algorithm.

The non-quadratic variational regularization model involves two hyperparameters. The smooth properties of reconstructed images depend strongly on their value and therefore they must be accurately determined, excluding an empirical estimation. We use a stochastic approach, based on the *Maximum Likelihood estimator*. The regularizing model is interpreted as a Markov Random Field (MRF). There are two difficulties with the estimation:

- The probability to be maximized is non concave.
- To compute the derivative of the likelihood function according to parameters, we need to sample from the prior and posterior image densities. Due to the convolution, it is impossible to directly sample images from the posterior law.

We propose in this paper to use a stochastic method to estimate the parameters, so we first make a statistical interpretation of the regularization contstraint, in a Bayesian framework. Then we explain why we choose the Maximum Likelihood estimator, and we focus on its gradient computation. We use a Monte Carlo method, which needs to sample images from the prior and posterior laws. Sampling is achieved by transforming these laws into Gaussian laws: this is possible by using a half quadratic extension of the φ-function as Geman & Yang [13], and by working in the cosine transform (DCT) space. We study some problems raised by sampling. Then we detail the MCMCML (Markov Chain Monte Carlo Maximum Likelihood) estimation algorithm, and study its convergence.

2 Problem statement

In this paper, X, Y and N are vectors of dimension $(N_x N_y)$, made from the pixels of an image in a lexicographic order. We denote $X_{i,j}$ the value of the pixel at column i and line j.

The degradation model is represented by the equation $Y = H\mathcal{X} + N$, where Y is the observed data, and $\mathcal{X}$ the original image. N is the additive noise and is supposed to be Gaussian, white and stationary. H is the convolution operator. The PSF h is positive, symmetric with respect to lines and columns, and verifies the Shannon property. H is a block-circulant matrix generated by h.

The noise standard deviation σ and the PSF h are known (see fig.6).

The regularized solution is computed by minimizing the energy

$$U(X) = ||Y - HX||/2\sigma^2 + \Phi(X) \tag{1}$$

where Φ is defined by $\Phi(X) = \lambda^2 \sum_{ij} \left\{ \varphi\left(\frac{(D_x X)_{ij}}{\delta}\right) + \varphi\left(\frac{(D_y X)_{ij}}{\delta}\right) \right\}$

λ and δ are the *hyperparameters*, D_x and D_y are first derivative discrete operators on columns and lines. φ is called a *potential function*, which has to exhibit boundary preserving properties [5]. φ is a symmetric, positive and increasing function, with a quadratic behaviour near 0 to smooth isotropically homogeneous areas and linear or sub-linear behaviour near ∞ to preserve high gradients (edges). We use the "hyper surfaces" convex function $\varphi(u) = 2\sqrt{1 + u^2} - 2$ [5].

2.1 Hyperparameter choice

The quality of the reconstructed solution depends on the hyperparameter choice. λ is a parameter which weights the regularization term versus the data term. Too high values of λ yield oversmooth solutions, and too small values lead to noisy images. The δ parameter is related to a threshold below which the gradients (due to the noise) are smoothed, and above which they are preserved. A high value of this threshold filters the edges as well as the noise, that yields over-regularized images. On the other hand, a small δ value provides insufficient noise filtering.

An empirical hyperparameter choice is very difficult, because there are two degrees of freedom. That is why we propose to do an automatic estimation to choose the optimal hyperparameters.

2.2 Bayesian interpretation of restoration problems

We propose to use a statistical estimator to choose the optimal hyperparameters. So we need a stochastic interpretation of the criterion defined by equation (1). In a Bayesian framework, the posterior probability of an image X observing the data Y is given by

$$P(X \mid Y) = P(Y \mid X) P(X) / P(Y) \tag{2}$$

where $P(Y)$ is a constant w.r.t. X. In this paper we write $P(X)$ instead of $P(x = X)$, where x denotes the multidimensional $(N_x N_y)$ random variable. To evaluate equation (2), we first have to express the probability of obtaining a corrupted data Y from an original image X. $P(Y \mid X)$ follows the distribution of the Gaussian noise, independent of X:

$$P(Y \mid X) = \frac{1}{K_\sigma} e^{-\frac{||N||^2}{2\sigma^2}} = \frac{1}{K_\sigma} e^{-\frac{||Y - HX||^2}{2\sigma^2}} \tag{3}$$

where K_σ is a normalizing constant. $P(Y \mid X)$ is called the likelihood of X.

$P(X)$ is the prior probability, i.e. the prior knowledge of the reconstructed solution, which defines the regularization model.

X follows a Gibbsian distribution:

$$P(X) = \frac{1}{Z} e^{-\Phi(X,\lambda,\delta)} \qquad (4)$$

Then X is a Markov Random Field (MRF), according to the Hammersley-Clifford theorem.

Finally, the posterior probability is

$$P(X \mid Y) = \frac{1}{Z_Y} e^{-\frac{\|Y - HX\|^2}{2\sigma^2} - \Phi(X,\lambda,\delta)} \qquad (5)$$

Z and Z_Y are normalizing terms (partition functions), which depend on the hyperparameters (λ, δ):

$$Z = \int_\Omega e^{-\Phi(X,\lambda,\delta)} \, dX$$
$$Z_Y = \int_\Omega e^{-\|Y - HX\|/2\sigma^2 - \Phi(X,\lambda,\delta)} \, dX$$

Ω is the state space (set of $N_x \times N_y$ size images with real pixels, hence Ω is of dimension $[-m, m]^{N_x N_y}$ where m is a fixed bound). These integrals are well-defined because the state space is bounded.

Minimizing $U(X)$ in (1) is equivalent to maximixe the posterior probability in (5) (MAP criterion). Therefore, the optimization can be done by either deterministic or stochastic algorithms. The former usually needs the unicity of the energy minimizer, but the latter (like simulated annealing [22]) works better with multiple energy minima. The considered energy is convex so we use a deterministic algorithm.

3   Hyperparameter estimation

3.1   Introduction

Hyperparameter estimation is a difficult problem, but is needed when using a parametric restoration algorithm, whose results' quality strongly depend on their value.

At first, we present a few estimators generally used for parameter estimation. We then explain why we prefer to use the last one, although it is very hard to implement.

If we consider the prior probability $P(X \mid \lambda, \delta)$, the ML estimator on λ and δ can be expressed as

$$(\hat{\lambda}, \hat{\delta}) = \arg\max_{\lambda,\delta} P(X \mid \lambda, \delta) = \arg\max_{\lambda,\delta} \frac{1}{Z} e^{-\Phi(X,\lambda,\delta)} \qquad (6)$$

Hyperparameters estimated this way make sense w.r.t. the *complete data* case, i.e. not corrupted by blur and noise. This is the case for a classification problem, as seen in [9] or [25].

For image reconstruction, hyperparameters depend on the type of degradation (variance of noise and size of H). This dependence is impossible if they are estimated from the original image (supposing that we know it).

Lakshmanan and Derin [23] proposed to use the following criterion:

$$(\hat{X}, \hat{\lambda}, \hat{\delta}) = \arg \max_{X, \lambda, \delta} P(X, Y \mid \lambda, \delta) \tag{7}$$

which involves the joint law of X and Y. To get the optimum the authors proposed to use the Generalized Maximum Likelihood (GML) algorithm. This criterion allows to use an approximate optimization technique, consisting of alternate maximizations on $\hat{X}$ and $(\hat{\lambda}, \hat{\delta})$. This method is suboptimal, but is fast and simplifies the problem. More recently, the same type of criterion has been used in [7], and in [21]. It is equivalent to the MAP estimate for X when λ and δ are fixed, and the parameters reduce to the ML estimator in the complete data case. In fact, the ML estimator is applied to the solution reconstructed with the current $(\hat{\lambda}, \hat{\delta})$. The convergence of this algorithm is not guaranteed and degenerated solutions can be found. As in the previous case, sampling is only needed from the prior model, without respect to the data. Therefore, this estimator does not seem to be appropriate to our problem, that is why we have chosen the following criterion.

3.2 The Maximum Likelihood Estimator

L. Younes [31] [32] proposed to use the ML estimator computed with the probability of the observed data knowing the λ and δ parameters:

$$(\hat{\lambda}, \hat{\delta}) = \arg \max_{\lambda, \delta} P(Y \mid \lambda, \delta) \tag{8}$$

As Y is a fixed observation, this probability depends only on the hyperparameters. To calculate (8), the joint distribution $P(X, Y)$ is integrated on X, then Bayes law is used to reduce to prior and posterior distributions. We finally obtain:

$$P(Y \mid \lambda, \delta) = \int_{\Omega} P(Y \mid X, \lambda, \delta) \, P(X \mid \lambda, \delta) \, dX \tag{9}$$

which leads to $P(Y \mid \lambda, \delta) = Z_Y \, / \, Z \, K_\sigma$, where Z and Z_Y are respectively the partition functions related to the prior and posterior distributions (see section 2.2), and $K_\sigma = \int_{\Omega} P(Y \mid X, \lambda, \delta) \, dX$.

The main difficulty of parameter estimation comes from these partition functions, which depend on (λ, δ) but are impossible to evaluate. So we optimize this criterion without explicitly computing Z and Z_Y, as we only compute its derivatives, by using a gradient descent algorithm.

The criterion is redefined as $J(\lambda, \delta) = -\log P(Y \mid \lambda, \delta)$ which yields

$$J(\lambda, \delta) = \log Z_Y(\lambda, \delta) - \log Z(\lambda, \delta) + \text{const.} \tag{10}$$

where the constant (w.r.t. λ and δ) is $\log K_\sigma$.

3.3 Gradient estimation

To optimize the log-likelihood (10), we should evaluate its derivatives w.r.t. the hyperparameters. Refer to [20] for more detailed calculus. These calculus are based on the following property:

PROPERTY 1 (LOG Z DERIVATIVE)
Let Z denote the partition function related to the prior distribution $P(X \mid \lambda, \delta) = Z^{-1} e^{-\Phi(X)}$ *and* $\theta = (\lambda, \delta)$; *then we have*

$$\frac{\partial \log Z}{\partial \theta^i} = -E_{X \sim P(X \mid \theta)} \left[\frac{\partial \Phi(X \mid \theta)}{\partial \theta^i} \right] \tag{11}$$

where $E[\,]$ *is the expectation of* X *w.r.t.* $P(X \mid \theta)$.

All the derivative evaluations are done by estimation of expectations, which could be approximated by the empirical mean estimate

$$E_{X \sim P(X)} \left[\frac{\partial \Phi(X)}{\partial \theta^i} \right] \simeq \frac{1}{N} \sum_{X^k \sim P(X)} \frac{\partial \Phi(X^k)}{\partial \theta^i} \tag{12}$$

where X^k is the kth vector of the chain (X^n) and is sampled from the law $P(X \mid \lambda, \delta)$. N is the number of samples X^k. The first derivative of the criterion (10), for each component of θ, is given by

$$\nabla J_{\theta^i} = E_Y \left[\frac{\partial \Phi}{\partial \theta^i} \right] - E \left[\frac{\partial \Phi}{\partial \theta^i} \right] \tag{13}$$

which exhibits two types of expectation:

- $E[\,]$ expectation for the prior law $X \sim P(X) = Z^{-1} e^{-\Phi(X)}$
  without respect to the data (section 3.4)
- $E_Y[\,]$ expectation for the posterior law $X \sim P_Y(X) = Z_Y^{-1} e^{-\|Y - HX\|^2 - \Phi(X)}$
  with respect to the data (section 3.4).

In order to use a gradient-type descent algorithm, we need to compute (13) for various values of θ. So we need to sample from prior and posterior distributions to minimize the $-$log-likelihood. In the following section, we present an efficient method to do this.

3.4 The sampling problem

Sampling from the posterior density Sampling from the posterior density is intractable by means of classical algorithms such as Gibbs sampler [10] or Metropolis [24] due to the large support of the PSF, inducing a large neighborhood for the conditional probability. We use the idea introduced by Geman & Yang [13] to derive a simulated annealing algorithm for MAP estimation of X. Classical methods generally use an integer bounded space for pixels, and a finite state space. Our algorithm works on real bounded pixels, which is compatible with the reconstruction algorithm. Some samples are presented in figure 1.

The idea is to *diagonalize* the convolution operator, which is equivalent to replacing the convolution by a point to point multiplication, in the *frequency space*. The difficulty is to diagonalize at the same time the non quadratic regularization term. That is made possible by using the half-quadratic form of φ first developed in [13]:

$$\varphi(u) = \inf_{b \in \mathbb{R}} \left[(u-b)^2 + \psi(b) \right] \tag{14}$$

Two auxiliary variable fields B^x and B^y (vectors of the same size as X) are defined, so that $\Phi(X) = \inf_{B^x, B^y} \Phi^*(X, B^x, B^y)$, with

$$\Phi^*(X, B^x, B^y) = \lambda^2 \sum_{ij} \left\{ \left(B_{ij}^x - \frac{D_x X_{ij}}{\delta} \right)^2 + \psi(B_{ij}^x) + \left(B_{ij}^y - \frac{D_y X_{ij}}{\delta} \right)^2 + \psi(B_{ij}^y) \right\}$$

This half-quadratic expansion has been used in [1,5] for image recovery, by alternate minimizations w.r.t. X and B^x, B^y in order to compute the regularized solution, defining a convergent algorithm.

We assume that $P(X)$ is the marginal of $P(X, B^x, B^y)$:

$$\int_{\Omega_B \times \Omega_B} P(X, B^x, B^y) \, dB^x dB^y = P(X) \quad \text{where } \Omega_B = \mathbb{R}^{N_x N_y}. \tag{15}$$

In fact, $P(X)$ defined by (15) does not correspond to the $P(X)$ previously defined in (4) and (1) with the φ-function, but is an approximation involving a function $\tilde{\varphi}$. Indeed, $\varphi(u)$ is the minimum of $(b-u)^2 + \psi(b)$, but there is no reason for $P(X)$ to be the marginal of a joint density of X, B^x, B^y built with the half-quadratic expansion of φ.

$$\text{Let define } \tilde{\varphi} \text{ by } \int_{-\infty}^{+\infty} e^{-\lambda^2 [(b-u)^2 - \psi(b)]} \, db = e^{-\lambda^2 \tilde{\varphi}(u)}$$

This is always possible because the integral is positive, and it is convergent due to the linear behaviour of ψ at $+\infty$ and to the symmetry of ψ. We have computed a graph of $\tilde{\varphi}$ by a numerical method. There is so little difference between the graphs of φ and $\tilde{\varphi}$, that in practice we can use φ in the reconstruction algorithm, even if the estimated parameters correspond to a $\tilde{\varphi}$ regularizing function.

$$P(X, B^x, B^y) \propto e^{-\frac{||Y - HX||^2}{2\sigma^2} - \Phi^*(X, B^x, B^y)} \tag{16}$$

This allows Φ to be quadratic w.r.t. X if B^x and B^y are fixed, and the distribution to be "half-Gaussian" in the sense that the law of X given B^x, B^y is *Gaussian*, and the variance-covariance matrix is diagonalized by a Fourier transform (see [13]). Conversely, B^x and B^y are conditionally *independent* given X, so they are sampled in a single pass.

We modify the method proposed in [13] by using a DCT, which allows us to fulfill the symmetric boundary conditions, so we avoid producing artefacts on the borders of the image [20]. This is equivalent to using images 4 times larger, symmetrized w.r.t. rows and columns, but we process real pixels, unlike the original algorithm. Due to the symmetry in the frequency domain the number of processed pixels remains constant $(N_x N_y)$.

Let us denote H^4 the $(2N_x)^2 (2N_y)^2$ block-circulant matrix obtained by extension of H. This is the convolution operator applied on the $2N_x \times 2N_y$ symmetrized images. h^4 is the generator of H^4, composed of the first column of H^4. The Fourier transform $\mathcal{F}[h^4]$ of h^4 gives the eigenvalues of H^4. In the same way we define d_x^4 and d_y^4, generators of D_x^4 and D_y^4, derivative operators on the symmetrized images.

ALGORITHM 1 (POSTERIOR MODIFIED GEMAN-YANG)
We first compute $\mathcal{F}[h^4]$, $DCT[Y]$ and W defined by
$$W = \left(\frac{1}{2\sigma^2} |\mathcal{F}[h^4]|^2 + \frac{\lambda^2}{\delta^2} \left(|\mathcal{F}[d_x^4]|^2 + |\mathcal{F}[d_y^4]|^2 \right) \right)^{-1},$$
where $\mathcal{F}$ stands for the Fourier transform.
Set $X^0 = \hat{X}$ (reconstructed image with initial parameters).

To obtain X^{n+1} from X^n, repeat until the convergence criterion is satisfied:

- Sample B^x, B^y w.r.t. the gradients of X^n, using the law
  $P_g(b) \propto \exp\left(\lambda^2 \left[b(2g/\delta - b) - \psi(b) \right] \right)$;
- Generate an R image, whose pixels follow the Gaussian law $N(0, 1/2)$;
- Sample X^{n+1} in the frequency space:
  $X^{n+1} = DCT^{-1} \left\{ W \left(\frac{\lambda^2}{\delta} DCT \left[D_x^t B^x + D_y^t B^y \right] + \frac{1}{2\sigma^2} \mathcal{F}[h^4] DCT[Y] \right) + \sqrt{W} R \right\}$

An image reconstruction scheme (estimation of X knowing Y, λ, δ) has been derived from this sampler [20], whose advantages (compared to classical ones) are both the speed and the respect of the symmetric boundary conditions.

Sampling from the prior model The classical Gibbs or Metropolis samplers could be used to sample from the prior model. But we prefer to use a prior modified Geman-Yang sampler, derived from the previous posterior sampler, to improve the computing time of our estimation method.

Let us consider in the same manner the augmented stochastic process $P(X, B^x, B^y)$. To sample from this distribution, we must modify the posterior algorithm 1 in the following way (because we do not take into account the observed data anymore): replace W by $W_0 = \frac{\delta^2}{\lambda^2} \left(|\mathcal{F}[d_x^4]|^2 + |\mathcal{F}[d_y^4]|^2 \right)^{-1}$, and the sample X^{n+1} is given by $DCT^{-1} \left[\frac{\lambda^2}{\delta} W_0 DCT \left[D_x^t B^x + D_y^t B^y \right] + \sqrt{W_0} R \right]$.

The differences between Gibbs and Geman-Yang prior samples can be seen in figure 2: the second ones are more space varying, because they are sampled in the Fourier space. The number of iterations is the same for both samplers (see discussions on convergence studies below).

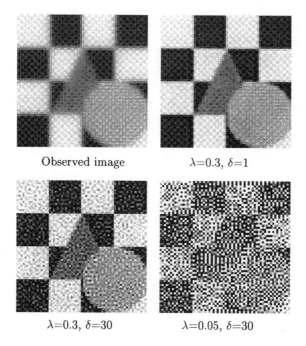

Observed image $\lambda=0.3$, $\delta=1$

$\lambda=0.3$, $\delta=30$ $\lambda=0.05$, $\delta=30$

Fig. 1. 64×64 samples of the posterior law (5)

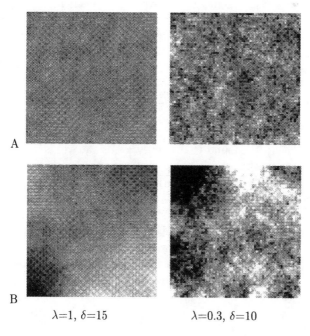

A

B

$\lambda=1$, $\delta=15$ $\lambda=0.3$, $\delta=10$

Fig. 2. (A) Gibbs and (B) modified Geman-Yang Prior samples, computed with 20 iterations

Convergence studies Some theoretical studies concerning Markov Chain convergence can be found in [26], [27] and [29]. We essentially retain that the possibility of transition between two states of the Ω state space is a sufficient convergence condition. For the augmented stochastic process $P(X, B^x, B^y)$, we ensure that for every realization of the random fields X and B^x, B^y, the transition probabilities $P(X \mid B^x, B^y)$ and $P(B^x, B^y \mid X)$ are strictly positive. The modified Geman and Yang methods we present herein are correct sampling methods, because $P(X, B^x, B^y)$ is an invariant measure as regards these two transitions. Indeed,

$$\forall\, X, X', B, B' \in \Omega :$$
$$P(X, B).P(X' \mid B) = P(X', B).P(X \mid B) \tag{17}$$
$$P(B, X).P(B' \mid X) = P(B', X).P(B \mid X)$$

To evaluate the speed of convergence of the algorithms we can only do experimental studies, because the Ω state space is not finite. The M first samples of the Markov Chain must be discarded, therefore a criterion must be chosen to determine M. In [4] are detailed several methods: some of them are based on the L^2 norm between the current and equilibrium distributions; others use the rejection rate of the Metropolis sampler. We prefer a faster method, which consists in using a plot of the energy $\Phi(X)$, and a preliminary study to determine the number of iterations which will be used during the estimation step.

A comparative convergence speed study has been made to prove the efficiency of our prior sampling method versus classical samplers. To do this, we estimate the energy of prior samples generated by Metropolis, by Gibbs and by the modified Geman-Yang sampler. Then we look how this energy is varying with the number of iterations, and we determine the number of iterations which is necessary to reach the equilibrum distribution. Although the state spaces of the samplers are not the same (real pixels for Geman-Yang and integer pixels for Gibbs and Metropolis), inducing a little difference between the generated images, we conclude that the modified Geman-Yang sampler is more interesting than the two others. We found that energy needs the same number of iterations (from 5 to 10) to reach a stable value, but the speed of each iteration depends much on the chosen method. There is a little difference between Gibbs and Geman-Yang samples (see fig. 2 for an illustration), Geman-Yang ones showing larger structures than the others. This difference has been detected over a large set of samples, and seems to disappear after a few hundreds of iterations, so the convergence seems to be slower for the Gibbs sampler. Classical samplers successively explore the pixels; our method is much faster, because the pixels are processed simultaneously in the frequency space, which enables long distance interactions at once.

We also studied the variation of the convergence speed with the hyperparameter value, and concluded that a phase transition [14] exists. As illustrated in figure 3, the energy of samples shows a great variation around $\lambda \simeq \lambda_c$ for a fixed δ and a nonconvex φ-function, that changes the aspect of the generated images. In the transition area, we remark that 10 more iterations are necessary to ensure convergence of the chain. It means the sampler oscillates between two kinds of states (constant for $\lambda > \lambda_c$ and noisy for $\lambda < \lambda_c$), before reaching an equilibrum state. For a non convex φ-function, the energy plot exhibits a discontinuity at λ_c, that is not the case for convex functions (like the one we use).

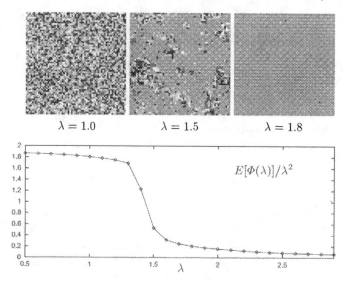

Fig. 3. Gibbs samples and plot of their Φ energy divided by λ^2, $\delta = 5$, with a nonconvex Geman & McClure φ-function

The main difficulty of convergence is due to the initialization. Practically, the speed of convergence depends strongly on the initial image. Obviously the best choice for the prior sampler is a constant image, whereas random initialization need thousands of iterations to produce the same result. Indeed, prior samples are generally small fluctuations around this constant. Therefore the best choice seems to be the maximizer of the sampled probability. In the same manner, the posterior sampler must be initialized by $\hat{X}$, the reconstructed image with the λ and δ hyperparameters. As seen in fig. 1, posterior samples are fluctuations around $\hat{X}$.

3.5   MCMCML algorithm

To optimize the likelihood criterion (10) we use a descent method.

ALGORITHM 2 (MCMCML)

- *Initialization: The ratio λ/δ corresponds to the best Wiener filter and $\delta = 7\sigma$.*
- *Compute $\hat{X}$: $\hat{X}$ is reconstructed from Y, with the current couple $\theta_n = (\lambda_n, \delta_n)$.*
- *Compute $E[\,]$ and $E_Y[\,]$ with θ_n: Generate 2 Markov Chains with the Modified Geman-Yang method, one sampled from the prior model, and another one sampled from the posterior distribution, and compute the respective empirical means (12).*

- *Iteration from θ_n to θ_{n+1} : $\theta_{n+1}^i = \theta_n^i - \alpha \nabla J_{\theta^i}(\theta_n^i) \left(\frac{\nabla J_{\theta^i}(\theta_n^i + \Delta\theta^i) - \nabla J_{\theta^i}(\theta_n^i)}{\Delta\theta^i} \right)^{-1}$
  where $\alpha < 1$ and $\Delta\theta^i$ is a step used to compute the second derivative.*
- *Stopping criterion: we stop the algorithm if $\frac{|\nabla J_{\theta^i}|}{\theta^i} < \epsilon$.*

Estimation and restoration processes are simultaneous, as the modified Geman-Yang sampler is initialized with $\hat{X}$.

Convergence and accuracy If we compute the Hessian matrix (see [20] for details) we see that this matrix depends on the data and the hyperparameters, so that the criterion cannot generally be concave w.r.t. θ. As experimental studies have shown, the criterion is *locally quadratic* w.r.t. θ close to the optimum. This enables to use a Newton-Raphson descent method to optimize θ. The second derivative is needed, and we evaluate it in an empirical way. To ensure the convergence we must choose α smaller than 1 if θ is far from the optimum.

The estimation accuracy only depends on the ϵ threshold and the expectation's accuracy, which is given by the size and the number of the samples needed for computing the means. An ϵ value corresponding to 3% estimation error or less has no visible effects on the reconstruction result. This accuracy is reached in less than 5 iterations for the image shown in figure 6.

The images shown in fig. 6 and the degradation model are provided by the French Space Agency (CNES); they simulate the future SPOT 5 satellite images. The algorithm presented herein has been successfully employed by Alcatel Space Industries to reconstruct real satellite data provided by the French defense agency. These images are confidential, so they cannot be shown here. For the CNES 512×512 image, the computation time is 13 s for estimation (made on a 64×64 area) and 12.6 s for reconstruction (Sun Ultra 1, 167 MHz).

Study of the algorithm: multiple solutions For each initial δ we can find a value of λ for which (λ, δ) is optimal. Figure 4 shows sets of local minima that cancel the criterion derivatives. So for each δ an optimal $\hat{\lambda}$ could be found. For large δ, the model is nearly quadratic, that reduces to a single λ/δ parameter model: this graph is also asymptotically linear.

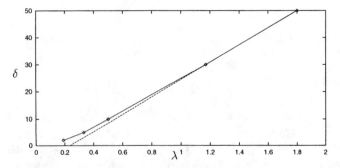

Fig. 4. Sets of local minima $(\hat{\lambda}, \hat{\delta})$ hyperparameters which cancel the log-likelihood derivatives ("hyper surfaces" φ-function), the dashed line corresponds to $\varphi(u) = u^2$

Then we fix the δ parameter, thus we only estimate λ by computing the ∇_λ likelihood derivative w.r.t. λ. The δ value is chosen in order to penalize the gradients due to noise and to preserve edges.

Good results are obtained with a fixed $\delta \simeq 7\sigma$. Changing δ does not influence the global SNR, but locally changes the restoration quality. High δ values improve edge reconstruction, while low values enable more efficient noise filtering in homogeneous areas (see [20]).

As we can see in figure 5 the SNR is optimal for the estimated $\hat{\lambda}$ value. The SNR is defined by a mean square difference between original and reconstructed images.

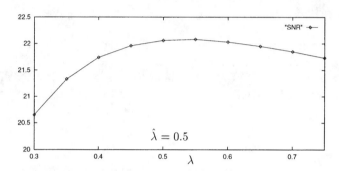

Fig. 5. SNR with λ varying around $\hat{\lambda}$ ($\delta = 10$)

Other possible algorithms The Generalized Stochastic Gradient (proposed by Younes [31], [32]) is another possible estimation method. Two Markov chains, (X_1^n) and (X_2^n), are sampled from prior and posterior laws. Compared to our method, the same criterion derivative is used, weighted by a factor to control convergence, therefore the solution is the same. The essential difference is that in the method proposed in [31] a *single* sample is needed for each descent step and all samples are kept, without trying to reach the equilibrum distribution.

A version of this algorithm was used in [21] and [33], where the posterior samples were replaced by the $\hat{X}$ restored image with the current hyperparameters, which avoids sampling the posterior law, and is equivalent to the Lakshmanan & Derin estimator mentionned in section 3.1.

4   Conclusion

The proposed algorithm is automatic and fast, and ables to simultaneously reconstruct images and estimate the parameters.

As we actually do not estimate δ because of the multiplicity of $(\hat{\lambda}, \hat{\delta})$, it seems necessary to introduce a priori knowledge of hyperparameters if we want the optimal couple to be unique. A possible choice of the $P(\lambda, \delta)$ law could be linked to the probability to obtain (λ, δ) for a large set of images.

If we want to get qualitatively better reconstructed solutions, we should take into account higher order derivatives. Second order models as in [13] seem to give better results. But this automatically introduces more parameters to be estimated, so more gradient criteria must be evaluated. That does not change the nature of the algorithm.

The chosen model is homogeneous: we assume the same (λ, δ) couple is convenient for the entire (large size) image. A better choice would be to divide the image into small areas on which we estimate the hyperparameters. Indeed, homogeneous areas would be better reconstructed if processed separately, because the λ value estimated on them is higher than the one which corresponds to edge areas. We are currently working on this problem.

Acknowledgements

The authors would like to thank the French Space Agency (CNES) for providing the image of Nîmes (SPOT 5 simulation), Marc Sigelle, from ENST, Paris, and Xavier Descombes, from INRIA, Sophia Antipolis, for interesting discussions, and Simon Wilson, from Trinity College, Dublin, for his kind remarks.

References

1. G. Aubert, L. Vese, *A variational method in image recovery*, SIAM J. Numer. Anal., Vol 34, No 5, pp 1948-1979, Oct. 1997.
2. R. Azencott, *Image analysis and Markov fields*, in Int. Conf. on Ind. and Appl. Math., SIAM Philadelphia, 1988.
3. J. Besag, *Spatial Interaction and Statistical Analysis of Lattice Systems*, A. Roy. Stat. Soc. Series B, Vol 36, pp 721-741, 1974.
4. S. P. Brooks, G. O. Roberts, *Assessing Convergence of Markov Chain Monte Carlo Algorithms*, Univ. of Cambridge, may 1997.
5. P. Charbonnier, L. Blanc-Féraud, G. Aubert, M. Barlaud, *Deterministic edge-preserving regularization in computed imaging*, IEEE Trans. Image Proc., Vol 6, No 2, pp 298-311, Feb. 1997.
6. B. Chalmond, *Image Restoration using an Estimated Markov Model*, Signal Processing, 15, pp 115-129, 1988.
7. F. Champagnat, Y. Goussard, J. Idier, *Unsupervised deconvolution of sparse spike train using stochastic approximation*, IEEE Trans. on Image Processing, Vol 44, No 12, Dec. 1996.
8. G. Demoment, *Image Reconstruction and Restoration: Overview of Common Estimation Structures and Problems*, IEEE Trans. on ASSP, Vol 37, No 12, pp 2024-2036, Dec. 1989.
9. X. Descombes, R. Morris, J. Zerubia, M. Berthod, *Estimation of Markov random field prior parameters using Markov Chain Monte Carlo Maximum Likelihood*, INRIA Research Report No 3015, Oct. 1996, and IEEE Trans. on Image Proc. to be published in 1999.
10. S. Geman, D. Geman, *Stochastic relaxation, Gibbs distributions, and the Bayesian restoration of images*, IEEE Trans. Patt. Anal. Machine intell., Vol 6, No 6, pp 721-741, Nov. 1984.
11. D. Geman, *Random fields and inverse problems in imaging*, Springer-Verlag, Berlin, 1990.

12. D. Geman, G. Reynolds, *Constrained restoration and recovery of discontinuities*, IEEE Trans. Patt. Anal. Machine intell., Vol 16, No 3, pp 367-383, Mar. 1992.
13. D. Geman, C. Yang, *Nonlinear image recovery with half-quadratic regularization and FFTs*, IEEE Trans. Image Proc., Vol 4, No 7, pp 932-946, Jul. 1995.
14. H.-O. Georgii, *Gibbs Measures and Phase Transitions*, Gruyter - Studies in Mathematics, Vol 9, 1988.
15. C. Geyer, E. A. Thompson, *Constrained Monte Carlo Maximum Likelihood for dependent data*, J. R. Statist. Soc. B, Vol 54, No 3, pp 657-699, 1992.
16. C. Geyer, *On the convergence of Monte Carlo Maximum Likelihood calculations*, J. R. Statist. Soc. B, Vol 56, No 1, pp 261-274, Nov. 1994.
17. W. R. Gilks, S. Richardson, D. J. Spiegelhalter, *Markov Chain Monte Carlo in practice*, Chapman & Hall, 1996.
18. J. Hadamard, *Lectures on Cauchy's Problem in Linear Partial Differential Equations*, Yale University Press, New Haven, 1923.
19. W. Hastings, *Monte Carlo sampling methods using Markov chains and their applications*, Biometrika, No 57, pp 97-109, 1970.
20. A. Jalobeanu, L. Blanc-Féraud, J. Zerubia, *Estimation d'hyperparamètres pour la restauration d'images satellitaires par une méthode "MCMCML"*, INRIA Research Report No 3469, Aug. 1998.
21. M. Khoumri, L. Blanc-Féraud, J. Zerubia, *Unsupervised deconvolution of satellite images*, IEEE Int. Conference on Image Processing, Chicago, USA, 4-7 oct. 1998.
22. P. J. M. van Laarhoven, E. H. L. Aarts, *Simulated Annealing : theory and applications*, D. Reidel, 1987.
23. S. Lakshmanan, H. Derin, *Simultaneous parameter estimation and segmentation of Gibbs random fields using simulated annealing*, IEEE Trans. Patt. Anal. Machine intell., Vol 11, No 8, pp 799-813, Aug. 1989.
24. N. Metropolis, A. Rosenbluth, M. Rosenbluth, A. Teller, E. Teller, *Equation of state calculations by fast computing machines*, J. of Chem. Physics, Vol 21, pp 1087-1092, 1953.
25. R. Morris, X. Descombes, J. Zerubia, *Fully Bayesian image segmentation - an engineering perspective*, INRIA Research Report No 3017, Oct. 1996.
26. J. J. K. O Ruanaidh, W. J. Fitzgerald, *Numerical Bayesian methods applied to signal processing*, Statistics and Computing, Springer-Verlag, 1996.
27. C. Robert, *Méthodes de Monte Carlo par chaînes de Markov*, Economica, Paris, 1996.
28. M. Sigelle, *Simultaneous image restoration and hyperparameter estimation for incomplete data by a cumulant analysis*, Bayesian Inference for Inverse Problems, part of SPIE's Int. Symposium on Optical Science, Engineering and Instrumentation, Vol 3459, San Diego, USA, 19-24 jul. 1998.
29. M. A. Tanner, *Tools for statistical inference*, Springer Series in Statistics, Springer-Verlag, 1996.
30. A. N. Tikhonov, *Regularization of incorrectly posed problems*, Sov. Math. Dokl., Vol 4, pp 1624-1627, 1963.
31. L. Younes, *Estimation and annealing for Gibbsian fields*, Ann. Inst. Poincaré, Vol 24, No 2, pp 269-294, 1988.
32. L. Younes, *Parametric inference for imperfectly observed Gibbsian fields*, Prob. Th. Fields, No 82, pp 625-645, Springer-Verlag, 1989.
33. J. Zerubia, L. Blanc-Féraud, *Hyperparameter estimation of a variational model using a stochastic gradient method*, Bayesian Inference for Inverse Problems, part of SPIE's Int. Symposium on Optical Science, Engineering and Instrumentation, Vol 3459, San Diego, USA, 19-24 jul. 1998.

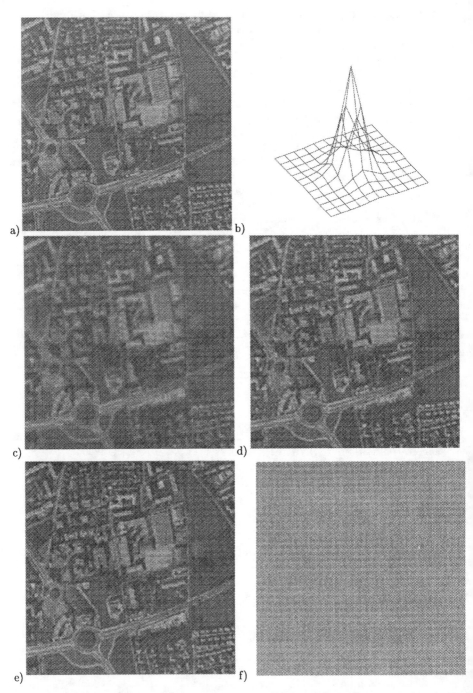

Fig. 6. a) 256×256 original image extracted from "Nîmes" © CNES, b) PSF (h) 11×11 pixels, c) observed image (blur and noise $\sigma = 1.35$, SNR=16.6 dB) , d) Wiener reconstruction (SNR=20.2 dB), e) reconstructed solution with the proposed algorithm, $\lambda = 0.5$ and $\delta = 10$ (SNR=22.1 dB), f) error image

Auxiliary Variables for Markov Random Fields with Higher Order Interactions

Robin D. Morris[1][2]

[1] RIACS, NASA Ames Research Center, MS 269-2, Moffett Field, CA 94035 USA.
[2] This work was performed while the author held a National Research Council-NASA Ames Research Associateship.
rdm@ptolemy.arc.nasa.gov
Tel: +1 650 604 0158 Fax: +1 650 604 3594

Abstract. Markov Random Fields are widely used in many image processing applications. Recently the shortcomings of some of the simpler forms of these models have become apparent, and models based on larger neighbourhoods have been developed. When single-site updating methods are used with these models, a large number of iterations are required for convergence. The Swendsen-Wang algorithm and Partial Decoupling have been shown to give potentially enormous speed-up to computation with the simple Ising and Potts models. In this paper we show how the same ideas can be used with binary Markov Random Fields with essentially any support to construct auxiliary variable algorithms. However, because of the complexity and certain characteristics of the models, the computational gains are limited.

1 Introduction

Markov Random Fields (MRFs) were introduced into the image processing literature in 1984 [3], and have since been widely used for many tasks, mainly in low level vision. Despite the increase in computational power that has become available, and the inherent parallelisation that can be applied to the computational algorithms, computation with MRF models and single-site updating algorithms (the usual forms of the Gibbs sampler and Metropolis-Hastings [5] algorithms) is time consuming. In statistical physics applications, where the aim is to simulate large interacting spin systems, the Swendsen-Wang (SW) algorithm [11] was developed to speed up the computation, especially at the critical point, when simulating the Ising [8] or Potts models. The Multi-Level Logistic model of [3] is just the Potts model, and so the SW algorithm is applicable in that case. However, it has become apparent that the Ising or Potts model does not capture the image characteristics that are important in segmentation tasks [10]. This has motivated the development of MRFs with longer range and more complex forms of interaction [1, 12] to model more adequately the structures present in typical segmentation imagery. The application of these models has proved successful, but single-site updating algorithms have proved computationally intensive, especially when there is a requirement to estimate the hyperparameters of these models [4]. Motivated by the success of the SW algorithm to dramatically speed up computation with the Ising or Potts model, in this paper we investigate the use of the auxiliary variable approach applied to MRFs with long range interactions, both in the form used in the SW algorithm, and the partial decoupling approach proposed in [6]. The SW algorithm has most benefit near the critical point of the Ising model. The auxiliary variable methods developed in this paper appear to be of limited benefit for the simulation of models with long-range interactions. This is likely to be because the models are being used well away from any critical regimes. Indeed the presence or absence of critical behaviour in models with long-range interaction has to be demonstrated on a model-by-model basis. Whether the long-range interaction model considered in this paper has a phase transition is currently unknown.

2 Auxiliary Variables for Markov Random Fields

2.1 The Swendsen-Wang algorithm and Partial Decoupling

The idea behind auxiliary variable methods is the following:

It is desired to simulate a distribution $\pi(\mathbf{x})$. Auxiliary variables $\mathbf{u}$ are introduced, with conditional distribution $\pi(\mathbf{u}|\mathbf{x})$. This gives a joint distribution $\pi(\mathbf{x}, \mathbf{u}) = \pi(\mathbf{u}|\mathbf{x})\pi(\mathbf{x})$, with the desired marginal distribution for $\mathbf{x}$ of $\pi(\mathbf{x})$. Simulation of this distribution is generally performed by alternately updating $\mathbf{u}$ and $\mathbf{x}$ – the idea being to define $\pi(\mathbf{u}|\mathbf{x})$ such that the updates cause rapid mixing. The realisations of $\mathbf{x}$ are those desired.

The Ising model is defined by

$$\pi(\mathbf{x}) \propto \exp\left(\beta \sum_{i \sim j} \mathrm{I}[x_i = x_j]\right) \tag{1}$$

where $i \sim j$ indicates nearest neighbour pairs.

For the SW algorithm the distribution $\pi(\mathbf{u}|\mathbf{x})$ is defined such that the u_{ij} are independent, and is

$$p(u_{ij}|\mathbf{x}) \propto \exp\left(-\beta \mathrm{I}[x_i = x_j]\right) \mathrm{I}[0 \leq u_{ij} \leq \exp\left(\beta \mathrm{I}[x_i = x_j]\right)] \tag{2}$$

where u_{ij} can be considered as a continuous 'bond' variable between the pixels x_i and x_j and $\mathrm{I}[\cdot]$ is the indicator function. This results in

$$\pi(\mathbf{x}|\mathbf{u}) \propto \prod_{i \sim j} \mathrm{I}[0 \leq u_{ij} \leq \exp\left(\beta \mathrm{I}[x_i = x_j]\right)] \tag{3}$$

What does this choice of distribution give us? Considering first $p(u_{ij}|\mathbf{x})$, for $u_{ij} > 1$ we must have $\exp\left(\beta \mathrm{I}[x_i = x_j]\right) > 1$, or equivalently, $x_i = x_j$. Thus $u_{ij} > 1$ constrains x_i and x_j to be in the same state. Conversely, if x_i and x_j are in the same state, what is the probability of $u_{ij} > 1$? From the conditional distribution in equation 2 we have that

$$p(u_{ij} > 1|x_i = x_j) = 1 - \exp(-\beta) \tag{4}$$

Since it is only important whether u_{ij} is greater or less than one we may think of the u_{ij} as binary bond variables. From equation 4 the bond variable is present between two pixels in the same state with probability $1 - \exp(-\beta)$. To sample $\mathbf{u}$ thus involves placing bonds between neighbouring pixels of the same state with probability $1 - \exp(-\beta)$ and omitting bonds between neighbouring pixels of differing states.

Once the bonds are in place, the conditional distribution $\pi(\mathbf{x}|\mathbf{u})$ says that all configurations where bonded pixels are of the same state are equally probable. Thus to update $\mathbf{x}$ we form clusters of connected pixels and assign to all pixels of the cluster the same state, chosen uniformly from the allowed states. This scheme allows potentially large clusters of pixels to change state at each iteration, allowing the Markov chain to explore the distribution freely.

In the discussion above we have assumed that the parameter β in the Potts model is positive, inducing clustering. However, the SW algorithm is still applicable if β is negative. In this case a similar argument gives that neighbours in different states are constrained to remain in different states with probability $1 - \exp(\beta)$. This forms 'clusters' where neighbouring sites in the cluster must be in different states.

Partial decoupling was introduced [6, 7] to overcome some problems with the basic form of the SW algorithm when simulating systems with data. In this case growing the clusters without taking any account of the data can be unhelpful, as the clusters are unlikely to reflect the structure in the data. A modification to the conditional distribution of the auxiliary variables allows the data to be taken into account when forming the clusters. Later in this paper, we will use this modification to systematically reduce the strength of the constraints introduced by the complexity of the higher order MRF models.

The SW algorithm forms clusters which are coloured independently. The idea behind partial decoupling is to reduce the probability that bonds will be placed, with the consequence that the clusters will not be independent, and so the colouring of the clusters themselves will have to be updated using a Markov chain Monte Carlo (MCMC) scheme [6].

Instead of the definition of $p(u_{ij}|\mathbf{x})$ in equation 2 above, we now define

$$p(u_{ij}|\mathbf{x}) \propto \exp(-\delta\beta\mathrm{I}[x_i = x_j])\mathrm{I}[0 \leq u_{ij} \leq \exp(\delta\beta\mathrm{I}[x_i = x_j])] \qquad (5)$$

where δ is a fixed constant between zero and one. This results in

$$\pi(\mathbf{x}|\mathbf{u}) \propto \exp\left(\sum_{i \sim j}(1 - \delta)\beta\mathrm{I}[x_i = x_j]\right) \times \prod_{i \sim j}\mathrm{I}[0 \leq u_{ij} \leq \exp(\delta\beta\mathrm{I}[x_i = x_j])] \quad (6)$$

So now we form clusters (or enforce the dissimilarity of neighbours, in the case of negative β) by bonding pixels with probability $1 - \exp(-\delta\beta)$. However, the clusters thus formed are not independent, and their colouring must be updated conditionally on the pixels neighbouring the cluster. Thus cluster $\mathcal{I}$ takes colour

k, conditional on its neighbors, $\mathcal{N}(\mathcal{I})$ with probability

$$\pi(k|\mathcal{N}(\mathcal{I})) \propto \exp\left(\sum_{i\sim j:i\in\mathcal{I},j\in\mathcal{N}(\mathcal{I})} (1-\delta)\beta I[k=x_j]\right) \tag{7}$$

and is updated using, for example, the Gibbs sampler or the Metropolis-Hastings algorithm.

In [7] the δ's were chosen to split the lattice up into regular blocks. In [6] the data was used to set the values of δ to encourage clusters supported by the data. In this paper we will use the δ's to avoid overconstraining the possible updates, and to reduce the complexity introduced by negative β's.

2.2 The 'chien' model

In its original formulation the 'chien' model [1] was defined as a binary MRF, where the potential function considered a 5×5 neighbourhood, and 3×3 cliques. For a clique of size 3×3 there are 512 configurations. When symmetries are removed, this reduces down to 51 classes. These classes are shown in figure 1. By considering the energy associated with lines and edges, the 51 parameters, c_1 to c_{51}, in figure 1 were reduced down to functions of three parameters, representing boundary length (e), line length (l) and noise (n). The reader is referred to [1] for full details of this model.

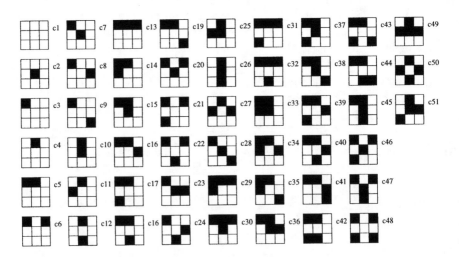

Fig. 1. The 51 classes of clique considered in the 'chien' model

2.3 Auxiliary variables for the 'chien' model

Conventionally, the probability of a configuration of an MRF is written in the following form

$$p(\mathbf{x}) = \frac{1}{Z} \exp\left(\sum_c V_c(\mathbf{x})\right) \tag{8}$$

That is, the total energy of the configuration is made up of a sum over all the cliques of the potentials associated with each clique configuration. In the Ising/Potts model these cliques are the neighbouring pairs, and the configurations the state of homogeneity of these pairs. In the 'chien' model the cliques are 3×3 blocks, and the configurations are those shown in figure 1. For the Ising/Potts model, writing the energy in the form

$$p(\mathbf{x}) \propto \exp\left(\sum_{i \sim j} \beta_{ij} \mathrm{I}[x_i = x_j]\right) \tag{9}$$

indicates how to introduce auxiliary variables to induce clusters with either reduced or eliminated dependency between the clusters. To introduce similar auxiliary variables for the 'chien' model, we must write the pdf for the 'chien' model in a similar form.

The pdf for the 'chien' model can be written as

$$p(\mathbf{x}) \propto \exp\left(\sum_j v(a_j, b_j, c_j, d_j, e_j, f_j, g_j, h_j, i_j)\right) \tag{10}$$

where the pixels are labeled as shown in figure 2, the sum over j is over all sites in the image and $v(\cdot)$ is the value given by the classification in figure 1. In subsequent equations we will drop the j subscript to simplify the notation.

a	b	c
d	e	f
g	h	i

Fig. 2. Labeling of the pixels in a clique of the 'chien' model

Fig. 3. Sample from the 'chien' model, $e = 0.8, l = 1.5, n = 1.6$

The energy of a clique can be written, using this labeling, as

$$v(a, b, c, d, e, f, g, h, i) =$$

$$I[a = e]I[b = e]I[c = e]I[d = e]I[f = e]I[g = e]I[h = e]I[i = e]c_1$$
$$+ I[a = e]I[b = e]I[c = e]I[d = e]I[f = e]I[g = e]I[h = e](1 - I[i = e])c_3$$
$$+ I[a = e]I[b = e]I[c = e]I[d = e]I[f = e]I[g = e](1 - I[h = e])I[i = e]c_4$$
$$+ I[a = e]I[b = e]I[c = e]I[d = e]I[f = e]I[g = e](1 - I[h = e])(1 - I[i = e])c_5$$
$$\vdots$$
$$+ (1 - I[a = e])(1 - I[b = e])I[c = e](1 - d[d = e])(1 - I[f = e])$$
$$I[g = e]I[h = e]I[i = e]c_{35}$$
$$\vdots$$
$$+ (1 - I[a = e])(1 - I[b = e])(1 - I[c = e])(1 - I[d = e])$$
$$(1 - I[f = e])(1 - I[g = e])(1 - I[h = e])(1 - I[i = e])c_2 \quad (11)$$

where the c's are the clique parameters from figure 1. A computer algebra package can be used to expand this expression, and then to reduce it into a simple form, where each entry is a product of terms of the form $I[\cdot = \cdot]$, for example $I[a = e](c_7 - c_2)$ and $I[a = e]I[b = e]I[c = e]I[g = e]I[h = e](-c_2 + 3c_7 + 2c_{10} - 3c_{15} + c_{18} - 2c_{20} - 3c_{23} - c_{26} - c_{28} + c_{30} - c_{32} - c_{34} + 2c_{37} + 2c_{38} - c_{40} - c_{42} - c_{44} + 3c_{45} + c_{46} + c_{47})$ etc. In fact, terms involving all pairs, triples, 4-tuples, 5-tuples, 6-tuples, 7-tuples, 8-tuples and the single 9-tuple are present in this representation (some with a coefficient of zero). This enables us to write the energy in the form

$$v(a, b, c, d, e, f, g, h, i) =$$

$$I[a = e]\beta_{\{ae\}} + I[b = e]\beta_{\{be\}} + \cdots$$
$$+ I[a = e]I[b = e]\beta_{\{abe\}} + I[a = e]I[c = e]\beta_{\{ace\}} + \cdots$$
$$+ I[a = e]I[b = e]I[c = e]\beta_{\{abce\}} + \cdots$$
$$+ I[a = e]I[b = e]I[c = e]I[d = e]\beta_{\{abcde\}} + \cdots$$
$$+ I[a = e]I[b = e]I[c = e]I[d = e]I[f = e]\beta_{\{abcdef\}} + \cdots$$
$$+ I[a = e]I[b = e]I[c = e]I[d = e]I[f = e]I[g = e]\beta_{\{abcdefg\}} + \cdots$$
$$+ I[a = e]I[b = e]I[c = e]I[d = e]I[f = e]I[g = e]I[h = e]\beta_{\{abcdefgh\}} + \cdots$$
$$+ I[a = e]I[b = e]I[c = e]I[d = e]I[f = e]I[g = e]I[h = e]I[i = e]\beta_{\{abcdefghi\}}$$
$$(12)$$

In this representation, the distribution can be written as

$$\pi(\mathbf{x}) \propto \prod_k \exp \left(\sum_j I_k[a, b, c, d, e, f, g, h, i]\beta_{\{a,b,c,d,e,f,g,h,i\}} \right) \quad (13)$$

where $I_k[\cdot]\beta_{\{\cdot\}}$ are the terms in equation 12, k being the index of the term. This is now in a form which makes application of the extension to the SW algorithm in [2] clear.

We introduce a set of independent auxiliary variables, $\mathbf{u}_k$, corresponding to each of the terms in the representation of equation 12, with conditional distribution

$$\pi(u_k(j)|\mathbf{x}) = \exp\left(-I_k[\cdot]\beta_{\{.\}}\right)I[0 \leq u_k(j) \leq \exp\left(I_k[\cdot]\beta_{\{.\}}\right)] \qquad (14)$$

That is, for each term $I_k[\cdot]\beta_{\{.\}}$ we have a set of auxiliary variables, $\mathbf{u}_k$ as in the standard SW algorithm. Each element of $\mathbf{u}_k$, $u_k(j)$ is independent and drawn from a uniform distribution $U[0, \exp(I_k[a_j, b_j, \ldots]\beta_{\{.\}})]$, resulting in a conditional distribution for $\mathbf{x}$ of

$$\pi(\mathbf{x}|\mathbf{u}) = \prod_j \prod_k I[u_k(j) \leq \exp(I_k[\cdot]\beta_{\{.\}})] \qquad (15)$$

So we now have that $\mathbf{x}|\mathbf{u}$ is uniformly distributed, provided that the constraints introduced by the particular realisation of the $\mathbf{u}$ variables are satisfied. These constraints now form 'bonds' between groups of up to 9 pixels at a time, constraining the groups to be in the same state, for $\beta_{\{.\}} > 0$, or groups of up to 9 pixels are constrained to not all be in the same state, for $\beta_{\{.\}} < 0$. An analogous procedure can be performed for any binary MRF.

We can divide the terms into two classes, those where the coefficient $\beta_{\{.\}}$ is greater than zero, and those where it is less than zero. The update procedure is then as follows

1. For each term in the expansion of equation 11 with $\beta_{\{.\}} > 0$, for each site j, if all the pixels concerned are in the same state, constrain them to be in the same state in the next iteration with probability $1 - \exp(-\beta_{\{.\}})$.
2. For each term in the expansion with $\beta_{\{.\}} < 0$, for each site j, if all the pixels concerned are *not* in the same state, constrain them to not all be in the same state in the next iteration with probability $1 - \exp(\beta_{\{.\}})$.
3. Choose any random colouring which satisfies the constraints from steps 1 and 2.

The constraints generated in step 1 (type 1 constraints) are easily dealt with – there are only two ways that a group of n pixels can all be homogeneous in a binary MRF. Thus we can use all of the constraints with $\beta_{\{.\}} > 0$, irrespective of how many pixels are included in that term, to form clusters of pixels in a similar manner to the SW algorithm, where all the pixels in a cluster must be in the same state after the update.

The constraints from step 2 (type 2 constraints) are more problematic – there are $2^n - 2$ ways a group of n pixels can not all be in the same state in a binary MRF. The update strategy used was that known as 'generate-and-test' [9]. This heuristic method works well when either the density of the constraints is low (such that there are many configurations that satisfy all the constraints and finding one is relatively easy), or when the density of constraints is high, when there are very few solutions, but local changes that satisfy the constraints will almost always move towards one of the few global solutions.

The 'generate-and-test' approach results in the following algorithm.

1. Start from a random configuration that satisfies the type 1 constraints
2. Go through the list of type 2 constraints until one is not satisfied
3. To satisfy this constraint
   (a) flip the state of one of the clusters involved in this constraint
   (b) if the constraint is still not satisfied, un-flip this cluster, and flip another until the constraint is satisfied
4. Go to step 2 until all the constraints are satisfied

Because the constraints are generated from the current colouring of the pixels, we know that there is at least one colouring which satisfies all the constraints in stage 2. (However, this colouring is not of interest to us; the whole point of constructing the auxiliary variable algorithm is to find a recolouring that is significantly different from the current colouring.) Irreducibility is, however, guaranteed, as there is a non-zero probability of no constraints being placed, resulting in the x variables being independent, and so any state can be reached in one update.

Figure 3 shows a sample, generated by the single-site Gibbs sampler, from the 'chien' model, with the parameters being $e = 0.8, l = 1.5, n = 1.6$, after 10,000 iterations. These parameters were chosen to give a sample which shows regions together with fine structure. Starting from a random initial image (each pixel is black or white with probability 0.5), figure 4 shows the clusters induced by the type 1 constraints for this set of parameter values – the bond-graph shows how the pixels are constrained, and the cluster map shows how the bonds divide the image up into groups (there are actually 109 regions in this image). Clearly with this density of type 1 constraints finding a colouring which satisfies the type 2 constraints will be easy. It will, however, be very similar to the initial colouring. This motivates the use of the partial decoupling approach – the δ's can be chosen to reduce the density of type 1 constraints.

If the initial state is all of one colour the situation is worse – almost every pixel will be bonded into one region, and the sampler will be essentially immobile.

2.4 Partial decoupling for the 'chien' model

The density of the type 1 constraints lead to an almost immobile algorithm in the previous subsection. Here we consider how partial decoupling may help the mobility of the sampler by reducing the density of the type 1 constraints and by eliminating the type 2 constraints. This also results in an easier update algorithm.

To derive the partial decoupling algorithm, the same representation of the energy as in equation 12 is used. However, the conditional distribution of the u_k's is now

$$\pi(u_k(j)|\mathbf{x}) \propto \exp\left(I_k[\cdot]\delta_{\{\cdot\}}\beta_{\{\cdot\}}\right) I[0 \leq u_k(j) \leq \exp\left(I_k[\cdot]\delta_{\{\cdot\}}\beta_{\{\cdot\}}\right)] \qquad (16)$$

where $\delta_{\{\cdot\}}$ is the factor associated with each of the auxiliary variables. This enables the influence in the clustering and anti-clustering of each term in equation

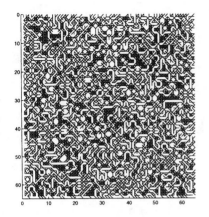

Fig. 4. Bond graph (left) and region map (right) from the type 1 constraints for the SW type algorithm (see text)

12 to be controlled. In practise this enables the complexity of the update algorithm caused by the terms with $\beta_{\{\cdot\}} < 0$ to be eliminated, by choosing $\delta_{\{\cdot\}} = 0$ for these terms. In the experiments described below, the same value of δ was used for all the terms with $\beta_{\{\cdot\}} > 0$.

With this set of auxiliary variables, the conditional distribution of $\mathbf{x}|\mathbf{u}$ is now

$$\pi(\mathbf{x}|\mathbf{u}) \propto \prod_j \left[\prod_{k:\beta_{\{\cdot\}}<0} \exp\left(I_k[\cdot]\beta_{\{\cdot\}}\right) \times \prod_{k:\beta_{\{\cdot\}}>0} \exp\left(I_k[\cdot](1 - \delta_{\{\cdot\}})\beta_{\{\cdot\}}\right)\right.$$

$$\left. \times \prod_{k:\beta_{\{\cdot\}}>0} I[u_k(j) \leq \exp(I_k[\cdot]\delta_{\{\cdot\}}\beta_{\{\cdot\}})] \right] \quad (17)$$

The final term of this equation is the cluster constraints – when updating the $\mathbf{u}_k : \beta_{\{\cdot\}} > 0$, the pixels are bonded with probability $1 - \exp(-\delta_{\{\cdot\}}\beta_{\{\cdot\}})$, and this terms says that all the pixels in a cluster must be the same colour.

The first two terms give the distribution of the colours of the clusters. From them we can easily derive the conditional distribution for the colour of each cluster, given its neighbours. For computational purposes, it is convenient to transform the representation back into the form of equation 10, except that now the potentials of the configurations have been modified by the inclusion of the $(1 - \delta_{\{\cdot\}})$ terms. This allows simpler computation when computing the probabilities of the allowed colours for the regions when implementing the Gibbs sampler.

This results in a form of block update algorithm. The weakened cluster constraints form regions, the size and shape of which is a function of the model. These regions are then recoloured with probabilities which reflect the cluster formation process and the model.

Figure 5 shows the bond-graph and the corresponding cluster map for the application of the partial decoupling algorithm to an initial random image. Clearly the density of the bonds is much reduced from figure 4, so the algorithm should be more mobile. The colouring of the clusters is updated conditionally on the cluster's neigbours, using the Gibbs sampler. Figure 6 shows the initial image and the result after one iteration. The algorithm clearly moves rapidly towards the equilibrium distribution. However, further updates using the Partial Decoupling algorithm rapidly move towards an almost uniform image – even with reduced bonding strength, the number of possible ways the pixels in a uniform region can be bonded results in most of them being joined into one cluster, and then the conditional update will preferentially re-colour the regions to eliminate edges. The algorithm is thus of limited applicability – it moves rapidly towards the equilibrium distribution, but moves slowly once it has converged.

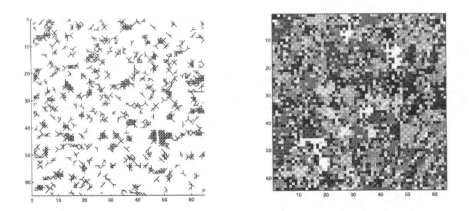

Fig. 5. Bond graph (left) and region map (right) for the Partial Decoupling algorithm ($\delta = 0.0067$)

Fig. 6. Initial random image (left) and image after one iteration of the Partial Decoupling algorithm (right)

3 Conclusions

We have shown how to construct an auxiliary variable algorithm for the MRF known as the 'chien' model, and explained how this method may be used with an essentially arbitrary binary MRF. Because of the strength of interaction in the model, however, introducing a full set of auxiliary variables results in a sampler which moves very slowly. Reducing the set of auxiliary variables, and reducing the influence of those included, enables an algorithm to be constructed which is a form of block-update algorithm. This moves rapidly from a random start point towards the equilibrium distribution, but then moves into one of the modes of the distribution, and becomes immobile.

The SW algorithm for the Ising model shows most spectacular improvement at the critical point, when the correlation length becomes infinite. The 'chien' model with parameters corresponding to the characteristics of real images does not seem to exhibit this behaviour, and so auxiliary variables are of less benefit. Whether the 'chien' and other higher order models do show critical behaviour for some parameter values is an open problem. If they do, then the algorithm described in this paper should be very useful for simulating the models in those behaviour regimes.

References

1. X. Descombes, J.F. Mangin, E. Pechersky, and M. Sigelle. Fine structures preserving model for image processing. In *Proc. 9th SCIA 95 Uppsala, Sweden*, pages 349–356, 1995.
2. R.G. Edwards and A.D. Sokal. Generalization of the Fortuin-Kasteleyn-Swendsen-Wang representation and Monte Carlo algorithm. *Physical Review Letters*, pages 2009–2012, 1988.
3. S. Geman and D. Geman. Stochastic relaxation, Gibbs distributions and the Bayesian restoration of images. *IEEE Trans PAMI*, 6(6):721–741, November 1984.
4. C. J. Geyer and E.E. Thompson. Constrained Monte Carlo Maximum Likelihood for dependent data. *JRSS - B*, 54(3):657–699, 1992.
5. W.K. Hastings. Monte Carlo sampling methods using Markov Chains and their applications. *Biometrika*, 57:97–109, 1970.
6. D.M. Higdon. Auxiliary variable methods for Markov chain Monte Carlo with applications. *Journal of the American Statistical Association*, 93:585-595, June 1998
7. M. Hurn. Difficulties in the use of auxiliary variables in Markov chain Monte Carlo methods. *Statistics and Computing*, 7:35–44, 1997.
8. E. Ising. . *Zeitschrift Physik*, 31:253, 1925.
9. S. Minton, M.D. Johnston, A.B. Phillips, and P. Laird. Solving large-scale constraint satisfaction and scheduling problems using a heuristic repair method. In *Proceedings of AAAI*, 1990.
10. R.D. Morris, X. Descombes, and J. Zerubia. The Ising/Potts model is not well suited to segmentation tasks. In *Proceedings of the IEEE Digital Signal Processing Workshop*, September 1996.
11. R.H. Swendsen and J-S. Wang. Nonuniversal critical dynamics in Monte Carlo simulations. *Physical Review Letters*, 58:86–88, 1987.
12. H. Tjelmeland and J. Besag. Markov Random Fields with higher order interactions. *Scandinavian Journal of Statistics*, 25:415-433, 1998.

Unsupervised Multispectral Image Segmentation Using Generalized Gaussian Noise Model

P. Rostaing, J.-N. Provost and C. Collet

Groupe de Traitement du Signal, Ecole Navale, Lanvéoc-Poulmic,
BP 600 - 29240 Brest-Naval, France.
email : name@ecole-navale.fr
Tel: +33 2 98 23 40 19
Fax: +33 2 98 23 38 57

Abstract. This paper is concerned with hierarchical Markov Random Field (MRF) models and their applications to multispectral image segmentation. We present an extension of the classic Gaussian model for the modelization of the data likelihood based on a Generalized Gaussian (GG) model, requiring a "shape parameter". In order to obtain an unsupervised multispectral image segmentation, we develop a two step algorithm. In the first step, we estimate the parameters associated with a causal Markovian model (on a quad-tree [1]) and a generalized Gaussian modeling for the data-driven term, by using an Iterative Conditional Estimation (ICE algorithm [16]). One of the originality of this paper consists in explicitly decorrelate the multispectral observations during the estimation step on a quad-tree structure. A second step gives the segmentation map obtained with the estimated parameters, according to the Modes of Posterior Marginals (MPM) estimator. The main motivation of the paper is to extend the variety of noise models which results of the distribution mixture on multispectral images. Some results on synthetic and SPOT images validate our approach.

Keywords : correlated sensors, Generalized Gaussian, multispectral segmentation, quad-tree, parameter estimation.

1 Introduction

One of the main interest of MRF modeling coupled with a Bayesian formulation consists in establishing an explicit link between observation field and label field jointly to the introduction of contextual and *a priori* information [9]. Nevertheless, non-causal MRF models are known to yield iterative and computational intensive segmentation algorithms [12]. Besides, the problem of *unsupervised* Markovian segmentation is complex. The main difficulty is that the estimation of parameters is required for the segmentation, while one or several segmentations are usually required for parameter estimation [13]. In this paper, we work

---

[1] Acknowledgments: The authors thank P. Pérez and P. Bouthemy for fruitful comments and suggestions.

on a hierarchical MRF attached to the nodes of a quad-tree. The specific structure of these models results in an attractive causality property through scale [9], which allows the design of exact non-iterative inference algorithms [12], similar to those employed in the framework of Markov chain models [8].

We consider a couple of random fields $W = (X, Y)$, with $Y = \{Y_s, s \in S\}$ the field of observations located on a lattice S of N sites s, and $X = \{X_s, s \in S\}$ the label field. Each of the Y_s takes its value in $\Lambda_{obs} = \{0, ..., 255\}^C$ where C stands for the number of spectral bands, *i.e.*, $Y_s = \left[Y_s^{(1)}, ..., Y_s^{(C)}\right]^T$ is a vector of size C. Moreover, each X_s takes its value in $\Lambda_{label} = \{\omega_1, ..., \omega_K\}$ where K is the number of classes. The distribution of (X, Y) is defined, firstly, by $P_X(x)$, the distribution of X supposed to be stationary and Markovian in scale, and secondly, by the site-wise conditional data likelihoods $P_{Y_s|X_s}(y_s|x_s)$, which depend on the concerned class label x_s. If the data are assumed to be independent *conditionally* on the labeling process X, one gets

$$P_{Y|X}(y|x) = \prod_{s \in S} P_{Y_s|X}(y_s|x) = \prod_{s \in S} P_{Y_s|X_s}(y_s|x_s) \qquad (1)$$

In real life, labeled samples are usually not available and we have to estimate the parameter from unlabeled samples, *i.e.*, the label field X is hidden. In statistics, the problem is well known as the incomplete data problem. The observations Y correspond to the *incomplete data* whereas W constitutes the *complete data*. Prior distribution $P_X(x)$ depends on some parameter vector Φ_x, while data likelihood $P_{Y|X}(y|x)$ depends on another parameter vector Φ_y. Both of them has to be estimated. Joint and posterior distributions $P_{Y,X}(y, x) = P_X(x)P_{Y|X}(y|x)$ and $P_{X|Y}(x|y) \propto P_X(x)P_{Y|X}(y|x)$ thus depend on $\Phi = \{\Phi_x, \Phi_y\}$. This will be made explicitly when necessary, *i.e.*, denoting posterior distribution as $P_{X|Y}(x|y, \Phi)$.

If we note $f_i(y_s) = P_{Y_s|X_s}(y_s|x_s = \omega_i)$, one considers generally [7, 5, 6] that the general expressions of $\{f_1, ..., f_K\}$ are known and depend on a parameter set Φ_y. The general case often studied is the Gaussian mixture $\mathcal{N}(\mu_i, \sigma_i^2)$ which is totally described by $\Phi_y = \{(\pi_i, \mu_i, \sigma_i^2) \forall i \in [1, K]\}$ where π_i represents the proportions of each class in the mixture. But in the general case, $f_i(y_s)$ is not exactly and accurately known and one needs to introduce a more general shape for the site-wise likelihood.

The Generalized Gaussian model introduced here for the data-driven term, has proved successful in capturing the variety of the noise laws present in the distribution mixture of synthetic multispectral images. For example, as shown in [14], in the specific case of sonar images, the generalization of the conditional likelihood probabilities (from Rayleigh law to Weibull law) increases the adequation between observations and modelization thanks to the introduction of a "shape parameter". The number of corresponding parameters is indeed incremented, but the shape parameter allows to best fit observations and model.

Unfortunately, in reality, most multispectral observations are correlated [10] and in this case, the data driven term expression is not always known. In the Gaussian case, sensors can be independent or not because the mathematical expression of the Gaussian likelihood $P_{\mathbf{Y}_s|X_s}(\mathbf{y}_s|x_s)$ is nevertheless well known. Moreover, as mentioned above, the conditional probabilities are not always necessarily Gaussian in practice. In this paper, we propose to generalize the method presented in [8] to the case of possibly non-Gaussian correlated sensors (conditioned to the random process X) in a quad-tree structure. To do that, we take the following way [17] :

- each sensor is decorrelated from the others by using a Cholesky transform noted A_i for the ith class,
- then, the components of $\mathbf{Z}_s = A_i\mathbf{Y}_s$ are assumed to be independent conditionally to $X_s = \omega_i$:

$$P_{\mathbf{Z}_s|X_s}(z_s|x_s) = P(z_s^{(1)}, ..., z_s^{(C)}|x_s) = \prod_{c=1}^{C} P_{Z_s^{(c)}|X}(z_s^{(c)}|x_s). \quad (2)$$

The approach we choose to solve the unsupervised MRF-based segmentation problem consists in having a two step process. First, a parameter estimation step is conducted to infer both the noise model parameters and the hierarchical MRF model parameters. In this paper, the parameter estimation method is based on the ICE algorithm [16]. Let us notice that, with a similar approach, the Stochastic Expectation Maximization (SEM) algorithm [4][3] aims at determining the Maximum Likelihood estimates of the parameter Φ_y by making use of simulation of the missing data. This way requires the use of the ML estimators of the GG modeling given in [15]. Nevertheless, the ICE approach is more general, because we keep the choice of the estimator of Φ_y. Then, a second step (segmentation step) is devoted to the segmentation itself, using the values of the estimated parameters. At the end, the final label field corresponding to the segmentation is processed by using the Modes of Posterior Marginals (MPM) estimator [12].

This paper is organized as follows. In Section 2, we detail the Generalized Gaussian modelization and the algorithm for sensor decorrelation. Section 3 presents estimation and classification steps on the proposed hierarchical Markovian model in the multispectral case. Experimental results obtained on synthetic scenes are reported in Section 4. Then, we conclude with some perspectives.

2 Noise model

2.1 Generalized Gaussian noise model

The noise model considered here is referred to as Generalized Gaussian (GG) noise and is obtained by generalizing the Gaussian density [11] to represent different degrees of exponential decay

$$P_Z(z) = [2\Gamma(1/p)]^{-1}\eta(p)\, p\, \exp\left[-(\eta(p)|z-\mu|)^p\right] \quad (3)$$

where $\Gamma(.)$ is the gamma function, p is a positive parameter governing the rate of decay ($p = 1$ for Laplacian noise, $p = 2$ for Gaussian noise, $p > 8$ for nearly uniform noise), μ is the mean, and[2]

$$\eta(p) \triangleq \left[\frac{\Gamma(3/p)}{\sigma^2 \Gamma(1/p)} \right]^{1/2} \qquad (4)$$

where σ^2 is the variance. The shape of the GG probability density function (pdf) (3) is illustrated in Fig.1 for some values of the parameter p. We note that for small values of p (*i.e.*, $p < 2$), probability density functions have heavier tail than those of the Gaussian densities which produce an impulsive random sample. This noise model is well fitted to some physical data properties. For example, according to Algazi and Lerner [2], densities representative of certain atmospheric impulse noises can be obtained by picking $0.1 < p < 0.6$.

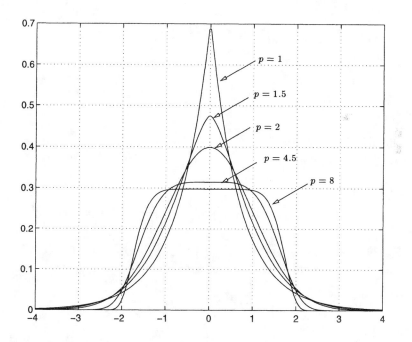

Fig. 1. Generalized Gaussian probability density function with $\sigma^2 = 1$, $\mu = 0$ and for severals values of p.

Statistical properties of the maximum likelihood estimators of the Generalized Gaussian density function are considered in [15]. The maximum likelihood estimators of the GG parameters (*i.e.*, mean, variance and shape parameter) can

---

[2] The symbol $\triangleq$ stands for "equals by definition"

be used to derived the SEM algorithm from one realization of the random field X according to the *a posteriori* probability. However, in the following, estimators based on the the empirical moments are derived and implemented because they lead to easier (*i.e.*, less time computational) estimators. Hence, the ICE algorithm is then performed (see subsection 3.3).

2.2 Moments based estimators of the GG parameters

Let us consider N random samples z_i, $i = 1, \ldots, N$, assumed to be independent and identically distributed (iid) from the pdf given in (3). Three parameters are required to characterize the GG pdf from the samples of z_i. First, the empirical mean $\widehat{\mu}$ is performed

$$\widehat{\mu} = \frac{1}{N} \sum_{i=1}^{N} z_i \tag{5}$$

Secondly, the estimator $\widehat{\sigma}^2$ of σ^2, based on the second-order moment, is simply the empirical variance

$$\widehat{\sigma}^2 = \frac{1}{N} \sum_{i=1}^{N} (z_i - \widehat{\mu})^2 \tag{6}$$

and finally, the estimation of the shape parameter p is based on the estimation of the centered fourth-order moment $\mu(4)$ of the GG pdf given by the general relation

$$\mu(n) = E[(Z - E[Z])^n] = \frac{\Gamma(\frac{n+1}{p})}{\Gamma(1/p)\eta(p)^n} \quad \text{for } n \text{ even.} \tag{7}$$

By substituting (4) in (7) for $n = 4$ and using estimators of mean $\widehat{\mu}$, variance $\widehat{\sigma}^2$ and centered fourth-order moment $\widehat{\mu}(4)$, we obtain the following relation

$$\widehat{\mu}(4) = \widehat{\sigma}^4 \frac{\Gamma(5/p)\Gamma(1/p)}{\Gamma(5/p)^2} \triangleq \widehat{\sigma}^4 h(p) \tag{8}$$

where

$$\widehat{\mu}(4) = \frac{1}{N} \sum_{i=1}^{N} (z_i - \widehat{\mu})^4 \tag{9}$$

Eq.(8) can be numerically solved[3] in order to obtain the estimator $\widehat{p}$ of the "shape" parameter p.

---

[3] Eq.(8) gives a unique estimate because of the monotone decreasing property of function $h(p)$ for $p > 0$.

2.3 Correlated non-Gaussian mixture model

In order to validate the use of GG marginal distribution for image segmentation, algorithm will be, firstly, performed on synthetic image and secondly, on SPOT image. In the following, the generation of synthetic images is presented. The resulting correlated non-Gaussian mixture model will be considered for the image segmentation algorithm.

The model used for the correlated non-Gaussian data observed by the sensors is presented in Fig.2. An independent zero-mean random vector $Z_s = [Z_s^{(1)}, Z_s^{(2)}, \ldots, Z_s^{(C)}]^T$, conditionally to X_s, with any marginal distributions $P_{Z_s^{(c)}|X_s}(z_s^{(c)}|x_s = \omega_i) = g_i^c(z_s^{(c)})$ (for $c = 1, \ldots, C$), and unit diagonal covariance matrix, is passed through a spatial linear filter L_i of length C to produce a correlated non-Gaussian vector conditionally to X_s, with a dependence structure generated by L_i. The mean of each sensor $\mu_i^{y,(c)}$ for $c = 1 \ldots C$ is finally added to obtain the correlated non-Gaussian vector $Y_s = [Y_s^{(1)} \ Y_s^{(2)} \ldots Y_s^{(C)}]^T$, conditionally to X_s.

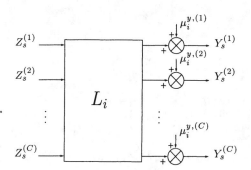

Fig. 2. Correlation model of an array of C sensors. L_i is a $C \times C$ mixture triangular matrix for the ith class.

The Fig.3 presents simulated images for marginals modeled as Generalized Gaussian pdf with the correlation structure presented in Fig.2. The channel $Y^{(c)}$ for $c = 1, 2$ and 3 is plotted by resizing the dynamic on 256 grey levels.

The ground truth image is the same as Pieczynski et-$Al.$ in [17]. The noise parameters are chosen such that :

- means and standart deviations are supposed to be the same for each channel ($i.e.,$ $\mu_i^y = \mu_i^{y,(1)} = \mu_i^{y,(2)} = \mu_i^{y,(3)}$ and $\sigma_i^y = \sigma_i^{y,(1)} = \sigma_i^{y,(2)} = \sigma_i^{y,(3)}$), correlation coefficients are supposed to be the same for cross channels ($i.e.,$ $\rho_i^y = \rho_i^{y,(1,2)} = \rho_i^{y,(1,3)} = \rho_i^{y,(2,3)}$)
- the "shape" parameters are also supposed to be the same for each channel ($i.e.,$ $p_i^z = p_i^{z,(1)} = p_i^{z,(2)} = p_i^{z,(3)}$)
- values of the above parameters for the two classes ($i = 1$ and 2) are set to : $p_1^z = 0.5$, $p_2^z = 4$, $m_2^y - m_1^y = 2\sigma_1^y = 2\sigma_2^y$ and $\rho_1^y = \rho_2^y = 0.8$.

Modeling the noise process as Fig.2 suggests the reversed model : the dependence structure of this model is straightforward, but solving for the marginals $g_i^c(z_s^{(c)})$ is more difficult. In order to circumvent this problem, Pieczynski *et-Al.* [17] suppose that marginals necessarily belong to a set of probability density functions. In this paper, we assume the independence of GG distributions $g_i^c(z_s^{(c)})$ which are representative of a large variety of pdf governing by the rate of decay parameter.

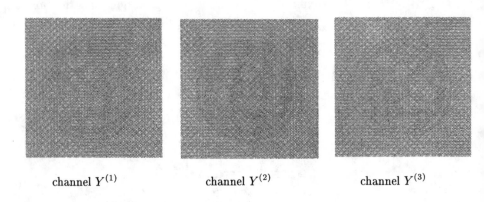

channel $Y^{(1)}$ channel $Y^{(2)}$ channel $Y^{(3)}$

Fig. 3. Non-Gaussian correlated observed image Y acording to the noise model presented in Fig.2 with GG marginals $g_i^c(z_s^{(c)})$.

The observed random vector Y_s, conditionally to $X_s = \omega_i$, described by its positive-definite symmetric covariance matrix, Σ_i^y ($i = 1, \ldots, K$), admits the Crout (Cholesky) factorization

$$\Sigma_i^y = L_i\, L_i^T \tag{10}$$

where L_i is a $C \times C$ unique lower triangular matrix. Then, the correlated vector Y_s can be transformed into an uncorrelated vector

$$Z_s = A_i Y_s \qquad \text{with } A_i = L_i^{-1}. \tag{11}$$

Directly from (11), we have $\mu_i^z = A_i\mu_i^y$ and the pdf's of Z_s and Y_s are related via

$$P_{Y_s|X_s}(y_s|x_s = \omega_i) = |A_i| P_{Z_s|X_s}(A_i y_s|x_s = \omega_i)$$

$$= |A_i| \prod_{c=1}^{C} g_i^c(\underbrace{[A_i y_s]^{(c)}}_{z_s^{(c)}}). \tag{12}$$

where $|.|$ denotes the determinant.

The following section presents the estimation and segmentation steps based on a hierarchical Markovian approach.

3 Hierarchical Markovian Approach

3.1 Introduction

In order to perform an unsupervised multispectral image segmentation, we should use Markov chains as proposed in [8]. Nevertheless, this approach needs a transformation of the image into a one dimensional set [18], which decreases the wealth of 2D representation. In this paper, we prefer a more general framework for image segmentation using a hierarchical Markovian modelization, based on the quad-tree [12]. Furthermore, as explained in section 2, we will consider Generalized Gaussian family as likelihood model.

3.2 Properties on the Quad-tree

We consider two sets of random variables : the labels $X = (X_s)_{s \in S}$ and the observed data $Y = (Y_s)_{s \in S}$. Both variables are indexed by S, a set of pixels on a quad-tree (*cf.* Fig 4a). $S = \{S^n\}_{n=0,...,R} = \{S^0, ..., S^R\}$ where R is the number of levels and $S^R = r$ is the root of the tree [12]. In our application, Y is defined only for $n = 0$. s^- is the unique parent of a pixel s and its four children will be designated by $s+ = \{t : s = t^-\}$ (*cf.* Fig. 4b). In the following, X^n and Y^n stand for $(X_s)_{s \in S^n}$ and $(Y_s)_{s \in S^n}$ respectively. Under Markovian assumption of causality in scale, *i.e.*, $P\left(X^n | X^k, k > n\right) = P\left(X^n | X^{n+1}\right)$ we assume that the transition probabilities can be factorized as :

$$P\left(X^n | X^{n+1}\right) = \prod_{s \in S^n} P\left(X_s | X_{s-}\right)$$

and that the observation model has for expression :

$$P\left(Y | X\right) = \prod_{s \in S} P\left(Y_s | X_s\right)$$

As a consequence, we deduce that (X, Y) is Markovian on the quad-tree [12] with

$$P\left(X, Y\right) = P\left(X_r\right) \prod_{s \in S} P\left(Y_s | X_s\right) \prod_{s > r} P\left(X_s | X_{s-}\right)$$

3.3 The segmentation algorithm

We note the prior root probabilities π_i and the prior parent-child transition probabilities a_{ij}:

$$\pi_i = P_{X_r}\left(x_r = \omega_i\right) \tag{13}$$

$$a_{ij} = P_{X_s | X_s}\left(x_s = \omega_j | x_{s-} = \omega_i\right) \tag{14}$$

In the following, we will use the *a posteriori* probabilities:

$$\xi_s(i) = P_{X_s | Y}\left[x_s = \omega_i | y\right] \tag{15}$$

$$\Psi_s(i, j) = P_{X_s, X_s^- | Y}\left[x_s = \omega_j, x_{s-} = \omega_i | y\right] \tag{16}$$

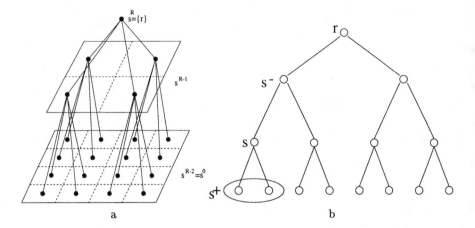

Fig. 4. a) The quad-tree structure b) Notations on a dyadic tree

These probabilities are estimated with the "Forward-Backward" algorithm [12, 10] which allows the non iterative and exact estimation of the marginal *a posteriori* distribution. The different parts of the algorithm are the following :

A) Initialization step Before to start the algorithm, we need to initialize some parameters :

$$\Phi^{[0]} = \left\{ \Phi_x^{[0]} = \{\pi_i, a_{ij}\}, \Phi_z^{[0]}(c) = \{\mu_i^{z,(c)}, \sigma_i^{z,(c)}, p_i^{z,(c)}\} \right\} \qquad (17)$$

with $\forall (i,j) \in [1, K]$ and $c \in [1, C]$

$$A_i^{[0]} = \begin{bmatrix} 1 & 0 & 0 \\ 0 & 1 & 0 \\ 0 & 0 & 1 \end{bmatrix} \qquad (18)$$

$$Z^{[0]} = Y \qquad (19)$$

(a) Data model parameters $\Phi_z^{[0]}(c) = \left\{ \mu_i^{z,(c)}, \sigma_i^{z,(c)}, p_i^{z,(c)} = 2 \right\}$ with $c \in [1, K]$ are obtained by Fuzzy K-means clustering technique [1] on image $Z^{[0]}$, generalized for multispectral images.

(b) Prior model parameters $\Phi_x^{[0]}$ are initialized by $\pi_i = \frac{1}{K}$, $a_{ij_{i \neq j}} = \frac{1}{2(K-1)}$ and $a_{ii} = \frac{1}{2}$.

Z is the image in the decorrelated space and $\Phi_z(c)$ for $c = 1, \dots, C$ are the conditional parameters in this space. In the following, conditional laws in Y,

$$f_i(y_s) = P_{Y_s | X_s}[y_s | x_s = \omega_i]$$

will be deduce from Generalized Gaussian conditional laws in Z,

$$g_i^{(c)}(z_s^{(c)}) = P_{Z_s^{(c)} | X_s}[z_s^{(c)} | \omega_i] = GG\left(z_s^{(c),[q]}, \mu_i^{z,(c),[q]}, \sigma_i^{z,(c),[q]}, p_i^{z,(c),[q]} \right) \qquad (20)$$

by

$$f_i^{[q]}(\boldsymbol{y}_s) = \prod_{c=\{R,G,B\}} g_i^{(c)}\left(z_s^{(c),[q]}\right) \times \left|A_i^{[q]}\right| \tag{21}$$

where A_i is defined in Eqs.10 and 11. $[q]$ represents the qth iteration.

B) Iterative Conditional Estimation step   For the kth iteration :

(a) Bottom-up pass on the quad-tree allows the estimation of $P_{X_s|Y_{\geq s}}(x_s|y_{\geq s})$
and $P_{X_s|X_{s-},Y_{\geq s}}(x_s|x_{s-},y_{\geq s})$. Particularly, for the root we have $s = r$ and
$P_{X_r|Y_{\geq r}}(x_r|y_{\geq r}) = P_{X_r|Y}(x_r|y)$.

(b) Top-down pass provides one realization of $X = x^{[q]}$ by random drawing
according to $P_{X_r|Y}(x_r|y)$ and $P_{X_s|X_{s-},Y_{\geq s}}(x_s|x_{s-},y_{\geq s})$.

(c) Updating of prior parameters:

$$a_{ij}^{[q+1]} = \frac{\sum_{s>r} \delta(x_{s-}^{[q]},\omega_i)\delta(x_s^{[q]},\omega_j)}{\sum_{s>r} \delta(x_{s-}^{[q]},\omega_i)} \tag{22}$$

$$\pi_i^{[q+1]} = \delta(x_r^{[q]},\omega_i) \tag{23}$$

where $\delta(x_s,\omega_i) = 1$ if $x_s = \omega_i$ and 0 elsewhere.

(d) Computation of $\Sigma_i^{y,[q]}$ according to the empirical moments.

(e) $A_i^{[q]}$ is calculated with the Cholesky decomposition such as

$$A_i^{[q]} \Sigma_i^{y,[q]} A_i^{[q]\,t} = I$$

(f) Determination of $\boldsymbol{Z}^{[q]}$:

$$\boldsymbol{Z}_s^{[q]} = A_i^{[q]}\boldsymbol{Y}_s, \forall s \in S^0 \text{ and } x_s^{[q]} = \omega_i$$

(g) Calculation of $\Phi_z^{[q+1]}$ for evaluation of Eq. (21).

(h) $q \leftarrow q+1$. If $q < q_{max}$, we repeat the iterative estimation step, using $\Phi^{[q+1]}$.

C) Classification step   the segmentation is performed according to the MPM criterion:

$$\hat{X}_s = \arg\max_{\omega_i \in \Lambda_{eq}} P_{X_s|Y}(x_s = \omega_i|y) = \arg\max_{\omega_i \in \Lambda_{eq}} \xi_s(i) \tag{24}$$

4   Images segmentation results

In this section, we present some results obtained on synthetic multispectral images presented in Fig.3 and one SPOT image.

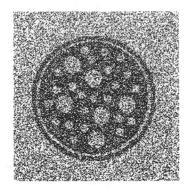 

(a) Gaussian correlated noise model   (b) non-Gaussian correlated noise model

Fig. 5. Segmented image on a quad-tree structure with Gaussian and non-Gaussian correlated noise

4.1 Synthetic multispectral image

The segmentation taking into account the GG model is best fitted to recover the real picture (Fig. 5b). The estimated parameters are close to the true parameters used to generate the multispectral correlated pictures. The obtained values for the estimation of the "shape" parameters are $\hat{p}_1^y = .55$ and $\hat{p}_2^y = 3.8$ again $p_1^y = .5$ and $p_2^y = 4$ for the true values. The percentage of error of bad labeled pixels is 30% in the Gaussian case and 20% in the GG case.

Other simulation results allow us to assert that :

- if the parameter p is far from 2 (particularly for small values, *i.e.*, $p < 2$) then the GG model is well adapted in comparison with the Gaussian one,
- the interest of the GG model appears for strongly noised images.

4.2 SPOT image

Fig.6 presents the three channels (infrared $Y^{(1)}$ (R), green $Y^{(2)}$ (G), blue $Y^{(3)}$ (B)) of a SPOT image which is represented with RGB colors in Fig. 7. The resulting segmentation using the GG noise model is presented in RGB colors on Fig.8.

Six classes are considered. Table 1. presents the estimated "shape" parameters for each class. It is interesting to note that this parameter is different from 2 on these real data and allows to best fit GG model to data instead of Gaussian model. However, by using the classic multispectral Gaussian model, a very close segmentation to Fig.8 is obtained. It is probably because the image doesn't look like strongly noisy.

5 Conclusion

This paper presents a hierarchical MRF modeling to segment multispectral correlated images. The interest of our approach consists in the modeling of the data

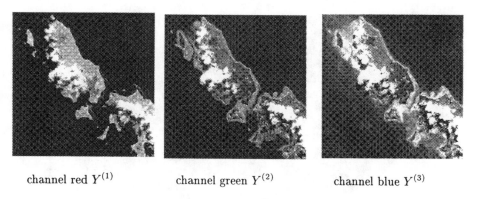

| channel red $Y^{(1)}$ | channel green $Y^{(2)}$ | channel blue $Y^{(3)}$ |

Fig. 6. Observed SPOT image Y for each channel.

class $n^o i$	1	2	3	4	5	6
channel R	1.28	5.64	1.01	3.42	1.68	2.26
channel G	0.71	1.81	0.97	4.31	2.30	2.63
channel B	0.48	2.88	1.73	2.35	1.99	1.19

Table 1. Estimation of the shape parameter

likelihood, which is based on a generalization of Gaussian pdf. Using a shape parameter, this modeling allows to capture the variety of the noise laws present in the distribution mixture. As our goal is to make the segmentation unsupervised, of course the number of parameter increases, even if the shape parameter allows to best fit observations and model. This is why we chose an Iterated Conditionnal Estimation based on the empirical moments, estimated on one X random sample at each iteration, for computational cost considerations. The parameter estimation step is also computed according to a ICE approach on a quad-tree and takes into account, explicitly, the correlation between sensors. Then, the segmentation step is based on MPM criterion.

This approach has been validated on synthetic correlated multispectral pictures, as for the estimation step than for the segmentation one. More precisely, we have shown that the Generalized Gaussian modeling improves the performance of the segmentation algorithm, in the presence of impulsionnal noise.

A   Generalized Gaussian generating procedure

In this appendix we present a generating procedure for the independent identically distributed (iid) random variable z_i $(i = 1, \ldots, N)$ modeled according to the generalized Gaussian distribution.

Fig. 7. Observed multispectral SPOT image

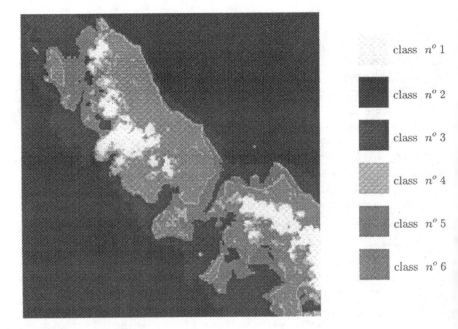

class n° 1

class n° 2

class n° 3

class n° 4

class n° 5

class n° 6

Fig. 8. Segmented image on a quad-tree structure with non-Gaussian correlated noise model.

The procedure starts by considering a random variable X distributed according to the gamma pdf :

$$P_X(x) = [\Gamma(1/p)]^{-1} x^{1/(p-1)} \exp(-x) u(x) \qquad (25)$$

where $u(.)$ is the unit step function and p is the positive parameter governing the rate of decay of the GG distribution to be generated. The generation of a gamma random variable is easily performed by exploiting its reproductive property (see, for example, the algorithm G-2 of [19]).

The second step of the procedure considers the nonlinear transformation

$$Y = \frac{1}{\eta(p)} X^{1/p} \qquad (26)$$

where $\eta(p)$ is defined in (4). It follows immediately that the pdf of Y is given by

$$P_Y(y) = [\Gamma(1/p)]^{-1} \eta(p)\, p \, \exp\left[-(\eta(p)|y|)^p\right] u(y) \qquad (27)$$

Finally, a random variable with GG pdf is easily generated multiplying Y by a random variable taking only equiprobable values ± 1.

References

1. X. Descombes A. Lorette and J. Zérubia. Extraction des zones urbaines fonde sur une analyse de la texture par modlisation markovienne. Rapport interne, INRIA - Sophia-Antipolis, 1998.
2. V. R. Algazi and R. M. Lerner. Binary detection in white non-Gaussian noise. Technical Report DS-2138, Lab., Lexington, MA, 1964.
3. G. Celeux and J. Diebolt. L'algorithme SEM : un algorithme d'apprentissage probabiliste pour la reconnaissance de mélange de densités. *Revue de statistiques appliquées*, 34(2):35–52, 1986.
4. G. Celeux and J. Diebolt. A random imputation principle : the stochastic EM algorithm. *INRIA research report*, (901):1–19, septembre 1988.
5. R. Chellappa and R.L. Kashyap. Statistical inference in gaussian markov random field models. *IEEE Computer Society Conference on Pattern Recognition and Image Processing*, pages 77–80, juin 1982.
6. Y. Dong, B. Forster, and A. Milne. Segmentation of radar imagery using gaussian markov random field models and wavelet transform techniques. *IEEE International Geoscience and Remote Sensing Symposium Proceedings (IGARSS-97)*, 4:2054–2056, 1997.
7. R. J. Elliott and L. Aggoun. Estimation for discret Markov random fields observed in gaussian noise. *IEEE Transaction on Information Theory*, IT-40(5):739–742, September 1994.
8. N. Giordana and W. Pieczynski. Estimation of generalized multisensor hidden Markov chains and unsupervised image segmentation. *IEEE Transactions on Pattern Analysis and Machine Intelligence*, 19(5):465–475, 1997.
9. C. Graffigne, F. Heitz, P. Pérez, F. Prêteux, M. Sigelle, and J. Zerubia. Hierarchical Markov random field models applied to image analysis : a review. In *SPIE Neural Morphological and Stochastic Methods in Image and Signal Processing*, volume 2568, pages 2–17, San Diego, 10-11 July 1995.

10. P. Pérez J.-M. Laferté, F. Heitz and E. Fabre. Hierarchical statistical models for the fusion of multiresolution image data. In *Proc. Int. Conf. Computer Vision*, pages 908–913, Cambridge, USA, June 1995.

11. M. Kanefsky and J. B. Thomas. On polarity detection schemes with non-Gaussian inputs. *J. Franklin Inst.*, 280:120–138, 1965.

12. J.-M. Laferté, P. Pérez, and F. Heitz. Discrete markov image modeling and inference on the quad-tree. Technical Report 1198, INRIA, July 1998.

13. M. Mignotte, C. Collet, P. Pérez, and P. Bouthemy. Unsupervised segmentation applied on sonar images. In *Proc. International Workshop EMMCVPR'97 : Energy Minimisation Methods in Computer Vision and Pattern Recognition (Springer editor)*, volume 1223, pages 491–506, Venice, Italy, May 1997.

14. M. Mignotte, C. Collet, P. Pérez, and P. Bouthmy. Three-class markovian segmentation of high resolution sonar images. *Journal of Computer Vision and Image Understanding*, submitted.

15. T. T. Pham and R. J. P. deFigueiredo. Maximum likelihood estimation of a class of non-gaussian densities with application to l_p deconvolution. *IEEE Trans. on Acoustics, Speech and Signal Processing*, 37(1):73–82, january 1989.

16. W. Pieczynski. Champs de Markov cachés et estimation conditionnelle itérative. *Traitement du signal*, 11(2):141–153, 1994.

17. W. Pieczynsky, J. Bouvrais, and C. Michel. Unsupervised bayesian fusion of correlated sensors. In *First International Conference on Multisource-Multisensor Information Fusion*, Las Vega, Nevada, USA, June 6-9 1998.

18. J.A. Provine and R.M. Rangayyan. Lossless compression of Peano scanned images. *Journal of Electronic Imaging*, 3(2):176–181, 1994.

19. R.Y. Rubinstein. *Simulation and Monte Carlo Method*. NY: Wiley, 1987.

Adaptive Bayesian Contour Estimation: A Vector Space Representation Approach*

José M. B. Dias

Instituto de Telecomunicações,
Instituto Superior Técnico,
1049-001 Lisboa, PORTUGAL
bioucas@lx.it.pt

Abstract. We propose a vector representation approach to contour estimation from noisy data. Images are modeled as random fields composed of a set of homogeneous regions; contours (boundaries of homogeneous regions) are assumed to be vectors of a subspace of $L^2(T)$ generated by a given finite basis; B-splines, Sinc-type, and Fourier bases are considered. The main contribution of the paper is a smoothing criterion, interpretable as *a priori* contour probability, based on the *Kullback distance* between neighboring densities. The *maximum a posteriori probability* (MAP) estimation criterion is adopted. To solve the optimization problem one is led to (joint estimation of contours, subspace dimension, and model parameters), we propose a *gradient projection* type algorithm. A set of experiments performed on simulated an real images illustrates the potencial of the proposed methodology

1 Introduction

Boundary estimation/detection plays a key role in image analysis/understanding, pattern recognition, computer vision, computer graphics, and computer-aided animation. Although the approaches to contour estimation are numerous, most of them share the same spirit: contours are obtained through the maximization of objective functions composed of a *prior term*, that favors contours with some attributes (e.g., continuity, smoothness, elasticity, and rigidity), and a *data term*, that measures the adjustment to data.

As in many other fields, different aspects of contour estimation have been addressed either under the energy-minimization framework or under the Bayesian framework.

1.1 Energy-Minimization Framework

Under the energy viewpoint, data and prior terms are interpretable as external energy, which attracts the contour to the desired features, and internal energy (e.g., due to contour *tension* and *rigidity*), respectively. This perspective was introduced in the original work of Kass [16], where the concept of *snake* (or active

---

* This work was supported by the Portuguese PRAXIS XXI program, under project 2/2.1.TIT/1580/95.

contour, or deformable model) was put out: "A snake is an energy-minimizing spline guided by internal constraint forces and influenced by image forces that pull it towards features such as lines and edges".

Since its introduction, the initial concept of active contour has been modified and improved in order to adapt it to different image classes and to overcome some of its drawbacks; namely, snake attraction by artifacts, snake degeneration, convergence and stability of the deformation process, myopia (i.e., use of image data only along the contour neighborhood), initialization, and model parameters estimation. References [3], [5], [11], [28] are illustrative examples of approaches to solve common problems with different snake techniques;

1.2 Bayesian Framework

Under the Bayesian viewpoint, the objective function referred above and its data and prior terms are interpretable as the *posterior* contour probability, the *likelihood function* associated to the observation mechanism, and the contour prior probability, respectively; since the sought contour maximizes the posterior probability, it is interpretable as the *maximum a posteriori* (MAP) estimate.

In many imaging problems (e.g., medical imaging, synthetic aperture radar, synthetic aperture sonar) the likelihood function can be derived from the knowledge of the generation mechanism [7], [10], rather than from other heuristic and common sense arguments. A statistical framework is therefore, in these cases, the correct choice.

Relevant advantages of the Bayesian approach are the following:

(a) it allows to include prior knowledge about the parameters to be estimated in a model-based fashion;

(b) it supplies an adequate framework for dealing with nuisance parameters (e.g., noise power, parameters distributions, blur coefficients).

1.3 Prior and Contour Representation

Contour representation and prior term, say p_c, are close issues that have received great attention, regardless of the viewpoint. In snake-type approaches the term p_c is, typically, of the form

$$p_c(\mathbf{c}) = \int R(\mathbf{c}(t)) \, dt, \tag{1}$$

where $R(\mathbf{c})$ measures the smoothness of the contour $\mathbf{c}$. Usually R is the combination of norms of different derivatives [16]. In the Bayesian approach, Markov random fields have been used as a way of modelling contour smoothness [7], [10], [13].

The prior contour information can be imposed by appropriate selection of function p_c and/or by introducing constraints on the set of admissible contours. For example, functional R in (1) can be tailored in order to have continuous derivative contours. Another possibility is to find the solution in a constrained space;

one can assume, for example, that contours belong to a parametrized family; i.e., $c(t) = c(t; \alpha)$, where α is defined in a given set Θ. This is the case of the deformable parametrized models/templates (e.g., Fourier [11], spline [2], and wavelet descriptors [4]).

1.4 Proposed Approach

Herein we address contour estimation under the Bayesian setting. We assume that images are piece-wise homogeneous random fields, and that contours are the boundaries of open connected sets.

Likelihood function

The likelihood function is derived from the image generation mechanism. We assume that pixels within each homogeneous region are independent samples of a selected random variable. For example, coherent amplitude images (e.g., ultrasound and synthetic aperture radar and sonar images) are Rayleigh distributed [26], X-ray images are very well approximated with a Gaussian distribution [19], and nuclear and confocal microscopic images are Poisson distributed [22].

We take as hypothesis that the random variables associated with the image pixels are independent, i.e., we assume the so-called *conditional independence property* [14]. In an image system, this is a correct assumption if the *resolution volumes* contributing to different pixels are disjoint. This is approximately the case in most acquisition systems, since there is no information gain in acquiring extremely correlated neighboring data.

Prior

We assume that contours belong to a finite-dimensional subspace spanned by a given vector basis. Smoothness properties of contours are closely related to those of basis vectors and to the subspace dimension K [6].

Roughly, the basis dimension determines the *frequency content* of contours. What should then be a suitable subspace dimention K? From the error projection point of view, K should be as large as possible. However, as K increases the subspace becomes less constrained and the estimated contours more *noisy*.

We tackle the estimation of the subspace dimension by assuming that contours $c(t; \alpha, K)$ are random, with probability density function of the form

$$p_c(c(t; \alpha, k)) = p_K(k). \tag{2}$$

The density p_K is chosen to be a decreasing function of K, thus favoring smooth contours. The exact structure of p_K is derived with basis on the estimate goodness.

The *maximum a posteriori probability* (MAP) estimation criterion is adopted. To solve the optimization problem one is led to, a *gradient projection* type algorhitm is used.

Dealing with contours as subspace elements is very appealing, namely due to the following:

(a) it is a parametrized approach: given a subspace basis, a natural parametriza-
tion is the set of basis coefficients, which are normally much smaller than
the basis dimension;

(b) given a generic contour $c \in S$, the closest contour of c in a subspace of S is
given by the projection of c onto this subspace.

Fourier descriptors [11], B-splines [2], and wavelets [4] have already been
proposed in the field of contour estimation. However, only work [11] explores the
vector space perspective; namely, it introduces a *minimum description length*
(MDL) [20] type principle for the determination the subspace dimension.

The MDL criterion, as applied in [11], is a smoothing criterion depending
only on the subspace dimension K. The smoothing criterion herein proposed,
besides depending on K, depends also on the Kullback distance between neighboring densities. This modification plays a key rule in assuring that estimated
space dimension is, to a great extent, independent of the neighboring densities
parameters.

The mais contribution of this work are the following:

(a) the study of the adequacy of subspaces generated by B-splines, Sinc-type,
and Fourier bases to the smoothness contour modeling;

(b) the introduction of a criterion for the subspace dimension estimation based
on the estimate robustness;

(c) the proposal of a complete adaptive scheme that iteratively estimates the
contour, the distribution parameters, and the subspace dimension.

The paper is organized as follows. Section 2 addresses aspects of contour representation using B-splines, Sinc-type and Fourier, bases. Section 3 proposes two
algorithms for contour estimation: the first assumes that the subspace dimension
is know; the second estimates the subspace dimension jointly with the contours.
Finally Section 4 presents results obtained with real data.

2 Contour Representations and Subspaces

Contours are closed periodic curves $c(t) = \{x(t), y(t)\}$, such that $c(t) = c(t+T)$.
For notational convenience, assume that contours are defined in the complex
plane C, and, therefore, $x(t)$ and $y(t)$ are the real and imaginary parts of c,
respectively.

We assume that $c \in L^2(T)$ (contour power over its period is finite). Since
$L^2(T)$ is a separable Hilbert space [17], then there exist bases $\{\varphi_n(t)\}$, for
$n = 0, 1, \ldots$, in $L^2(T)$, such that each vector $c \in L^2(T)$ is given by the linear combination

$$c(t) = \sum_{n=0}^{\infty} \alpha_n \varphi_n(t). \tag{3}$$

For an orthoganal basis $\{\varphi_n(t)\}$, coefficients α_n are unique and given by

$$\alpha_n = (c, \varphi_n) \equiv \frac{1}{T} \int_0^T c(t) \varphi_n^*(t) \, dt. \tag{4}$$

Equality (3) is to be understood as a limit in norm.

In the proposed approach, contours are chosen to be elements of the subspace $S_K \equiv \text{span}\,(\varphi_0, \ldots, \varphi_{K-1})$ generated by the vector basis $\{\varphi_n(t)\}$, for $n = 0, 1, \ldots, K - 1$. Each contour is then written as

$$c(t; \boldsymbol{\alpha}) = \sum_{n=0}^{K-1} \alpha_n \varphi_n(t). \tag{5}$$

where $\boldsymbol{\alpha} \equiv \{\alpha_0, \ldots, \alpha_{K-1}\}$.

The contour smoothness constraint is enforced by adequate selection of the basis $\{\varphi_n\}$ and of the subspace dimension K. In this work we consider B-spline [6], Sinc-type, and Fourier bases.

2.1 B-splines Basis

Spline functions are piecewise polynomials [6], which have been widely used to represent contours and surfaces in computer graphics, computer vision, and signal and image processing [8], [9], [25], [24]. Given the set of so-called *knots* $\{t_0 < t_1 < \ldots < t_k\} \subset \Re$, a m-order spline is a piece-wise polynomial function defined on $[t_0, t_k]$, which are C^{m-1} continuous on $[t_m, t_{k-m}]$. Given a knot sequence, the set of all splines which are C^{m-1} continuous on $[t_m, t_{k-m}]$ is a linear space of dimension $(k - n)$. The family of so-called B-splines functions, generated by the Cox-deBoor recursion [6], is a basis for this linear space. For equispaced knots, the B-spline are named *uniform*, and given by

$$\mathcal{B}_i^m(t) = \mathcal{B}_0^m(t - iT_s), \tag{6}$$

where index i denotes the i-th basis element, $T_s = t_{i+1} - t_i$, and

$$\mathcal{B}_0^m(t) = \underbrace{\mathcal{B}_0^0(t) * \mathcal{B}_0^0(t) * \ldots \mathcal{B}_0^0}_{m \text{ convolutions}}, \tag{7}$$

with

$$\mathcal{B}_0^0 = \begin{cases} 1 & t_i \leq t < t_{i+1} \\ 0 & otherwise. \end{cases}$$

Since we are interested in representing periodic curves, splines and their B-spline basis must be modified accordingly. For this purpose, define $\{\tilde{t}_n, n \in \mathcal{Z}\}$, with $\tilde{t}_n = t_{n \bmod k}$, as the periodic extension of the knot sequence $\{t_0 < t_1 < \ldots < t_k\}$ [12]. The basis functions $\tilde{\mathcal{B}}_i^m(t)$ are now periodic extensions of $\mathcal{B}_i^m(t)$, with period $T = t_k - t_0$, given by

$$\tilde{\mathcal{B}}_i^m(t) = \sum_{n=-\infty}^{\infty} \mathcal{B}_{i+nT}^m(t). \tag{8}$$

When using the spline representation, we assume that contours are elements of the subspace $S_K \equiv \text{span}\,(\tilde{\mathcal{B}}_0^m, \tilde{\mathcal{B}}_1^m, \ldots, \tilde{\mathcal{B}}_{K-1}^m)$ and, therefore, C^{m-1} continuous; the degree of smoothness is enforced by the subspace dimension K: as the subspace dimension increases, contours becomes less constrained.

B-splines exhibit local control: when representing curves as linear B-spline combinations, modifying a coefficient causes only a small part of the curve to change. This leads to simple an effective algorithms for computing displacements of the active contour under the influence of image forces.

In all examples herein presented we use $m = 3$. The spline contours are therefore C^2 continuous. This a common choice in vision and computer graphics [9]. Nevertheless, the concepts to be presented apply to any m-order spline.

2.2 Sinc-type Basis

A natural way to impose smoothness is to constrain the curves to be F-bandlimited (i.e., having maximum frequency F). The set of F-bandlimited curves of finite energy is a linear space; the sequence $\{S_n(t)\} = \{S_0(t - nT_s)\}$, where $T_s = 1/2F$ and

$$S_0(t) = \sqrt{2F} \frac{\sin 2\pi Ft}{2\pi Ft} \equiv \sqrt{2F} \operatorname{sinc}(2Ft), \tag{9}$$

is an orthonormal space basis. The projection of a curve $c(t)$ on $S_n(t)$ is exactly $c(nT_s)$ (see, e.g., [27]).

In order to adapt basis elements $S_n(t)$ to periodic curves, one should have $|S_0(t)| \simeq 0$ for $|t| \geq T/2$. Since $S_0(t)$ goes to zero as $|t|^{-1}$, for $|t| \to \infty$, this might not be fulfilled, if $K = T/T_s$ is too small. To overcome this difficulty, we replace the basis function (9) with

$$S_0(t) = \sqrt{2F} \frac{\sin 2\pi Ft}{2\pi Ft} \frac{\cos(2\pi Ft\rho)}{1 - (4\rho Ft)^2}. \tag{10}$$

Basis (10) is the impulse response of a raised cosine filter with a roll-off factor ρ [15], which goes to zero as $|t|^{-3}$, for $|t| \to \infty$. Seting, for example, $\rho = 0.4$, we can take, for most practical purposes, $S_0(t) \simeq 0$ for $|t| > 3T_s$.

The basis $\{S_n(t)\} = \{S_0(t - nT_s/(1+\rho))\}$, with $S_0(t)$ given by (10) generates the space of $F(1 + \rho)$-bandlimited functions. Therefore the smoothness of space elements is enforced by selecting F. Since the sampling interval $T_s/(1 + \rho)$ must be equal to T/K (i.e., an integer number of bases over T), the relation between F and K is

$$F = \frac{1}{2(1 + \rho)} \frac{K}{T}. \tag{11}$$

The periodic extension $\tilde{S}_0^m(t)$ of $S_0^m(t)$, with period T, is given by

$$\tilde{S}_i^m(t) = \sum_{n=-\infty}^{\infty} S_{i+nT}^m(t). \tag{12}$$

When using the bandlimited representation, we assume that contours are elements of the subspace $S_K = \operatorname{span}(\tilde{S}_0, \tilde{S}_1^m, \ldots, \tilde{S}_{K-1})$; the degree of smoothness is enforced by choosing K, which determines the maximum content frequency of contours according to (11).

As the B-spline basis, also the Sinc-type function (11), exhibits local control: the energy of $\tilde{S}_i(t)$ is concentrated at $t = iTs$.

2.3 Fourier Basis

The Fourier orthonornal basis is, probably, the representation most often used for periodic functions. In this representation the basis elements are given by

$$\mathcal{F}_n(t) = e^{j\frac{2\pi}{T}nt} \qquad n \in \mathcal{Z}. \tag{13}$$

With the Fourier representation, the most natural way of imposing smoothness is to restrict $L^2(T)$ to the finite subspace S_K generated by $\{\mathcal{F}_n(t)\}$, for $n = -K+1, \ldots, 0, \ldots, K-1$. The generated contours are then given by

$$c(t) = \sum_{n=-K+1}^{K-1} \alpha_n e^{j\frac{2\pi}{T}nt}. \tag{14}$$

Subspace S_K is obtained by filtering $L^2(T)$ elements with an ideal low-pass filter of cut frequency K/T. As in the Sync-type basis, the smoothness degree is enforced by selecting the maximum contour content frequency. Contrarily to the B-spline and Sync-type representations, the Fourier basis does not exhibit local control.

2.4 Contour Sampling and Fitting

Due to the discrete nature of digital images, one frequently faces the problem of finding, in a given subspace S_K, the *closest* element of a set of discrete points. In other words, given $\mathbf{c} \equiv \{c(t_0), c(t_1), \ldots, c(t_{N-1})\}$, find $\hat{\mathbf{c}} \in \mathcal{C}^N$, such that

$$\hat{\mathbf{c}} = \arg \min_{\mathbf{w}(w),\ w \in S_K} \|\mathbf{c} - \mathbf{w}\|, \tag{15}$$

where $\mathbf{w} \equiv \{w(t_0), w(t_1), \ldots, w(t_{N-1})\}$.

In the fitting problem at hand, the set $\{t_0, t_1, \ldots, t_{N-1}\}$ and the period T are not known. Herein we take $T = N$ and $t_i = i$, for $i = 0, 1 \ldots, N-1$, which is termed the *uniform assignment* strategy.

Define matrix $\mathbf{B}$ such that

$$[\mathbf{B}]_{ij} \equiv \tilde{\varphi}_j(t_i), \qquad i = 0, 1, \ldots, N-1, \qquad j = 0, 1, \ldots, K-1, \tag{16}$$

where $\tilde{\varphi}_j$ is one of the basis functions (8), (12), or (13).

In terms of matrix $\mathbf{B}$, minimization (15) is written as

$$\hat{\mathbf{c}} = \arg \min_{\mathbf{w} \in \mathcal{R}(\mathbf{B})} \|\mathbf{c} - \mathbf{w}\|, \tag{17}$$

where $\mathcal{R}(\mathbf{B})$ stands for the span generated by the columns of $\mathbf{B}$ (notation $\mathbf{B}_k$, when used, stresses that $k = \dim(\operatorname{span}(\mathbf{B}_k))$. Using the Euclidian norm, and assuming that $K \leq N-1$, the projection (17) is given by

$$\hat{\mathbf{c}} = \mathbf{B}\mathbf{B}^{\#}\mathbf{c}, \tag{18}$$

with

$$\mathbf{B}^{\#} \equiv (\mathbf{B}^H\mathbf{B})^{-1}\mathbf{B}^H \tag{19}$$

being the pseudoinverse matrix of $\mathbf{B}$ [21]. Matrix $\mathbf{B}^{\#}$ also solves the following problem:

$$\boldsymbol{\theta} = \arg \min_{\boldsymbol{\alpha} \in C^K} \|\mathbf{c} - \mathbf{B}\boldsymbol{\alpha}\| \tag{20}$$

$$= \mathbf{B}^{\#}\mathbf{c}, \tag{21}$$

being, therefore, $\hat{\mathbf{c}}$ also given by $\mathbf{c}^{\#} = \mathbf{B}\boldsymbol{\theta}$.

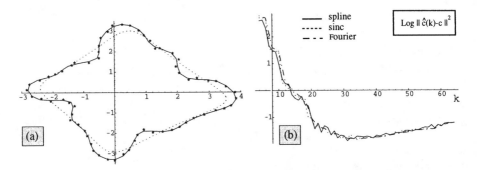

Fig. 1. (a) Projection of a noisy contour onto the subspace generated by Fourier basis. Stars represent the discrete contour to be projected, whereas doted and solid lines represent the projection onto $\mathcal{R}(\mathbf{B}_{10})$ and $\mathcal{R}(\mathbf{B}_{30})$, respectively. (b) Representation error for B-spline, Sinc-type, and Fourier bases.

Fig. 1(a) shows the projection of a hand traced contour contaminated with white noise, on the subspace generated by the Fourier basis. A complex zero-mean Gaussian random variable with standard deviation of 0.1 was added to each coordinate; stars represent the discrete contour to be projected; doted and solid lines represent the projections onto $\mathcal{R}(\mathbf{B}_{10})$ and $\mathcal{R}(\mathbf{B}_{30})$, respectively. Projection onto $\mathcal{R}(\mathbf{B}_{10})$ is clearly underfitted, while projection on $\mathcal{R}(\mathbf{B}_{30})$ is nearly optimum. This can be perceived from the error projection plotted on Fig. 1(b). The minimum error occurs, for the three representations, roughly at $K = 30$. For large values of K, the representation error increases, as the respective subspaces are now unable to smooth out the *high frequency components* of noise.

The similarity between the three representations, at least for the example presented, is evident. However, we would like to call attention to the following point: the representation error on the subspaces generated by Sinc-type and Fourier bases decreases until it reaches a minimum. This is not the case with the B-splines basis: the representation error, in this latter case, may increase, although slightly, with K. This behavior results from the non-nested structure of subspaces generated by the B-splines, whereas the subspaces generated by Sinc-type and Fourier bases are nested: the linear space of F-banlimited functions contains all subspaces of W-bandlimited functions with $W \le F$.

Subspaces having nested structure might be a desirable feature when the space dimension is unknown and it should be somehow estimated.

3 Image Generation Model

Let $\mathbf{c} = \{c_0, \ldots, c_{N-1}\}$ be the boundary of a connected region R_1 of the plane and R_2 the set of points not in R_1. Denote x_i as the image gray-level observed at i-th pixel, $\mathbf{x} = \{x_i\}$ as the set of image gray-levels, p_x as the gray-level density, and $\boldsymbol{\psi}_x = \{\boldsymbol{\psi}_1, \boldsymbol{\psi}_2\}$ as the density parameters (i.e., $p_x(x_i) = p_x(x_i|\boldsymbol{\psi}_1)$) for $i \in R_1$ and $p_x(x_i) = p_x(x_i|\boldsymbol{\psi}_2)$) for $i \in R_2$). Since we take as hypothesis that the image random variables, conditioned to the contour, are independent, it follows that

$$p_{\mathbf{x}|c}(\mathbf{x}|\mathbf{c}, \boldsymbol{\psi}_x) = \left(\prod_{i \in R_1} p_x(x_i|\boldsymbol{\psi}_1) \right) \left(\prod_{i \in R_2} p_x(x_i|\boldsymbol{\psi}_2) \right). \tag{22}$$

According to the proposed approach, contour $\mathbf{c}$ belongs to the subspace $\mathcal{R}(\mathbf{B}_K)$, being therefore given by $\mathbf{c} = \mathbf{B}_K \boldsymbol{\alpha}$, for $\boldsymbol{\alpha} \in \mathcal{C}^K$. Subscript K will occasionally be omitted.

3.1 Bayesian Approach to Contour Estimation

In accordance with the rationale already exposed, we assume that contours $\mathbf{c}(K) = \mathbf{c}(K, \boldsymbol{\alpha})$ are random vectors with probability density function given by

$$p_c(\mathbf{c}(k)) = p_K(k|\boldsymbol{\psi}_c), \tag{23}$$

where $\boldsymbol{\psi}_c$ denotes a parameter vector of p_K. Hence, the MAP estimate of the pair $(\mathbf{c}, K)$ is

$$(\hat{\mathbf{c}}, \hat{K}) = \arg \max_{k, \mathbf{c} \in \mathcal{R}(\mathbf{B}_k)} p_{\mathbf{x}|c}(\mathbf{x}|\mathbf{c}, \boldsymbol{\psi}_x) p_K(k|\boldsymbol{\psi}_c). \tag{24}$$

3.2 Known Space Dimension

Consider now that K is know. The MAP estimate (24) is, under this condition, simply the *maximum likelihood* (ML) contour estimate given by

$$\hat{\mathbf{c}} = \arg \max_{\mathbf{c} \in \mathcal{R}(\mathbf{B}_k)} L_{\mathbf{x}|c}(\mathbf{x}|\mathbf{c}, \boldsymbol{\psi}_x), \tag{25}$$

where $L_{\mathbf{x}|c}(\mathbf{x}|\mathbf{c}, \boldsymbol{\psi}_x) \equiv \log p_{\mathbf{x}|c}(\mathbf{x}|\mathbf{c}, \boldsymbol{\psi}_x)$ is the *loglikelihood function*.

To compute $\hat{\mathbf{c}}$, we implement an ascent type iterative algorithm that, in the t-th iteration, implements the following steps:

1. determine, in the unconstraint space $\mathcal{C}^N$, a contour displacement $\Delta\mathbf{c}^{(t)}$ that increases $L_{\mathbf{x}|c}(\mathbf{x}|\mathbf{c}, \boldsymbol{\psi}_x)$;
2. project $\mathbf{c}^{(t)} + \Delta\mathbf{c}^{(t)}$ onto the constrained subspace $\mathcal{R}(\mathbf{B})$, thus obtaining $\mathbf{c}^{(t+1)}$.

The displacement $\Delta\mathbf{c}^{(t)}$ is computed along orthogonal lines, as schematized in Fig. 2. Underlying this choice is the fact that the gradient of $L_{\mathbf{c}|c}(\mathbf{x}|\mathbf{c})$, computed with respect to $\mathbf{c}$, is orthogonal to the tangent vector $d\mathbf{c}/dt$ [23].

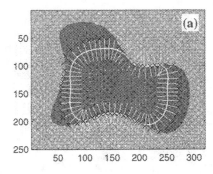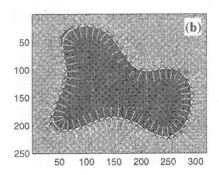

Fig. 2. Contour displacements computed along orthogonal lines. Crosses show the maximum of the loglikelihood function, along each orthogonal line. The doted line denotes the initial contour.

To prevent contours to be self intercepting, the orthogonal lines should be not too large. Work [23] proposes a technique for selecting long range orthogonal curves that do not intercept each other. Herein, however, we do not follow the mentioned technique, since it is not suited to our setting.

When the vector ψ_x is not known, we determine ML estimate of vector $(\mathbf{c}, \psi_x)$ according to

$$(\hat{\mathbf{c}}, \hat{\psi}_x) = \arg \max_{\mathbf{c} \in \mathcal{R}(\mathbf{B}), \psi_x} L_{\mathbf{x}|c}(\mathbf{x}|\mathbf{c}, \psi_x). \tag{26}$$

To compute $(\hat{\mathbf{c}}, \hat{\psi}_x)$, given by (26), the following iterative scheme is implemented:

Initialization: set $\mathbf{c}^{(0)}$, $\psi_x^{(0)}$, and δ
DO
   step 1: $\Delta\mathbf{c}^{(t)} = \arg \max_{\mathbf{u} \in \mathcal{O}(\mathbf{c}^{(t)})} L_{\mathbf{x}|c}(\mathbf{x}|\mathbf{c}^{(t)} + \mathbf{u}, \psi_x^{(t)})$
   where $\mathcal{O}(\mathbf{c}) \subset \mathcal{C}^N$ is the set of points defining
   orthogonal displacements to the contour $\mathbf{c}$
   step 2: $\mathbf{c}^{(t+1)} = \mathbf{c}^{(t)} + \mathbf{B}\Delta\mathbf{c}^{(t)}$
   step 3: $\psi_x^{(t+1)} = \arg \max_{\psi_*} L_{\mathbf{x}|c}(\mathbf{x}|\mathbf{c}^{(t+1)}, \psi_x)$
   step 4: $\Delta L = L_{\mathbf{x}|c}^{(t+1)} - L_{\mathbf{x}|c}^{(t)}$
While $|\Delta L| \geq \delta$.

Vector $\psi_x^{(t+1)}$ can be written in terms of regions R_1 and R_2 as

$$\psi_1^{(t)} = \arg \max_{\psi_1} \sum_{i \in R_1^{(t)}} L_{x|c}(x_i|\psi_1) \tag{27}$$

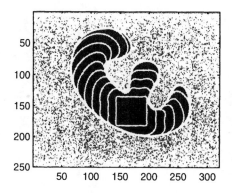

Fig. 3. Sequence of contour estimates produced by the proposed technique. The inner square represents the initial contour.

$$\psi_2^{(t)} = \arg\max_{\psi_1} \sum_{i \in R_2^{(t)}} L_{x|c}(x_i|\psi_2). \tag{28}$$

Expressions (27) and (28) depend on the particular structure of p_x. For example, for Gaussian densities with mean μ and variance σ^2, estimates of $\psi_i \equiv \{\mu_i, \sigma_i^2\}$, for $i = 1, 2$, are given by the sample mean and sample variance within the respective region.

Fig. 2, parts (a) and (b), displays estimates $\hat{c}^{(2)}$ and $\hat{c}^{(5)}$, respectively, of a Gaussian image with parameters $\{\mu_1 = 60, \sigma_1 = 15\}$ and $\{\mu_2 = 160, \sigma_2 = 30\}$. The boundary is obtained from a hand traced contour followed by projection onto $\mathcal{R}(\mathbf{B}_7)$.

Fig. 3 shows a sequence of contours estimates produced by the proposed algorithm. The long range nature of the external forces pulls the contour outwards as it was under an expansion force.

Fig. 4, part (a) and (b), displays two final estimates of Gaussian images with parameters $\psi_{(a)} = \{(\mu_1 = 80, \sigma_1 = 15), (\mu_2 = 160, \sigma_1 = 30)\}$ and $\psi_{(b)} = \{(\mu_1 = 100, \sigma_1 = 15), (\mu_2 = 100, \sigma_1 = 30)\}$. The estimated contours are nearly the true ones, even for image (b), which exhibits no contrast at all (i.e., $\mu_1 = \mu2$).

3.3 Unknown Space Dimension

Consider now that the space dimension is unknown and, consequently, it is also to be estimated jointly with the contour. Noting that $\mathbf{c} = \mathbf{c}(K)$, and according to expression (24), the MAP estimate of the space dimension is given by

$$\hat{K} = \arg\max_{k} \left\{ L_K(k|\psi_c) + \arg\max_{\mathbf{c} \in \mathcal{R}(\mathbf{B}_K)} L_{\mathbf{x}|c}(\mathbf{x}|\mathbf{c}, \psi_x) \right\}, \tag{29}$$

where $L_K(k|\psi_c) \equiv \log p_K(k|\psi_c)$. The estimated contour is the ML solution studied in the previous section, with the space dimension set to $\hat{K}$.

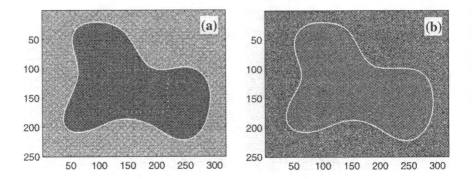

Fig. 4. Illustration of performance at very low contract: gray-levels in image (b) have the same mean value in both regions; in spite of this, the estimated contour is identical to the one estimated from image (a), which has high contrast.

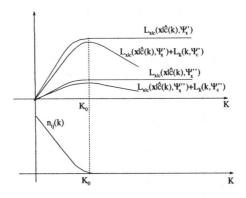

Fig. 5. Behavior of the loglikelihood function for two different vector parameters.

As in any Bayesian approach, the prior term must be specified. The first thought that could come to mind is to assume that $p_K(k|\psi_c)$ is uniformly distributed for $K_{min} \le K \le K_{max}$; the estimate of $(\mathbf{c}, K)$ would therefore be interpretable as a ML estimate. Unfortunately, this attempt would fail. The reason is the following: due the nested nature of parameter spaces $(\mathcal{R}(\mathbf{B}_K) \subset \mathcal{R}(\mathbf{B}_{K+1}))$, the loglikelihood function $L_{\mathbf{x}|c}(\mathbf{x}|\hat{\mathbf{c}}(\hat{K}), \psi_x)$ will be a monotonically (or at least nondecreasing) function of K, so it will reach its maximum at K_{max}.

The problem of choosing the order of competing models of different dimensions is termed a *model order selection* problem. Among the approaches that have been suggested to this problem, the *Akaike information criterion* (AIC) [1], and the *minimum description length* (MDL) [20] have gained popularity. Work [11], also on contour estimation, applies the MDL principle to derive the term $L_K(k|\psi_c)$ of (29).

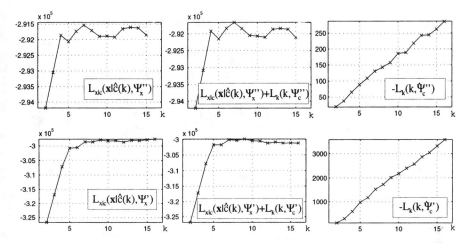

Fig. 6. Loglikelihood and prior behavior for the images shown in Fig. 7.

In this work we propose the prior

$$p_K(k|\psi_c) = \frac{1}{Z}e^{-\alpha k}, \qquad \alpha > 0, \tag{30}$$

where Z is a normalizing constant, and α is given by

$$\alpha = \frac{n_c}{4}\mu D(\psi_1, \psi_2), \tag{31}$$

where n_c is the number of contour pixels, $\mu \simeq 0.1$, and

$$D(\psi_1, \psi_2) \equiv D(\psi_1\|\psi_2) + D(\psi_2\|\psi_1), \tag{32}$$

is the *symmetric Kullback distance* [18] and $D(\psi_1\|\psi_2)$ the *Kullback distance* [18] between densities $p_x(x_i|\psi_2)$ and $p_x(x_i|\psi_2)$ given by

$$D(\psi_1\|\psi_2) = E_{\psi_1} \log \frac{p_x(x_i|\psi_1)}{p_x(x_i|\psi_2)}. \tag{33}$$

The derivation of prior (30) is out of the scope of this paper. We present, however, an informal justification. Aiming at this purpose, define

$$L_{c|\mathbf{x}}(k|\mathbf{x}, \psi) = L_K(k|\psi_c) + \arg\max_{c\in\mathcal{R}(\mathbf{B}_K)} L_{\mathbf{x}|c}(\mathbf{x}|c, \psi_x). \tag{34}$$

Define also the sets A_{12} and A_{21} containing pixel indexes wrongly classified: in the first case region 1 has been detected, whereas in the second case region 2 has been detected. Assume that the true dimension space is k_0 and introduce

$$\Delta L_{c|\mathbf{x}}(k) \equiv L_{c|\mathbf{x}}(k|\mathbf{x}, \psi) - L_{c|\mathbf{x}}(k_0|\mathbf{x}, \psi). \tag{35}$$

The difference $\Delta L_{c|\mathbf{x}}(k)$ can be written in terms of $A_{12}(k)$ and $A_{21}(k)$ as

$$\Delta L_{c|\mathbf{x}}(k) = \sum_{i \in A_{12}(k)} \log \frac{p_x(x_i|\psi_1)}{p_x(x_i|\psi_2)} +$$
$$\sum_{i \in A_{21}(k)} \log \frac{p_x(x_i|\psi_2)}{p_x(x_i|\psi_1)} + \Delta L_K(k), \qquad (36)$$

where

$$\Delta L_K(k) \equiv L_K(k|\psi_c) - L_K(k_0|\psi_c). \qquad (37)$$

A lengthy manipulation of (36), and a few weak assumptions, lead to

$$E\{\Delta L_{c|\mathbf{x}}(k)\|k_0\} = -n_{12}(k)D(\psi_1, \psi_2) + \Delta L_K(k), \qquad (38)$$

where $n_{ij}(k) \equiv \#A_{ij}(k)$ is the number of elements of A_{ij}.

The interpretation of (38) is clear: term $n_{12}(k)D(\psi_1, \psi_2)$ tends to zero as the the number of missclassified pixels tends to zero. The vanishing rate is proportional to the symmetric Kullback distance $D(\psi_1, \psi_2)$.

Fig. 5 schematizes the behavior of the loglikelihood function $L_{\mathbf{x}|c}(\mathbf{x}|\mathbf{c}, \psi_x)$ and of the prior term $L_K(k|\psi_c)$, for two vector parameters ψ_x' and ψ_x''. When $n_{ij}(k)$ approaches zero, $L_{\mathbf{x}|c}(\mathbf{x}|\mathbf{c}, \psi_x)$ approaches a constant. By adding an appropriate prior term $L_K(k, \psi_c')$ to the loglikelihood function, a maximum is obtained at $k = k_0$. For the second vector parameter ψ_x'', the increasing rate of $L_{\mathbf{x}|c}(\mathbf{x}|\mathbf{c}, \psi_x'')$ is slower than $L_{\mathbf{x}|c}(\mathbf{x}|\mathbf{c}, \psi_x')$. If the prior term $L_K(k, \psi_c')$ was used, a maximum would be obtained for $k < K_0$.

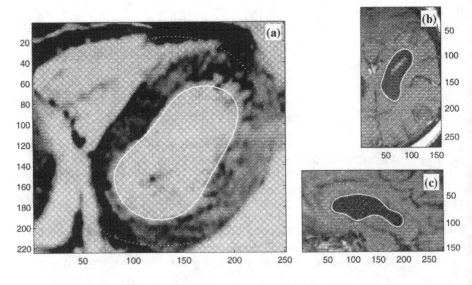

Fig. 7. Magnetic resonance images: (a) heart and (b,c) brain.

Fig. 6 displays the loglikelihood and prior behavior for images shown in Fig. 4: the left column plots data from part (b), while the right column plots data from part (a). For both cases the maximum of $L_{\mathbf{x}|c}+L_k$ is obtained for $k=7$. However, the loglikelihood function $L_{\mathbf{x}|c}(\mathbf{x}|\mathbf{c},\psi''_x)$ grows much slower than $L_{\mathbf{x}|c}(\mathbf{x}|\mathbf{c},\psi'_x)$, approximately by a factor of 12, determined in the interval $k \in \{2,3,4,5\}$. For the maximizer k to be the same, the prior term $L_K(k,\psi''_c)$ must grow slower than $L_K(k,\psi'_c)$ by the same factor. This is, with great approximation, what happens.

The symmetric Kullback distance, for Gaussian distributions, is given by

$$D(\psi_1,\psi_2) = \frac{(\sigma_1^2 - \sigma_1^2)^2 + (\sigma_1^2 + \sigma_1^2)(\mu_1 - \mu_2)^2}{2\sigma_1^2\sigma_2^2}. \tag{39}$$

Noting that the parameters associated with images displayed in Fig. 6 are $\psi''_x = \psi_x(a) = \{(\mu_1 = 80, \sigma_1 = 15), (\mu_2 = 160, \sigma_2 = 30)\}$ and $\psi'_x = \psi_x(b) = \{(\mu_1 = 100, \sigma_1 = 15), (\mu_2 = 100, \sigma_2 = 30)\}$, it follows that $D(\psi''_1,\psi''_2)/D(\psi'_1,\psi'_2) \simeq 10$, in accordance with the experimental results.

Fig. 7 shows estimated contours over real magnetic resonance images: (a) heart and (b,c) brain. The Gaussian model and the Fourier basis was used. The estimated space dimensions are 4, 6, and 7, which are in agreement with the contours frequency content.

We stress the methodology robustness with respect to image nonhomogeneities and poor contour initializations.

4 Concluding Remarks

This paper introduced a novel adaptive methodology to contour estimation from noisy images. The approach was Bayesian: images were modeled as as a set of homogeneous regions, in a statistical sense. Contours were assumed to be vectors of a subspace generated by a finite basis: B-splines, Fourier, and Sinc-type bases were studied. It was concluded that Fourier, and Sinc-type bases were better suited to the proposed technique due to its nesting property.

A relevant contribution of the paper was on the contour prior design. By parametrizing the *a priori* probability density function with the symmetric Kullback distance between densities of each homogeneous region, the proposed algorithm produces meaningful estimates.

The proposed scheme is completely adaptive; all model parameters are estimated jointly with the contour. Results obtained with simulated and real data show the adequacy of the proposed approach.

References

1. H. Akaike. A new look at statistical model identification. *IEEE Transactions on Automatic Control*, AC-19:716–723, 1974.
2. A. Amini, R. Curwen, and J. Gore. Snakes and splines for tracking non-rigid heart motion. In *Proc. European Conf. on Com. Vision - ECCV'96*, pages 251–261, Cambrige, 1996. Springer Verlag.

3. A. Amini, T. Weymouth, and R. Jain. Using dynamic programming for solving variational problems in vision. *IEEE Trans. on Pattern Analysis and Machine Intelligence*, PAMI-12(9):855–867, Sep. 1990.
4. G. Chuang and C. Kuo. Wavelet description of planar curves: theory and applications. *IEEE Trans. on Image Proc.*, 5:56–70, 1996.
5. L. Cohen and I. Cohen. Finite-element methods for active contour models and baloons for 2D and 3D images. *IEEE Trans. Pattern Anal. Machine Intell.*, 15:1131–1147, 1993.
6. C. de Boor. *A Practical Guide to Splines*. Springer Verlag, New York, 1978.
7. J. Dias and J. Leitão. Wall position and thickness estimation from sequences of echocardiograms images. *IEEE Transactions on Medical Imaging*, 15:25–38, Feb. 1996.
8. P. Dierckx. *Curve and Surface Fitting with Splines*. Oxford University Press, Oxford, 1993.
9. G. Farin. *Curves and Surface Fitting for Computer Aided Geometrical Design*. Oxford University Press, Oxford, 1993.
10. M. Figueiredo and J. Leitão. Bayesian estimation of ventricular contours in angiographic images. *IEEE Trans. Med. Imag.*, MI-11(3):416–429, September 1992.
11. M. Figueiredo, J. Leitão, and A. K. Jain. Adaptive b-splines and boundary estimation. In *Proc. of the IEEE Comp. Soc. Conf. on Com. Vision and Patt. Rec. - CVPR'97*, pages 724–730, San Juan (PR), 1997.
12. M. Flickner, J. Hafner, E. Rodriguez, and J. Sanz. Periodic quasi-orthogonal spline basis and applications to least square fitting of digital images. *IEEE Trans. on Image Proc.*, 5:71–88, 1996.
13. N. Friedland and D. Adam. Automatic ventricular cavity boundary detection from sequential ultrasound images using simulated annealing. *IEEE Transactions on Medical Imaging*, MI-8(4):344–353, December 1989.
14. S. Geman and D. Geman. Stochastic relaxation, Gibbs distribution and the Bayesian restoration of images. *IEEE Trans. Pattern Analysis and Machine Intelligence*, PAMI-6(6):721–741, Nov. 1984.
15. S. Haykin. *Communication Systems*. John Wiley & Sons, New York, 1983.
16. M. Kass, A. Witkin, and D. Terzopoulos. Snakes: Active countour models. *Intern. Journal of Comp. Vision*, 1:259–268, 1987.
17. E. Kreyszig. *Introduction to Functional Analysis with Applications*. John Wiley & Soons, New York, 1978.
18. S. Kullback. *Information Theory and Statistics*. Peter Smith, 1978.
19. A. Macovski. *Medical Imaging Systems*. Prentice-Hall, Englewood Cliffs, NJ, 1983.
20. J. Rissanen. Modeling by shortest data description. *Automatica*, 14:465–471, 1978.
21. L. Scharf. *Statistical Signal Processing. Detection, Estimation and Time Series Analysis*. Addison-Wesley, New York, 1991.
22. D. L. Snyder. *Random Point Process*. John Wiley & Sons, New York, 1975.
23. H. Tagare. Deformable 2-D template maching using orthogonal curves. *IEEE Trans. Medical Imaging*, 1:108–117, 1997.
24. M. Unser, A. Aldroubi, and M. Eden. B-spline signal processing: part i–efficient design and applications. *IEEE Trans. Signal Process.*, 41(2):834–847, 1993.
25. M. Unser, A. Aldroubi, and M. Eden. B-spline signal processing: part i–theory. *IEEE Trans. Signal Process.*, 41(2):821–833, 1993.
26. R.F. Wagner, M.F. Insana, and D.G.Brown. Statistical properties of radiofrequency and envelope detected signals with applications to medical ultrasound. *J. Opt. Soc. Am.*, 4:910–922, May 1987.
27. A. Zayed. *Advances in Shannon's Sampling Theory*. CRC Press, 1993.
28. S. Zhu and A. Yuille. Region competion: Unifying snakes, region growing, energy/Bayes/MDL for multi-band image segmentation. *IEEE Trans. Pattern Anal. Machine Intell.*, 18:884–900, 1996.

Adaptive Pixel-Based Data Fusion for Boundary Detection

Toshiro Kubota *

Intelligent Systems Laboratory, Department of Computer Science
University of South Carolina, Columbia, SC 29208, USA
kubota@cs.sc.edu, (803) 777-2404, (803) 777-3767 (Fax)

Abstract. It is of practical importance to fuse data obtained by multiple sensors for improving the performance of computer vision systems. This paper introduces an algorithm for pixel-based data fusion on the variational framework. An adaptive system fuses data effectively using a variational technique. Previously, we have introduced a technique to fuse gray-scale image and texture extracting features for segmenting an image with both textured and non-textured surfaces. This paper extends the study for more general multi-valued data and improve the previous algorithm in terms of the performance and speed.

Keywords: *segmentation, data fusion, color, texture, variational method*

1 Introduction

There has been an immense amount of interest in formulating various computer vision problems in the energy minimization framework[1,7]. One of such problems is segmentation or boundary detection [1]. Models such as *weak membrane model* (WMM) and *Mumford-Shah model* have gained much attention recently for segmentation under noisy circumstances [2,17,18]. It describes often conflicting terms of data fidelity and data smoothness along with a binary process to introduce discontinuities in the smoothness measure.

With this method, the objective becomes finding the global minimum of the energy function such as

$$E = \int \alpha\|s - g\|^2 + \mu\|\nabla s\|^2(1 - l) + \nu l \, dx \, dy . \tag{1}$$

where α, μ and ν are weight parameters and $\|\cdot\|$ is a L2-norm. The function $g(x, y)$ represents data collected by a sensor at the coordinate (x, y) and often referred to as *observation*, $s(x, y)$ is called *surface process* and represents a smooth

---

* Research partially supported by ONR Grant N00014-97-1-1163 and ARO Grant DAAH04-96-10326

[1] In this paper these two terms are considered to be interchangeable, although some claim that they should be treated differently as the former is region based and the latter is edge based.

surface constrained by the observation, $l(x, y)$ is a binary process and called *line process*. The line process represents existing/non-existing of a boundary at (x, y) with $l = 1/l = 0$, respectively. The first term is called *fidelity*, second term is *stabilizer* and the third term is *penalty*.

Through minimization of the energy function, the surface process maintains closeness to the observation, and develops into a smooth surface within a boundary where $l = 0$. The surface process is allowed to be discontinuous at boundaries where $l = 1$. The penalty term penalize having $l = 1$ so that it prevents l from being 1 everywhere. As ν increases, the number of boundary pixels in the result decreases.

The purpose of this paper is to extend this well-known model for multi-valued data where the data can be obtained from multiple sensors or from multiple feature extractors. We assume that each sensor or feature extractor is aligned so that no registration is required on the obtained data. This type of data fusion problems are often referred to as *pixel-based fusion* [16]. Thus, each pixel has a corresponding vector representation where each vector element is the value measured by each sensor or extracted by each feature extractor. Throughout the paper, each vector element is referred to simply as *feature*.

A key to a good boundary detection process is to detect discontinuities in spatially laid out data field. Each sensor or feature extractor is designed to capture only certain surface properties, thus can only detect certain surface boundaries. In order to build a reliable automated boundary detection system for general environments, it is important to have a mechanism of selecting a subset of sensors adaptively from region to region.

Previously, we have reported a fusion technique for boundary detection using gray-scale pixels and texture features. [10]. The technique combines the features by a set of normalized weights and it minimizes an energy term similar to WMM with respect to the weight to obtain a near optimal set of weights at each feature location. This paper gives another interpretation of the algorithm and proposes a new computational techniques. It also presents more extensive segmentation experiments on both synthetic and natural color images. The new technique is computationally cheaper than the previous one and gives better boundary representations in various experiments we have performed.

The paper is organized as follows. Section 2 describes our algorithm. It proposes an energy model based on the WMM and two different computational approaches for the minimization process. Section 3 provides various experimental results for color image segmentation. Section 4 provides experimental results for texture segmentation. Section 5 gives brief summary and conclusions.

2 Algorithm

2.1 Model I

The first model is the one reported in [10]. The main idea is to combine multiple features by normalized weights and to minimize the energy function (1) with respect to the weights to obtain a set of near-optimal weights. The energy is also minimized with s and l using some conventional method.

A rationale of this approach is to trust smooth data while disregard rough data assuming that the measurements be smooth within region boundaries. The quantity of λ or the amount of trust on the data is measured locally resulting in spatially variant adaptation of the system.

The idea of estimating parameters by minimizing an objective function is not new and has been studied in different perspectives [3, 20, 14]. In [20], the idea was applied for obtaining a set of compatibility coefficients for relaxation labeling problems. Region based segmentation algorithms often employ Estimation-Maximization (EM) strategy where segmentation (Estimation) and Maximum Likelihood parameter estimation (Maximization) are performed iteratively until convergence [3, 11]. Some heuristical approaches have been proposed as well. In [13], WMM incorporates Gabor features where the features are arranged in the Gabor 4D space (x, y, scale and orientation). An appropriate mixture is computed through diffusion in the 4D feature space. Our interest is to apply the idea of the relaxation labeling to WMM for boundary detection on multi-valued data field.

The weights are denoted as λ_i and satisfies the following constraints:

$$\sum_i \lambda_i = 1, \quad \lambda_i \geq 0. \tag{2}$$

Then the weighted features and surface processes are

$$\bar{g}_i = \lambda_i g_i , \tag{3}$$

$$\bar{s}_i = \lambda_i s_i . \tag{4}$$

The total energy of the model is

$$E = \sum_i \|\bar{s}_i - \bar{g}_i\|^2 + \mu_1 \|\nabla \bar{s}_i\|^2 (1 - l) + \nu l \tag{5}$$

If we assume

$$\frac{\delta \lambda_i}{\delta x}(1 - l) = \frac{\delta \lambda_i}{\delta y}(1 - l) = 0, \tag{6}$$

then

$$E = \sum_i \left\{ \lambda_i^2 \|s_i - g_i\|^2 + \lambda_i^2 \mu_1 \|\nabla s_i\|^2 (1 - l) \right\} + \nu l . \tag{7}$$

In order to ensure the assumption (6) to be valid, an additional constraint

$$\|\nabla \lambda_i\|^2 (1 - l) = 0 \tag{8}$$

is added.

It is desirable for the solution of the minimization process to be invariant to redundant features which can be either duplicates of already included features or features with no useful information such as white noise. A problem with the above model is that the fidelity and stabilizer are dependent on the number of features since $1/M \leq \sum_i \lambda_i^2 \leq 1$. Thus adding redundant features can reduce the influence of the fidelity and stabilizer terms relatively to the penalty. To achieve the invariance, ν can be made dependent on $\{\lambda_i\}$ as well by

$$\tilde{\nu} = \sum_i \lambda_i^2 \nu . \tag{9}$$

Now the objective is to minimize

$$E = \sum_i \lambda_i^2 \|s_i - g_i\|^2 + \lambda_i^2 \mu_1 \|\nabla s_i\|^2 (1 - l) + \tilde{\nu} l . \tag{10}$$

with

$$\sum_i \lambda_i = 1, \quad \lambda_i \geq 0, \quad \|\nabla \lambda_i\|^2 (1 - l) = 0 \tag{11}$$

Solving the Euler-Lagrange equation gives the update rule for s_i as

$$\frac{ds_i}{dt} \propto -\lambda_i^2 \|s_i - g_i\| + \nabla \cdot (\mu_1 \lambda_i^2 (1 - l) \nabla s_i) \tag{12}$$

The update rule for l using the mean field annealing is [6]

$$l = \frac{1}{1 + e^{(\tilde{\nu} - \mu_1 \sum \lambda_i^2 \|\nabla s_i\|^2)/T}} \tag{13}$$

where T is the temperature for the annealing process.

Initially, λ_i are all set to $1/M$. With some prior information available, this initial condition can be biased according to the prior, which may improve the system performance. The update process for λ_i takes two steps. First, λ_i is updated individually by minimizing

$$E = \sum_i \lambda_i^2 \|s_i - g_i\|^2 + \lambda_i^2 \mu_1 \|\nabla s_i\|^2 (1 - l) + \tilde{\nu} l + \mu_2 \|\nabla \lambda_i\|^2 (1 - l). \tag{14}$$

Its update rule is

$$\frac{d\lambda_i}{dt} \propto -\lambda_i \|s_i - g_i\|^2 - \mu_1 \lambda_i \|\nabla s_i\|^2 + \nabla \cdot (\mu_2 (1 - l) \nabla \lambda_i) - \lambda_i \nu l. \tag{15}$$

Second, $\{\lambda_i\}$ is normalized to meet the hard constraints of (2). In order to assure the positivity, any $\lambda_i < 0$ is set to 0 implying that s_i is not useful at the location. Then λs are normalized by

$$\lambda_i = \frac{\lambda_i}{\sum_j \lambda_j} \tag{16}$$

to ensure they sum up to 1. Thus the whole process of updating λ is similar to the probabilistic relaxation ([21]). A special care has to be taken when all λs are 0. In that case, they are set back to $1/M$ as the initial condition. However, this condition never happened in our experiments, mainly because the descent rate of λ_i is proportional to λ_i as seen in (15). As λ_i approaches to 0, the rate of its change decreases.

2.2   Model II

It can be seen easily that the energy equation, (10), can be rewritten as

$$E = \sum_i \lambda_i^2 E_i \qquad (17)$$

where E_i is the energy quantity (1) associated with the ith feature. The above equation suggests that λ can be interpreted as a normalized weight for E and not for s. The significance of this interpretation is that the dynamics of s_i is no longer dependent on λ_i and one can simply apply the pixel-based data fusion at the energy level instead of the feature level. The constraint (8) is no longer necessary for the formulation of (17). However, it enforces smoothness in λ and decreases the system's sensitivity to noise.

With this model, the update rule for s_i is

$$\frac{ds_i}{dt} \propto -\|s_i - g_i\| + \nabla \cdot (\mu_1(1 - l)\nabla s_i) \ . \qquad (18)$$

The update rules for l and λ_i are the same as (13) and (15), respectively.

Note that difference in Model I and II are not the energy model but the computation or minimization process. Both have the identical energy landscape as (10) and (17) are identical. Model I assigns a spatially variant diffusion speed for s_i where the speed is proportional to λ_i^2. On the other hand, Model II allows each surface process, s_i, to diffuse at the maximum speed independently from λ_i. Therefore, as will be demonstrated in the next section, Model I does not diffuse areas with high data fluctuations when another feature does not contain fluctuations in the region. This non-diffused areas will form separate regions by themselves as the energy minimization process converges.

As far as the amount of computation is concerned, this model is simpler than the first model since the update rule for s_i is simpler. With our implementation, Model II saved approximately 20% of computation time over Model I. It took approximately 300 and 400 seconds of CPU times to run 256x256 RGB color image segmentation on an SGI O_2 system with Model I and II, respectively. The same experiment took 220 seconds with WMM.

Model II can also be interpreted as the WMM energy with a modified feature gradient strength. With this interpretation, the gradient strength of the feature vector is computed as

$$\|\nabla s\|^2 = \sum_i \lambda_i \|\nabla s_i\|^2. \qquad (19)$$

Thus it can be considered as an adaptive gradient based segmentation technique. Similarly, $\{\lambda_i\}$ can be considered as a row of spatially variant linear transformation matrix.

Another alternative for the energy model is

$$E = \sum_i \lambda_i E_i. \qquad (20)$$

This is a different energy model from (10). With $\mu_2 = 0$ (i.e. without any spatial constraints on λ_i), (10) or (17) will yields

$$\lambda_i = \frac{E_i}{\sum_j E_j} \qquad (21)$$

as a solution. On the other hand, (20) will yield

$$\lambda_i = \begin{cases} 1 & if \ E_i = min_j E_j \\ 0 & otherwise \end{cases} . \qquad (22)$$

Thus, the former allows continuous values where $\lambda_i/\lambda_j = E_i/E_j$. The latter is a discrete system with a winner-take-all strategy. Both energy models produce very similar results in our experiments and (20) has a slightly less amount of computation compared to (17). The results presented in the following sections are produced by using the winner-take-all model (20).

3　Color image segmentation

The two algorithms described in the previous section are tested on RGB color images.

Problems of segmenting color images have been studied from various different perspectives and many algorithms have been proposed in literature. They can be categorized into 1) histogram based 2) clustering based [8, 23], and 3) Markov Random Field based techniques [15, 22, 19]. Techniques of the first class explicitly use histogram of each color component to separate foregrounds from the background. It is simple but does not work well on natural images. Techniques of the second class apply some clustering techniques in the transformed color space. Spatial information can be incorporated by adding pixel coordinates as features. Techniques of the third class incorporate both data proximity and spatial organization into Bayesian classification framework. Our scheme can be categorized into the third class as the WMM energy model can be linked to the Markov process through the Gibbs distribution [7].

Three test images were created for this experiments, each simulating a different type of noise sources. Figure 1 shows the test images. The columns in the figure show the luminance, R-band, G-band and B-band of RGB color images, respectively. The first one simulates noisy sensors in disjoint regions. The signal to noise ratio inside the defective regions is 0.425. The second image has noise everywhere on one of the sensors (Blue channel). It simulates a channel noise or the case where one of the sensor is ineffective in capturing information of the environment. The signal to noise ratio in the blue channel is 0.5. The third image has noise everywhere on all channels. It simulates system noise or a noisy environment. The signal to noise ratio of the image is 1.275. As seen in the figure, the luminance by itself does not contain enough information to extract boundaries reliably.

First, an input RGB image is normalized to make it zero mean and unit variance. This normalization allows the segmentation system to be independent of the contrast of the input image. Figure 2 is the results of applying boundary detection without updating λ_i. Thus, each feature is treated equally. It shows that the technique cannot extract the ground truth boundaries without picking up spurious edges. It performed well on the third case where all channels are equally noisy everywhere. In such cases, no weight adaptation is necessary.

Figure 3 is the results of applying the Model I to the test images. The technique could delineate the ground truth boundaries without picking up spurious edges. For the first image, it delineated the noisy regions as separate regions. This is acceptable since these "noise" could come from different surface characteristics of the regions. The technique, however, did need some adjustment on its parameters. Figure shows that $\nu = 0.03$ was too big for Image 2 but was right for Image 3.

Figure 4 is the results of applying the Model II to the test images. It delineated the ground truth boundaries in all test cases without any parameter tweaking. All the results were obtained with $\alpha = 0.05, \mu_1 = 1.0, \mu_2 = 10.0$, and $\nu = 0.01$.

The performance difference between Model I and II can be explained by examining the evolution of the surface processes. In Model I, the decent rate of s_i is proportional to λ_i^2 as described in Section 2. Within a noisy region where λ_i is small, s_i evolves very slowly. Eventually, this slow diffusion will allow the line process to form a boundary around the noisy region as the temperature, T in (13), decreases. The line process, however, does not form boundaries within the noisy region since the effective gradient strength, $\lambda_i \|\nabla s_i\|^2$, is small due to a small λ_i inside the region. Figure 5 compares the final surface processes for Model I and II.

We have applied the algorithm to natural images. First, Figure 6 gives the results of applying the WMM, Model I, and Model II energy minimization processes to the input image. The value of μ_2 is changed to 1.0 for this image since some regions such as the tail wing is smaller than regions in the previous experiments and the smoothness constraint on λ is relaxed. The results are very similar as the λ image does not show much variation in this case. The reason for this is that three color bands in natural images tend to have similar characteristics, and regions with high variation in one band tend to have high variations in the other two bands as well. Note that a bigger μ_2 will make the variation even smaller.

The amount of spurious edges within the terrain appear to be less for Model I and II than WMM. However, it is difficult to compare the quality of line processes quantitatively for natural images.

Next, the same algorithm is applied to a microscopic image of pituitary cells. Here, the objective is more clear that each cell needs to be separated from the background. Due to a high level of noise, α in 1 is lower to 0.25.

Although, the result of Model II appears to be less noisy, there is no significant improvements over the WMM.

4 Texture Segmentation

This section presents some experimental results of our algorithm for segmenting images comprising of both textured and non-textured regions. It is often the case that segmentation algorithms developed specifically for textured images do not precisely detect sharp boundaries between non-textured regions and those developed for non-textured images are too sensitive to texture edges and fail to detect meaningful boundaries for textured regions.

The idea here is to combine luminance value with features obtained by some texture feature detectors using the energy model, (17) so that the system can achieve a good trade-off between precise localization of non-textured regions and good classification of textured regions.

We have employed simple texture feature extractors, namely Laws texture metrics, for our experiments[12]. The Laws texture filters are separable pairs of 1D filters. Each 1D filter can be one of the following 5 filters:

$$L5 = \{ 1 \ \ 4 \ \ 6 \ \ 4 \ \ 1 \}/16,$$
$$E5 = \{-1 \ -2 \ \ 0 \ \ 2 \ \ 1\}/6,$$
$$S5 = \{-1 \ \ 0 \ \ 2 \ \ 0 \ -1\}/4,$$
$$W5 = \{-1 \ \ 2 \ \ 0 \ -2 \ \ 1\}/6,$$
$$R5 = \{ 1 \ -4 \ \ 6 \ -4 \ \ 1 \}/16.$$

We have used 3 pairs, L5-E5, E5-L5 and E5-E5, for our experiments. After the original image is filtered by the 3 separable filter pairs to produce 3 feature images, each feature is replaced by its absolute value and each feature image is smoothed with a concentric Gaussian filter with its standard deviation equal to 2 pixels. Finally, each texture feature is scaled by 2 to match the contrast of the original image.

This simple and crude texture features are combined with the original pixel value to form a feature vector with 4 elements at each pixel location.

Figures 8 gives results of WMM, Model I and II on two synthetic textured images. In both test cases, WMM could not detect ground-truth boundaries without picking up spurious texture edges. Both Model I and II localized boundaries between textured-non-textured and two non-textured regions. Due to simple texture feature extracting process, they could not reliably detect the boundaries between textured regions. We are currently experimenting various texture features for a more reliable discrimination process. Some of features under consideration are Gabor features[9, 4], Wavelet Fractal Signatures[5], and Fan filters[24].

5 Conclusion

The paper described an energy model and its minimization techniques for detecting region boundaries in a multi-valued data field. The energy model is a modification of WMM where a set of normalized weight, λ, participates in the minimization process. Two minimization techniques, (referred to as Model I and II throughout the paper), were described.

The technique is effective and improves the WMM when the noise characteristics of a multi-band image is different across the bands. It produces very similar result with the WMM when the noise characteristics are similar across the bands as demonstrated in our color image experiments.

With the same reasoning, the technique is effective on a feature vector image as each feature has different spatial variation.

The weight adaptation technique described in this paper does not take the signal level of each feature into account. For example, one sensor measures strong and uniform response in one region while the other sensor produced zero level signal in the same region. If the both responses are uniform across the region (i.e. $\nabla s = 0$) then the adaptation technique will weigh two sensors equally. However, intuitively, the sensor with the strong response should be trusted more than the other sensor. Thus, our future works include incorporating the measurement level into the adaptation criteria.

References

1. J. Besag. On the statistical analysis of dirty pictures. *J. Royal Statistical Soc., Ser. B*, 48:259–302, 1986.
2. A. Blake and A. Zisserman. *Visual Reconstruction*. MIT Press, Cambridge, MA, 1987.
3. C. Bouman and B. Liu. Multiple resolution segmentation of textured images. *IEEE Trans. Pattern Recog. and Machine Intel.*, 13(2):259–302, 1991.
4. J. G. Daugman. Two–dimensional spectral analysis of cortical receptive field profiles. *Vision Research*, 20:847–856, 1980.
5. F. Espinal, T. Huntsberger, B. Jawerth, and T. Kubota. Wavelet-based fractal signature analysis for automatic target recognition. *Optical Engineering*, 37(1):166–174, January 1998.
6. D. Geiger and F. Girosi. Parallel and deterministic algorithms from MRF's: Surface reconstruction. *IEEE Trans. Pattern Analysis and Machine Intel.*, 13(5):401–412, 1991.
7. S. Geman and D. Geman. Stochastic relaxation, Gibbs distribution, and the Bayesian restoration of images. *IEEE Trans. Pattern Analysis and Machine Intel.*, 6(6):721–741, 1984.
8. T.L. Huntsberger, C.L. Jacobs, and R.L. Cannon. Iterative fuzzy image segmentation. *PR*, 18:131–138, 1985.
9. A. K. Jain and F. Farrokhnia. Unsupervised texture segmentation using Gabor filters. *Pattern Recognition*, 24(12):1167–1186, 1991.
10. T. Kubota and T. Huntsberger. Adaptive pattern recognition system for scene segmentation. *Optical Engineering*, 37(3):829–35, 1998.
11. S. Lakshmanan and H. Derin. Simultaneous parameter estimation and segmentation of Gibbs random fields using simulated annealing. *IEEE Trans. Pattern Analysis and Machine Intel.*, 11:799–813, 1989.
12. K. I. Laws. *Textured Image Segmentation*. PhD thesis, University of Southern California, Institute for Robotics and Intelligent Systems, 1980.

13. T. S. Lee. A Bayesian framework for understanding texture segmentation in the primary visual cortex. *Vision Res.*, 35(18):2643–2657, 1995.

14. S.Z. Li. Parameter-estimation for optimal object recognition: Theory and application. *IJCV*, 21(3):207–222, February 1997.

15. J.Q. Liu and Y.H. Yang. Multiresolution color image segmentation. *PAMI*, 16(7):689–700, July 1994.

16. R. C. Luo and M. G. Kay. *Data fusion in robotics and machine intelligence*, chapter 2, page 47. Academic Press, 1992. Ed. A. Abidi and R. C. Gonzalez.

17. J-M. Morel and S. Solimini. *Variational Methods in Image Segmentation*. Birkhauser, Boston, MA, 1994.

18. D. Mumford and J. Shah. Optimal approximation by piecewise smooth functions and associated variational problems. *Communications on Pure and Applied Mathematics*, pages 577–685, 1989.

19. D.K. Panjwani and G. Healey. Markov random-field models for unsupervised segmentation of textured color images. *PAMI*, 17(10):939–954, October 1995.

20. M. Pelillo and M. Refice. Learning compatibility coefficients for relaxation labeling processes. *PAMI*, 16(9):933–945, September 1994.

21. A. Rosenfeld, R.A. Hummel, and S.W. Zucker. Scene labeling by relaxation operations. *SMC*, 6(6):420–433, June 1976.

22. E. Saber, A.M. Tekalp, and G. Bozdagi. Fusion of color and edge information for improved segmentation and edge linking. *IVC*, 15(10):769–780, October 1997.

23. T. Uchiyama and M.A. Arbib. Color image segmentation using competitive learning. *PAMI*, 16(12):1197–1206, December 1994.

24. A. B. Watson. The cortex transforms: rapid computation of simulated neural images. *Computer Vision, Graphics, and Image Processing*, 39:311–327, 1987.

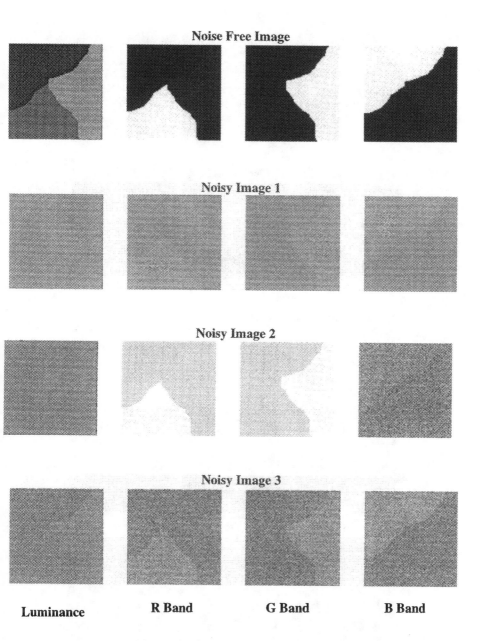

Luminance   R Band   G Band   B Band

Fig. 1. *Synthetic color test images.* 1st row: the original noise free image. 2nd row: disjoint noise image (SNR=0.425). 3rd row: noise everywhere in Blue band (SNR=0.5). 4th row: noise everywhere in all bands (SNR=1.275). 1st column: RGB images converted into gray scale. 2nd column: Red band. 3rd column: Green band. 4th column: Blue band.

184

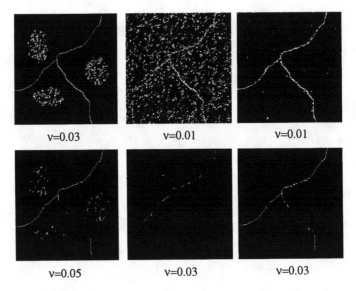

Fig. 2. *Results of boundary detection using WMM.* 1st column: the results for the noisy image 1 in Figure 1. 2nd column: the results for the noisy image 2 in Figure 1. 3rd column: the results for the noisy image 3 in Figure 1. $\alpha = 0.05$ and $\mu_1 = 1.0$. ν needed to be adjusted as shown.

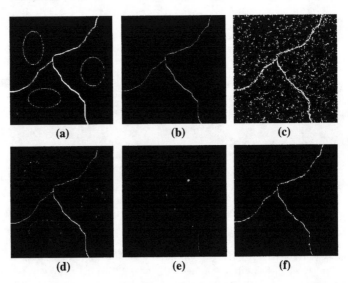

Fig. 3. *Results of boundary detection using Model I.* 1st column: the results for the Image 1 in Figure 1. 2nd column: the results for the Image 2 in Figure 1. 3rd column: the results for the Image 3 in Figure 1. $\alpha = 0.05$, $\mu_1 = 1.0$ and $\mu_2 = 10.0$. $\nu = 0.01$ for the 1st row and $\nu = 0.03$ for the 2nd row.

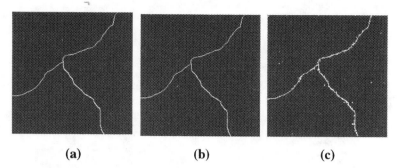

(a)   **(b)**   **(c)**

Fig. 4. *Results of boundary detection using Model II.* 1st column: the results for the noisy image 1 in Figure 1. 2nd column: the results for the noisy image 2 in Figure 1. 3rd column: the results for the noisy image 3 in Figure 1. All the results were obtained with the parameter set $\{\alpha = 0.05, \mu_1 = 1.0, \mu_2 = 10.0,$ and $\nu = 0.01\}$

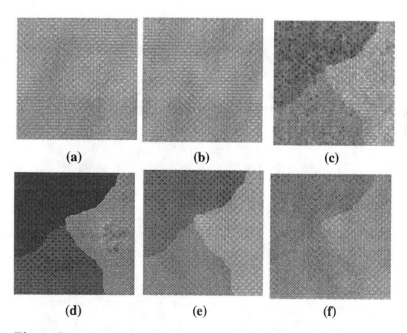

(a)   **(b)**   **(c)**

(d)   **(e)**   **(f)**

Fig. 5. *Comparison of surface process results.* The figure shows intermediate states of the surface process. The top row is the results of applying Model I on the test images shown in Figure 1. The bottom row is the results of applying Model II on the same test images. First, second and third columns are results of applying the algorithms on Image 1, Image 2 and Image 3 in Figure 1, respectively. Note that this figure only shows the luminance of the RGB images.

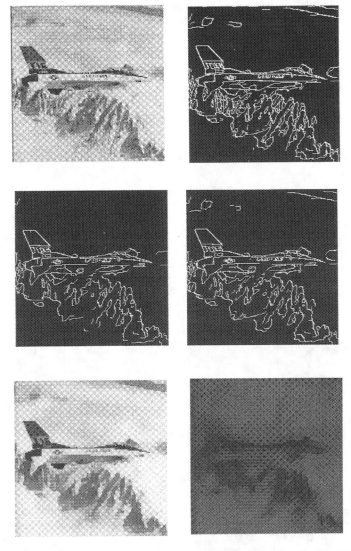

Fig. 6. *Results of boundary detection on a natural color image (Part 1).* Top-left: the luminance of the input color image. Top-right: the line process of WMM. Middle-left: the line process of Model I. Middle-right: the line process of Model II. Bottom-left: the luminance of the surface process. Bottom-right: the luminance of the lambda process. All the results are computed with $\{\alpha = 0.05, \mu_1 = 1.0, \mu_2 = 1.0,$ and $\nu = 0.01\}$

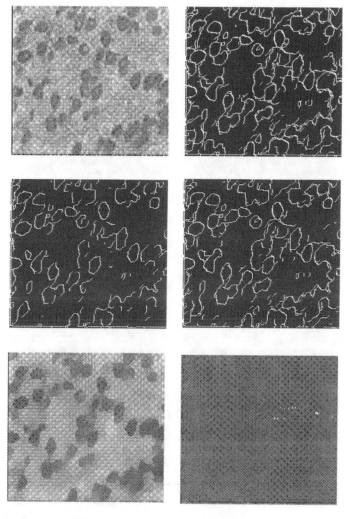

Fig. 7. *Results of boundary detection on a natural color image (Part 2).* Top-left: the luminance of the input color image. Top-right: the line process of WMM. Middle-left: the line process of Model I. Middle-right: the line process of Model II. Bottom-left: the luminance of the surface process. Bottom-right: the luminance of the lambda process. All the results are computed with $\{\alpha = 0.025, \mu_1 = 1.0, \mu_2 = 1.0,$ and $\nu = 0.01\}$

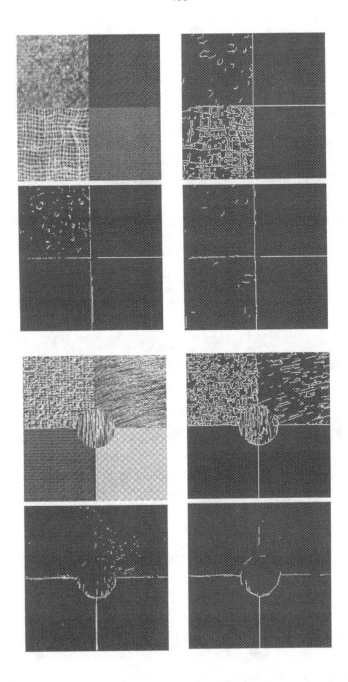

Fig. 8. *Results of boundary detection on textured image.* Two sets of results are shown here. Each set shows the original texture (top-left), the line process using WMM (top-right), the line-process using Model I (bottom-left), and the line-process using Model II (bottom-right). All the results are computed with $\{\alpha = 0.05, \mu_1 = 1.0, \mu_2 = 10.0, \text{ and } \nu = 0.01$

Bayesian A* Tree Search with Expected O(N) Convergence Rates for Road Tracking

James M. Coughlan and A.L. Yuille

Smith-Kettlewell Eye Research Institute, San Francisco, CA 94115, USA,
coughlan@ski.org

Abstract. This paper develops a theory for the convergence rates of A* algorithms for real-world vision problems, such as road tracking, which can be formulated in terms of maximizing a reward function derived using Bayesian probability theory. Such problems are well suited to A* tree search and it can be shown that many algorithms proposed to solve them are special cases, or variants, of A*. Moreover, the Bayesian formulation naturally defines a probability distribution on the ensemble of problem instances, which we call the **Bayesian Ensemble**. We analyze the Bayesian ensemble, using techniques from information theory, and mathematically prove expected O(N) convergence rates of inadmissible A* algorithms. These rates depend on an "order parameter" which characterizes the difficulty of the problem.

1 Introduction

Recently, it has become apparent [25] that a class of real world vision problems, formulated as Bayesian inference [18], can be solved using A* algorithms. This class includes tasks such as the detection and tracking of paths in noise/clutter, see figure (1). In particular, it was shown [25] that many of the algorithms used to solve these tasks (see, for example, [20],[2],[10],[12],[11]) could be interpreted as special cases, or variants, of A* algorithms. Incidently, a consequence of applying A* to Bayesian problems is that the prior probabilities, an essential component of the Bayesian approach, can be used to make stronger heuristic predictions than in standard A*, see [12],[28], which can result in improved performance.

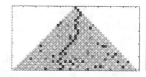

Fig. 1. The difficulty of detecting the target path in clutter depends, by our theory [27], on the order parameter K. The larger K the less computation required. Left, an easy detection task with $K = 0.8647$. Middle, a harder detection task with $K = 0.2105$. Right, an impossible task with $K = -0.7272$.

An advantage of expressing algorithms in a uniform framework, such as A*, is that it enables us to do theoretical and experimental comparisons between different algorithms to determine which ones are most effective. Moreover, one may hope to identify characteristics of the problem domain which determine the difficulty of the search tasks independently of the algorithm used. If so, then it may be possible to design optimally effective algorithms to solve the problems. These are the issues that we investigate in this paper. (See also, our related work [27]).

Broadly speaking, there are two strategies for evaluating the effectiveness of algorithms. The first is the worst case analysis used in much of computer science [9]. The second involves determining the convergence rates on typical problem situations (i.e. those which typically occur). This form of analysis requires having a probability distribution on the ensemble of problem instances. Karp and Pearl provided a fascinating analysis of binary tree A* search using this approach (see Chp. 5 [21],[15]). We argue that this second approach is of more relevance to the problems we are concerned with and so we will study it in our paper. Interestingly, however, there are some recent studies showing that order parameters exist for NP-complete problems and that these problems can be easy to solve for certain values of the order parameters [4],[24]. The connection between this approach and our own is a topic for further research.

We emphasize that the Bayesian formulation of our problems *naturally gives rise to a probability distribution on the ensemble of problem instances*, which we call the *Bayesian Ensemble*. This allows us to build on the foundations established by Karp and Pearl [21] to obtain expected convergence rates. Technically, our proofs involve adapting techniques from information theory, such as Sanov's theorem, which were developed to bound the probability of rare events occurring [7].

In particular, we formulate the problem of detecting target curves in clutter to be one of Bayesian inference [18]. This requires searching through a tree of possible target paths, see figure (2) for an illustration of this search task. This assumes that statistical knowledge is determined for the target and the clutter, as described in section (2). Such statistical knowledge has often been used in computer vision for determining optimization criteria to be minimized and techniques have been developed to learn it from real data [29]. We want to go one step further and use this statistical knowledge to determine good search strategies. In particular, we can prove that for certain curve and boundary detection algorithms we will obtain expected A* convergence rates by examining a number of nodes which is *linear in the size N of the problem*. In addition, the expected sort time per node is constant [6] (note that this does *not* necessarily imply that the expected time for the problem is linear in N). Moreover, our analysis helps determine important characteristics of the problem, similar to order parameters in statistical physics, which quantify the difficulty of the problem. These order parameters determine the constants in the convergence rate formulae and also determine the expected errors.

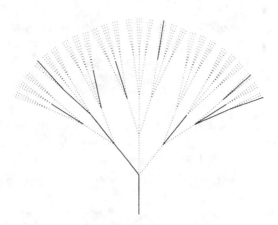

Fig. 2. A simulated road tracking problem where dark lines indicate strong edge responses and dashed lines specify weak responses. The branching factor is three. The data was generated by stochastic sampling using a simplified version of the models analyzed in this paper. In this sample there is only one strong candidate for the best path (the continuous dark line) but chance fluctuations have created subpaths in the noise with strong edge responses. The A* algorithm must search this tree starting at the bottom node.

As we will show, our convergence bounds become infinite at certain values of these order parameters. Is this an artifact of our proofs? Or is it a limitation of the A* search strategy? In related work [27], we prove instead that it corresponds to a fundamental difficulty with the problem. As proven in [27] similar order parameters characterize the difficulty of solving the problem *independently* of the algorithm employed. Moreover, at critical values of these order parameters there is a phase transition and the problem becomes insolvable. These fundamental bounds show that our proofs in this paper only break down as we enter the regime where the problem is unsolvable by any algorithm. The A* algorithm remains effective as we approach the critical value of the order parameters although, not surprisingly, the convergence rates get very slow.

The first section (2) of this paper describes the probabilistic formulation of road tracking that we use to prove our results. Section (3) introduces Sanov's theorem and illustrates how it can be applied to bound the probabilities of rare events. In section (4), we analyze the case of inadmissible heuristics (the case of admissible heuristics with pruning was analyzed in [26]). We conclude by placing this work in a larger context and summarizing recent extensions.

2 Mathematical Formulation of Road Tracking

Tracking curved objects in real images is an important practical problem in computer vision. We consider a specific formulation of the problem of road tracking from aerial images by Geman (D.) and Jedynak [12]. Their approach used a novel active search algorithm to track a road in an aerial photograph with empirical convergence rates of $O(N)$ for roads of length N. Their algorithm is highly effective for this application and is arguably the best currently available. In previous work [25], we showed that Geman and Jedynak's algorithm was a close approximation to A*. Other search algorithms such as Dijkstra and Dynamic Programming used in related visual search problems [20], [2],[10], [17]. [11],[5] can be shown to be special cases of A* [25].

Our approach assumes that both the intensity properties and the geometrical shapes of the target path (i.e. the edge contour) can be determined statistically. This path can be considered to be a set of elementary path segments joined together. We first consider the intensity properties along the edge and then the geometric properties.

The image properties of segments lying on the path are assumed to differ, in a statistical sense, from those off the path. More precisely, we can design a filter $\phi(.)$ with output $\{y_x = \phi(I(x))\}$ for a segment at point x so that:

$$P(y_x) = P_{on}(y_x), \quad if \text{ "}x\text{" } lies \ on \ the \ true \ path$$
$$P(y_x) = P_{off}(y_x), \quad if \text{ "}x\text{" } lies \ off \ the \ true \ path. \tag{1}$$

For example, we can think of the $\{y_x\}$ as being values of the edge strength at point x and P_{on}, P_{off} being the probability distributions of the response of $\phi(.)$ on and off an edge. The set of possible values of the random variable y_x is the *alphabet* with *alphabet size* M. See [12],[5] examples of distributions for P_{on}, P_{off} used in computer vision applications.

We now consider the geometry of the target contour. We require the path to be made up of connected segments $x_1, x_2, \ldots, x_N$. There will be a Markov probability distribution $P_g(x_{i+1}|x_i)$ which specifies prior probabilistic knowledge of the target. It is convenient, in terms of the graph search algorithms we will use, to consider that each segment x has a set of Q possible continuations. Following terminology from graph theory, we refer to Q as the *branching factor*. We will assume that the distribution P_g depends only on the relative orientations of x_{i+1} and x_i. In other words, $P_g(x_{i+1}|x_i) = P_{\Delta g}(x_{i+1} - x_i)$. An important special case is when the probability distribution is uniform for all branches (i.e. $P_{\Delta g}(\Delta x) = U(\Delta x) = 1/Q \ \forall \Delta x$).

By standard Bayesian analysis, the optimal path $X^* = \{x_1^*, \ldots, x_N^*\}$ maximizes the sum of the log posterior:

$$E(X) = \sum_i \log \frac{P_{on}(y_{(x_i)})}{P_{off}(y_{(x_i)})} + \sum_i \log \frac{P_{\Delta g}(x_{i+1} - x_i)}{U(x_{i+1} - x_i)}, \tag{2}$$

where the sum i is taken over all segments on the target. $U(x_{i+1} - x_i)$ is the uniform distribution and its presence merely changes the log posterior $E(X)$ by a constant value. It is included to make the form of the intensity and geometric terms similar, which simplifies our later analysis.

We will refer to $E(X)$ as the *reward* of the path X which is the sum of the *intensity rewards* $\log \frac{P_{on}(y_{(x_i)})}{P_{off}(y_{(x_i)})}$ and the *geometric rewards* $\log \frac{P_{\Delta g}(x_{i+1} - x_i)}{U(x_{i+1} - x_i)}$.

It is important to emphasize that our results can be extended to higher-order Markov chain models (provided they are shift-invariant). We can, for example, define the x variable to represent spatial orientation *and* position of a small edge segment. This will allow our theory to apply to models, such as snakes, used in recent successful vision applications [2], [12]. (It is straightforward to transform the standard energy function formulation of snakes into a Markov chain by discretizing and replacing the derivatives by differences. The smoothness constraints, such as membranes and thin plate terms, will transform into first and second order Markov chain connections respectively). Recent work by Zhu [33] shows that Markov chain models of this type can be learnt using Minimax Entropy Learning theory from a representative set of examples. Indeed Zhu goes further by demonstrating that other Gestalt grouping laws can be expressed in this framework and learnt from representative data.

Reward functions, such as equation (2), are ideally suited to A* graph/tree search algorithms [21],[23] and we will therefore analyze A* algorithms later in this paper, see section (4). As we will describe, A* searches the nodes – possible branches of the road/snake – which are most promising. The "goodness" $f(n)$ of a node n is $g(n) + h(n)$ where $g(n)$ is the reward to get to the node and $h(n)$ is a heuristic reward to get to the finish from n. The A* algorithm starts at the top of the tree and evaluates the child nodes (i.e. those connected to the top node by a single arc). These child nodes are placed in the *queue*. As the algorithm proceeds it selects the member of the queue with best evaluation, removes it from the queue, expands its children and enters them in the queue.

The evaluation of the nodes is based on the reward to reach it from the top node (i.e. the sum of the log posteriors) and on a heuristic reward based on anticipated future performance. More precisely, a path segment ending at x has a total reward $f(x) = g(x) + h(x)$ (note that the nonoverlapping path requirement implies that x determines a unique path to the initialization point). The choice of heuristic reward $h(x)$ is very important to the algorithm [21]. It can be proven that if $h(x)$ is an upper bound on the reward to get to the end of the path then A* is guaranteed to find the global maximum eventually. An A* algorithm whose heuristic satisfies this bound is called *admissible*. One that does not is called *inadmissible*. The problem is that admissible A* algorithms are guaranteed to find the best result but may do so slowly. By contrast, inadmissible $A*$ algorithms are often faster but may fail in certain cases.

Karp and Pearl [21] provided a theoretical analysis of convergence rates of A* search. They studied a binary tree where the rewards for each arc were 0 or 1 and were specified by a probability p. They then studied the task of finding the minimum reward path. This is an interesting task but it differs from ours in many respects. From our perspective, it resembles the task of finding the best path in the noise/clutter rather than detecting a true target in the presence of noise/clutter.

There are three elements to our proofs. The first is the use of Sanov's theorem to put exponential bounds on the probabilities of rare events – this theorem is described in section (3). The second is an onion peeling strategy to recursively explore the search tree, this is described more in [26]. The third is the summation of exponential series, generated by Sanov's theorem, which is described in more detail in [6].

3 Sanov's Theorem

This section introduces results from the theory of types [7] which we will use to prove our results. We will be particularly concerned with Sanov's theorem, which we state without proof later this section. To motivate this material we will apply it to the problem of determining whether a given set of measurements are more likely to come from a road or non-road but *without* making any geometrical assumptions about the likely shape of the road. The theorem assumes that we have an underlying distribution Q which generates a set of N independent identically distributed (i.i.d.) samples. From each sample set we can determine an empirical normalized histogram, or *type*, see figure (3). (This normalization ensures that the components of each type sums to one and hence can be interpreted as an empirical distribution). The law of large numbers states that these empirical histograms (when normalized) must become similar to the distribution Q as $N \mapsto \infty$. Sanov's theorem puts bounds on *how fast* the empirical histograms converge (in probability) to the underlying distribution. Thereby it puts bounds on the probability of rare events.

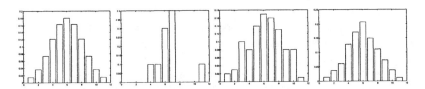

Fig. 3. Samples from an underlying distribution. Left to right, the original distribution, followed by histograms, or types, from 10, 100, and 1000 samples from the original. Observe that for small numbers of samples the types tend to differ greatly from the true distribution. But for large N the law of large numbers says that they must converge (with high probability).

More precisely, Sanov's Theorem states:

Sanov's Theorem. *Let $y_1, y_2, ..., y_N$ be i.i.d. from a distribution $Q(y)$ with alphabet size J and E be any closed set of probability distributions. Let $Pr(\phi \in E)$ be the probability that the type ϕ of a sample sequence lies in the set E. Then:*

$$\frac{2^{-ND(\phi^*||Q)}}{(N+1)^J} \leq Pr(\phi \in E) \leq (N+1)^J 2^{-ND(\phi^*||Q)}, \quad (3)$$

where $\phi^* = \arg\min_{\phi \in E} D(\phi||Q)$ is the distribution in E that is closest to Q in terms of Kullback-Leibler divergence, given by $D(\phi||Q) = \sum_{y=1}^{J} \phi(y) \log(\phi(y)/Q(y))$.

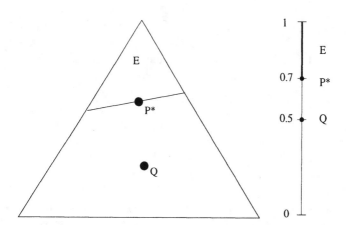

Fig. 4. Left, Sanov's theorem. The triangle represents the set of probability distributions. Q is the distribution which generates the samples. Sanov's theorem states that the probability that a type, or empirical distribution, lies within the subset E is chiefly determined by the distribution P^* in E which is closest to Q (in the sense of Kullback-Leibler). Right, Sanov's theorem for the coin tossing experiment. The set of probabilities is one-dimensional and is labelled by the probability $p(head)$ of tossing a head. The unbiased distribution Q is at the centre, with $P(head) = 1/2$, and the closest element of the set E is P^* such that $P^*(head) = 0.7$.

This is illustrated by figure (4). Intuitively, it shows that, when considering the chance of a set of rare events happening, we essentially only have to worry about the "most likely" of the rare events (in the sense of Kullback-Leibler divergence). Most importantly, it tells us that the probability of rare events falls off *exponentially* with the Kullback-Leibler divergence between the rare event (its type) and the true distribution. This exponential fall-off is critical for proving the results in this paper. Note that Sanov's theorem involves an *alphabet factor* $(N + 1)^J$. This alphabet factor becomes irrelevant at large N (compared to the exponential term). It does, however, require that the distribution Q is defined on a finite space, or can be well approximated by a quantized distribution on a finite space.

Sanov's theorem can be illustrated by a simple coin tossing example, see figure (4). Suppose we have a fair coin and want to estimate the probability of observing more than 700 heads in 1000 tosses. Then set E is the set of probability distributions for which $P(head) \geq 0.7$ $(P(head) + P(tails) = 1)$. The distribution generating the samples is $Q(head) = Q(tails) = 1/2$ because the coin is fair. The distribution in E closest to Q is $P^*(head) = 0.7, P^*(tails) = 0.3$. We calculate $D(P^*\|Q) = 0.119$. Substituting into Sanov's theorem, setting the alphabet size $J = 2$, we calculate that the probability of more than 700 heads in 1000 tosses is less than $2^{-119} \times (1001)^2 \leq 2^{-99}$.

In this paper, we will only be concerned with sets E which involve the rewards of types. These sets will therefore be defined by linear constraints on the types – in particular, constraints such as $\phi \cdot \alpha \geq T$, where $\alpha(y) =$

$\log(P_{on}(y)/P_{off}(y))$, $y = 1, ..., J$. (We define $\phi \cdot \alpha = \sum_{y=1}^{J} \phi(y)\alpha(y)$). This will enable us to derive results which will not be true for arbitrary sets E. We will often, however, be concerned with the probabilities that the rewards of samples from one distribution are greater than those from a second. It is straightforward to generalize Sanov's theorem to deal with such cases.

This leads to:

Theorem 1. *The probability that a sequence of samples from on-road has lower intensity reward than a sequence of samples from off-road is bounded below by* $(N + 1)^{-2J}2^{-2NB(P_{on},P_{off})}$ *and above by* $(N + 1)^{2J}2^{-2NB(P_{on},P_{off})}$, *where* $B(P_{on}, P_{off}) = -\log\{\sum_{y=1}^{J} P_{off}^{1/2}(y)P_{on}^{1/2}(y)\}$. *(N is the number of elements in each sequence of sample.)*

Proof. *This is a generalization of Sanov's theorem to the case where we have two probability distributions and two types. We define* $E = \{(\phi^{on}, \phi^{off}) : \phi^{off} \cdot \alpha \geq \phi^{on} \cdot \alpha\}$. *We then apply the same strategy as for the Sanov proof but applied to the product space of the two distributions* P_{on}, P_{off}. *This requires us to minimize:*

$$f(\phi^{off}, \phi^{on}) = ND(\phi^{off}||P_{off}) + ND(\phi^{on}||P_{on})$$

$$+\tau_1\{\sum_{y=1}^{J} \phi^{off}(y) - 1\} + \tau_2\{\sum_{y=1}^{J} \phi^{on}(y) - 1\} + \gamma\{\phi^{on} \cdot \alpha - \phi^{off} \cdot \alpha\}, \quad (4)$$

where the τ's and γ are Lagrange multipliers. The function $f(.,.)$ is convex in the ϕ^{off}, ϕ^{on} and the Lagrange constraints are linear. Therefore there is a unique minimum which occurs at:

$$\phi^{off*}(y) = \frac{P_{on}^{\gamma}(y)P_{off}^{1-\gamma}(y)}{Z[1-\gamma]}, \quad \phi^{on*}(y) = \frac{P_{on}^{1-\gamma}(y)P_{off}^{\gamma}(y)}{Z[\gamma]}, \quad (5)$$

subject to the constraint $\phi^{on} \cdot \alpha = \phi^{off} \cdot \alpha$. *The unique solution occurs when* $\gamma = 1/2$ *(because this implies* $\phi^{off*} = \phi^{on*}$ *and so the constraints are satisfied.) We define* $\phi_{Bh} = \phi_{\lambda^{-1}(1/2)} = P_{on}^{1/2}P_{off}^{1/2}/Z[1/2]$ *("Bh" is short for Bhattacharyya). We therefore obtain:*

$$(N + 1)^{-2J}2^{-N\{D(\phi_{Bh}||P_{off})+D(\phi_{Bh}||P_{on})\}} \leq Pr\{(\phi^{off}, \phi^{on}) \in E\}$$

$$\leq (N + 1)^{2J}2^{-N\{D(\phi_{Bh}||P_{off})+D(\phi_{Bh}||P_{on})\}}. \quad (6)$$

We define $B(P_{on}, P_{off}) = (1/2)\{D(\phi_{Bh}||P_{off}) + D(\phi_{Bh}||P_{on})\}$. *Substituting in for* ϕ_{Bh} *from above yields* $B(P_{on}, P_{off}) = -\log\{\sum_{y=1}^{J} P_{off}^{1/2}(y)P_{on}^{1/2}(y)\}$. *Hence result.*

This result tells us that the probability that an off-road sequence has higher reward than an on-road sequence falls off as $2^{-2B(P_{on},P_{off})N}$ for large N. We define the fall-off factor (i.e. the negative of the coefficient of N in the exponent) to be the *order parameter*. For this task, the order parameter is therefore $2B(P_{on}, P_{off})$ which is always positive (except for the degenerate case

$P_{on} = P_{off}$ for which it becomes zero). $B(P_{on}, P_{off})$ is just a measure of the distance between P_{on} and P_{off} and we will refer to it as the *Bhattacharyya distance* (because it is identical to the Bhattacharyya bound for Bayes error, see [22]). For this task, unlike tree search (see next section), there is no critical point and no phase transition.

4 Tree Search: A* and Inadmissible Heuristics

Our main result of this section is to prove convergence of A* algorithms with inadmissible heuristics. We prove that convergence is achieved with O(N) expected nodes opened and we put bounds on the expected errors of the solutions. It can also be proved that the expected search time per node is constant (i.e. independent of N) [6].

4.1 A* Convergence for the Bhattacharyya Heuristic

We now want to consider a traditional A* search strategy using a heuristic function but no pruning. In this section, we will formulate the problem for any heuristic and then obtain bounds for a special case, which we call the *Bhattacharyya heuristic* (again because it is directly related to the Bhattacharyya bound). In the following section, we will generalize our results of other inadmissible heuristics.

For a node W_M, at distance M from the start, we let $g(W_M)$ be the measured reward and $h(W_M)$ be the heuristic function. The A* algorithm proceeds by searching the node in the queue for which the combined reward $f(W_M) = g(W_M) + h(W_M)$ is greatest. How many nodes (or arcs) do we expect to search by this strategy? And what are the expected errors in our solutions?

The reward to reach W_M is just the reward of the log-likelihood data and prior terms along the path from the start to W_M. We define the *heuristic reward* $h(W_M) = (N - M)(H_L + H_P)$ where H_L and H_P are constants (H_L and H_P are heuristics for the likelihood and the prior respectively). As we have shown [6], convergence results can be proven for a range of values of H_L and H_P. In this paper, however, we will only consider special values H_L^*, H_P^* for which the analysis simplifies.

Observe that a path segment will be visited only if the reward to get to it (including its heuristic reward) is sufficiently high. More precisely, *if a segment n of a false path is searched then this implies that its reward is better than the reward of at least one point on the target path.* This is because the A* algorithm always maintains a queue of nodes to explore and searches the node segment with highest reward. The algorithm is initialized at the start of the target path and so an element of the target path will always lie in the queue of nodes that A* considers searching. Hence a node will never be explored if its reward is lower than all the rewards on the target path segments.

Since the length of all possible paths is constant we can ignore the constant factor $N(H_L + H_P)$ and the heuristic will then merely penalize path segments which have been tested. Then a false path of length n and a true path of length m will have effective rewards denoted by the random variables $S_{off}(n)$ and $S_{on}(m)$:

$$S_{off}(n) = \sum_{i=1}^{n}\{\log\frac{P_{on}(y_{x_i})}{P_{off}(y_{x_i})} - H_L\}_{off} + \sum_{i=1}^{n}\{\log\frac{P_{\Delta G}(x_{i+1} - x_i)}{U(x_{i+1} - x_i)} - H_P\}_{off},$$

$$S_{on}(m) = \sum_{i=1}^{m}\{\log\frac{P_{on}(y_{x_i})}{P_{off}(y_{x_i})} - H_L\}_{on} + \sum_{i=1}^{m}\{\log\frac{P_{\Delta G}(x_{i+1} - x_i)}{U(x_{i+1} - x_i)} - H_P\}_{on}(7)$$

where the subscripts off and on are used to denote false and true paths respectively (paths with a mixture of true and false segments will be dealt with later).

We now define types $\phi^{off}, \psi^{off}, \phi^{on}, \psi^{on}$ for false and true road samples with ϕ corresponding to the data and ψ to the prior. These types are normalized so that their components sum to 1, i.e. $\sum_{\mu=1}^{M} \phi_\mu = 1$, $\sum_{\nu=1}^{Q} \psi_\nu = 1$. The types will be computed for samples of variables lengths n, m. These lengths will be clear from the context so we will not label them explicitly (i.e. we will not use notation like ϕ_n to denote types taken from n samples).

Therefore we express the rewards of two sequences $S_{off}(n)$ and $S_{on}(m)$ by:

$$S_{off}(n) = n\{\phi^{off} \cdot \alpha - H_L\} + n\{\psi^{off} \cdot \beta - H_P\},$$
$$S_{on}(m) = m\{\phi^{on} \cdot \alpha - H_L\} + n\{\psi^{on} \cdot \beta - H_P\}, \tag{8}$$

where $\alpha(y) = \log(P_{on}(y)/P_{off}(y))$ and $\beta(\delta x) = \log(P_{\Delta g}(\delta x)/U(\delta x))$.

Recall that if a segment n of a false path is searched then its reward must be better than the reward of at least one subpath along the target path. This means that we should consider $Pr\{\exists m : S_{off}(n) \geq S_{on}(m)\}$. This, however, is hard to compute so we bound it above by $\sum_{m=0}^{\infty} Pr\{S_{off}(n) \geq S_{on}(m)\}$ (using Boole's inequality).

Our first result is Theorem 2, which is proven using Sanov's theorem (including the use of constrained optimization to find the fall-off coefficients) and results for the sums of exponential series. The main point of this result is to show that the chance of an off-road path having greater reward than *any* true road path falls off exponentially with the length of the off-road path.

We first define two *sub-order parameters* $\Psi_1 = D(\phi_{Bh}||P_{off}) + D(\psi_{Bh}||U)$ and $\Psi_2 = D(\phi_{Bh}||P_{on}) + D(\psi_{Bh}||P_{\Delta G})$. ($\psi_{Bh}$ is defined analogously to ϕ_{Bh} – see Theorem 1). These parameters will determine the convergence and error rates of the algorithm by means of the two functions:

$$C_1(\Psi) = \{\frac{1}{1 - 2^{-\{\Psi-\epsilon\}}} + \Xi(\epsilon, \Psi)\}, \ C_2(\Psi) = \{\frac{e^{-\{\Psi-\epsilon\}}}{(1 - e^{-\{\Psi-\epsilon\}})^2} + \hat{\Xi}(\epsilon, \Psi)\}, \tag{9}$$

where $\Xi, \hat{\Xi}$ are constants (independent of N) whose exact forms are given in [6].

It will be shown, in Theorem 3, that the order parameter for this problem is $K = \Psi_1 + \Psi_2 - \log Q$. This can be re-expressed as $2B(P_{on}.P_{off}) + 2B(P_{\Delta g}, U) - \log Q$ (observe the similarity with the result of Theorem 1). Note that this result depends on the search algorithm being A*. However, it has been shown that an identical order parameter is obtained [27] when analyzing whether the target can be detected by *any algorithm which computes the MAP estimate.*

Theorem 2. *The A* algorithm, using the Bhattacharyya heuristic* $H_L^* = \phi_{Bh} \cdot \boldsymbol{\alpha}$ *and* $H_P^* = \psi_{Bh} \cdot \boldsymbol{\beta}$*, gives:*

$$Pr\{S_{off}(n) \geq S_{on}(m)\} \leq \{(n+1)(m+1)\}^{2J+2Q} 2^{-(n\Psi_1 + m\Psi_2)}. \tag{10}$$

Moreover, the probability of a particular false path segment being searched falls off, to first order in n, as $C_1(\Psi_2)2^{-n\Psi_1}$ *where n is the number of segments by which this path segment diverges from the target path.*

Proof. *This first part of the proof is again a generalization of Sanov applied to product distributions, see Theorem 1. The new twist is that we have different length factors n and m and the heuristics (also we consider distributions on a four-dimensional product space instead of two dimensions). But for the Bhattacharyya heuristics this will make no difference. (More general heuristics are dealt with in [6]). Define:*

$$E = \{(\phi^{off}, \psi^{off}, \phi^{on}, \psi^{on}) : n\{\phi^{off} \cdot \boldsymbol{\alpha} - H_L^* + \psi^{off} \cdot \boldsymbol{\beta} - H_p^*\}$$
$$\geq m\{\phi^{on} \cdot \boldsymbol{\alpha} - H_L^* + \psi^{on} \cdot \boldsymbol{\beta} - H_p^*\}\}. \tag{11}$$

Applying the strategy from Theorem 1, we must minimize:

$$f(\phi^{off}, \psi^{off}, \phi^{on}, \psi^{on}) = nD(\phi^{off}\|P_{off}) + nD(\psi^{off}\|U)$$
$$+mD(\phi^{on}\|P_{on}) + mD(\psi^{on}\|P_{\Delta G}) + \tau_1\{\sum \phi^{off} - 1\} + \tau_2\{\sum \psi^{off} - 1\}$$
$$+\tau_3\{\sum \phi^{on} - 1\} + \tau_4\{\sum \psi^{on} - 1\}$$
$$+\gamma\{m\{\phi^{on} \cdot \boldsymbol{\alpha} - H_L^* + \psi^{on} \cdot \boldsymbol{\beta} - H_p^*\} - n\{\phi^{off} \cdot \boldsymbol{\alpha} - H_L^* + \psi^{off} \cdot \boldsymbol{\beta} - H_p^*\}\} \tag{12}$$

where the τ's and γ are Lagrange multipliers. As before, we know that this function $f(.,.,.,.)$ is convex so there is a unique minimum. Observe that $f(....)$ consists of four terms of form $nD(\phi^{off}\|P_{off}) + \tau_1\{\sum \phi^{off}\} - n\gamma\phi^{off} \cdot \boldsymbol{\alpha}$ which are coupled by shared constants. These terms can be minimized separately to give:

$$\phi^{off*} = \frac{P_{on}^\gamma P_{off}^{1-\gamma}}{Z[1-\gamma]}, \quad \phi^{on*} = \frac{P_{on}^{1-\gamma} P_{off}^\gamma}{Z[\gamma]}, \quad \psi^{off*} = \frac{P_{\Delta G}^\gamma U^{1-\gamma}}{Z_2[1-\gamma]}, \quad \psi^{on*} = \frac{P_{\Delta G}^{1-\gamma} U^\gamma}{Z_2[\gamma]}, \tag{13}$$

subject to the constraint given by equation (11).

As before, we see that the unique solution occurs when $\gamma = 1/2$. In this case:

$$\phi^{off*} \cdot \boldsymbol{\alpha} = H_L^* = \phi^{on*} \cdot \boldsymbol{\alpha}, \quad \psi^{off*} \cdot \boldsymbol{\beta} = H_P^* = \psi^{on*} \cdot \boldsymbol{\beta}. \tag{14}$$

The solution occurs at ϕ_{Bh}, ψ_{Bh} ($\phi_{\lambda^{-1}(1/2)}$ and $\psi_{\mu^{-1}(1/2)}$). Hence the first result.

We must now sum over m to obtain the bound for $P\{\exists m : S_{off}(n) \geq S_{on}(m)\}$. For large m, the alphabet terms are unimportant and we just need to sum the geometric series. However, we must add extra terms $\Xi(\epsilon, \Psi_2)$ to correct for the alphabet factors for small m, see [6]. Hence

$$Pr\{\exists m \; : \; S_{off}(n) \geq S_{on}(m)\} \leq (n+1)^{2J+2Q} C_1(\Psi_2) 2^{-n\Psi_1}. \qquad (15)$$

We can now state our main result about the convergence of A* using the Bhattacharyya heuristic. Our result, Theorem 3, builds on Theorem 2 by adding an onion peeling argument combined with the summation of exponential series. The key concept here is the onion-like structure of the tree representation, see figure (5). This structure allows us to classify all paths in terms of sets $F_1, F_2, F_3, \ldots$ which depend on where they branch off from the true path. Paths which are always bad (i.e. completely false) correspond to F_1. Paths which are good for one segment, and then go bad, form F_2 and so on. By peeling off segments it can be shown that results for F_1 can be readily adapted to $F_2, F_3, \ldots$.

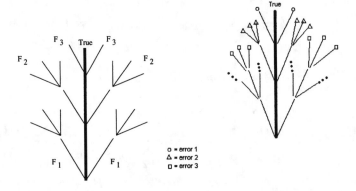

Fig. 5. Left: We can divide the set of paths up into N subsets $F_1, ..., F_N$ as shown here. Paths in F_1 are completely off-road. Paths in F_2 have one on-road segment and so on. Intuitively, we can think of this as an onion where we peel off paths stage by stage. Right: When paths leave the true path they make errors which we characterize by the number of false arcs. For example, a path in F_1 has error N, a path in F_i has error $N + 1 - i$.

Theorem 3. *Provided $\Psi_1 > \log Q$, the expected number of searches is $O(N)$ in the size of the problem and is bounded above by $C_1(\Psi_2)C_1(\Psi_1 - \log Q)N$. Moreover, the expected error in convergence is bounded above by $C_1(\Psi_2)C_2(\Psi_1 - \log Q)$, which is small, independent of the size N of the problem, and decays exponentially with $\Psi_1 - \log Q$. The order parameter $K = \Psi_1 + \Psi_2 - \log Q$.*

Proof. *We use the onion peeling strategy to express the expectation in terms of the expected number of nodes searched in $F_1, F_2, F_3 ..., F_N$. By the structure of our problem the expectations will be bounded by the same number for all F_i. Therefore the bound is linear provided the expectation for F_1 is finite. More precisely, we get $\sum_{i=1}^{N}\{1 + |F_i|\}$, where $|F_i|$ is the cardinality of F_i.*

Theorem 2 gives us a bound that a specific path of length n in F_1 will have higher reward than any subpath of the true path (a subpath must start at the beginning of the target path). We determine that the expected number of paths in F_1 of length n, with rewards higher than any subpath of the target path, is bounded above by $C_1(\Psi_2)(n+1)^{2J+2Q}Q^n 2^{-n\Psi_1}$, see equation (15), where $C_1(\Psi)$ is specified by equation (9). This can be summed over n again taking care with the alphabet factors, see [6] to obtain $C_1(\Psi_2)\{\frac{1}{1-2^{-(\Psi_1-\log Q)+\epsilon}} + \Xi(\epsilon,(\Psi_1-\log Q))\} = C_1(\Psi_1-\log Q)C_1(\Psi_2)$. This is finite provided $\Psi_1 > \log Q$. Our first result follows.

To put bounds on the expected errors of the algorithm we measure the error in terms of the expected number of off-road segments in the best path found by A^*. We use the onion peeling strategy again and consider the probability $Pr(n)$ that A^* will explore a path in F_{N+1-n} to the end, for any n, instead of proceeding along the true path. If this happens we will get an error of size n. The expected error can then be bounded above by $\sum_{n=1}^{\infty} Pr(n)n$.

We want to put an upper bound on $Pr(n)$. Observe that a path in F_{N+1-n} will be followed to the end only if its reward is greater than the heuristic reward along the true path, or the reward of one arc of the true path plus the heuristic reward for the remainder, or the reward for two true arcs plus the heuristic reward for the rest, and so on. We can apply Sanov to get probability bounds for these by using the constraints $n\{\phi^{off}\cdot\alpha+\psi^{off}\cdot\beta\} \geq m\{\phi^{on}\cdot\alpha+\psi^{on}\cdot\beta\}+(n-m)\{H_L^*+H_P^*\}$, where $m = 1,...,n$ is the number of arcs of the true path that are explored. These constraints, of course, are the same constraints $n\{\phi^{off}\cdot\alpha+\psi^{off}\cdot\beta-H_L^*-H_P^*\} \geq m\{\phi^{on}\cdot\alpha+\psi^{on}\cdot\beta-H_L^*-H_P^*\}$ which we used in Theorem 2 above. Therefore, by Boole's inequality,

$$Pr(n) \leq Q^n \sum_{m=0}^{\infty} \{(n+1)(m+1)\}^{2J+2Q} \times 2^{-\{n\Psi_1+m\Psi_2\}}. \qquad (16)$$

As before, we can sum the series with respect to m, see [6], to obtain:

$$Pr(n) \leq C_1(\Psi_2)(n+1)^{2J+2Q}2^{-n\{\Psi_1-\log Q\}}. \qquad (17)$$

The expected error is then bounded above by $\sum_{n=1}^{\infty} nPr(n)$. The dominant, exponential terms, can be summed (see [6]) yielding:

$$< Error > \leq C_2(\Psi_1 - \log Q)C_1(\Psi_2). \qquad (18)$$

5 Conclusion

Our analysis shows it is possible to track certain classes of image contours with linear expected node expansions (linear expected sorting time per node is shown in [6]). We have shown how the convergence rates, and the choice of A^* heuristics, depend on order parameters which characterize the problem domain. In particular, the entropy of the geometric prior and the Kullback-Leibler distance between P_{on} and P_{off} allow us to quantify intuitions about the power of geometrical assumptions and edge detectors to solve these tasks. Not surprisingly, the easiest target curves to detect are those for which the edge detector is most informative and the prior geometric knowledge most constraining. Our analysis allows us to quantify these intuitions. See [19] for analysis of the forms of P_{on}, P_{off} arising in typical images.

Our more recent work [27] has extended this work by showing that *similar order parameters can be used to specify intrinsic (algorithm independent) difficulty of the search problem and that phase transitions occur when these order parameters take critical values.* Fortunately, the proofs in this paper break down at closely related critical points. Therefore A* algorithms are an effective way to solve this problem in the regime for which it can be solved.

As shown in [25] many of the search algorithms proposed to solve vision search problems [20],[2], [12] are special cases of A* (or close approximations). We therefore hope that the results of this paper will throw light on the success of the algorithms and may suggest practical improvements and speed ups, see [5] for promising preliminary results.

Crucial to our analysis has been the use of Bayesian probability theory both to determine an optimization criterion for the problem we wish to solve *and* to define the *Bayesian ensemble* of problem instances. Analysis of the Bayesian ensemble led to the definition of order parameters which characterized the difficulty of the problem. It will be interesting to compare our results with those obtained by [4],[24] for completely different classes of problems and using different techniques. This is a topic for further research.

Acknowledgements

We want to acknowledge funding from NSF with award number IRI-9700446, from the Center for Imaging Sciences funded by ARO DAAH049510494, and from the Smith-Kettlewell core grant, and the AFOSR grant F49620-98-1-0197 to A.L.Y. Lei Xu drew our attention to Pearl's book on heuristics and we thank Abracadabra books for obtaining a second hand copy for us. We would also like to thank Dan Snow and Scott Konishi for helpful discussions as the work was progressing and Davi Geiger for providing useful stimulation. Also influential was Bob Westervelt's joking request that he hoped James Coughlan's PhD thesis would be technical enough to satisfy the Harvard Physics Department. David Forsyth, Jitendra Malik, Preeti Verghese, Dan Kersten, Suzanne McKee and Song Chun Zhu gave very useful feedback and encouragement. Finally, we wish to thank Tom Ngo for drawing our attention to the work of Cheeseman and Selman.

References

1. R. Balboa. PhD Thesis. Department of Computer Science. University of Alicante. Spain. 1997.

2. M. Barzohar and D. B. Cooper, "Automatic Finding of Main Roads in Aerial Images by Using Geometric-Stochastic Models and Estimation," *Proc. IEEE Conf. Computer Vision and Pattern Recognition*, pp. 459-464, 1993.
3. R. E. Bellman, *Applied Dynamic Programming*. Princeton University Press, 1962.
4. P. Cheeseman, B. Kanefsky, and W. Taylor. "Where the Really Hard Problems Are". In *Proc. 12th International Joint Conference on A.I.*. Vol. 1., pp 331-337. Morgan-Kaufmann. 1991.
5. J. Coughlan, D. Snow, C. English, and A.L. Yuille. "Efficient Optimization of a Deformable Template Using Dynamic Programming". In *Proceedings Computer Vision and Pattern Recognition. CVPR'98*. Santa Barbara. California. 1998.
6. J.M. Coughlan and A.L. Yuille. "Bayesian A* Tree Search with Expected O(N) Convergence Rates for Road Tracking". Submitted to *Artificial Intelligence*. 1999.
7. T.M. Cover and J.A. Thomas. **Elements of Information Theory**. Wiley Interscience Press. New York. 1991.
8. M.A. Fischler and R.A. Erschlager. "The Representation and Matching of Pictorial Structures". *IEEE. Trans. Computers*. C-22. 1973.
9. M.R. Garey and D.S. Johnson. **Computers and Intractability: A Guide to he Theory of NP-Completeness**. W.H. Freeman and Co. New York. 1979.
10. D. Geiger, A. Gupta, L.A. Costa, and J. Vlontzos. "Dynamic programming for detecting, tracking and matching elastic contours." *IEEE Transactions on Pattern Analysis and Machine Intelligence*, PAMI-17, March 1995.
11. D. Geiger and T-L Liu. "Top-Down Recognition and Bottom-Up Integration for Recognizing Articulated Objects". In *Proceedings of the International Workshop on Energy Minimization Methods in Computer Vision and Pattern Recognition*. Ed. M. Pellilo and E. Hancock. Venice, Italy. Springer-Verlag. May. 1997.
12. D. Geman. and B. Jedynak. "An active testing model for tracking roads in satellite images". *IEEE Trans. Patt. Anal. and Machine Intel*. Vol. 18. No. 1, pp 1-14. January. 1996.
13. U. Grenander, Y. Chow and D. M. Keenan, *Hands: a Pattern Theoretic Study of Biological Shapes*, Springer-Verlag, 1991.
14. D.W. Jacobs. "Robust and Efficient Detection of Salient Convex Groups". *IEEE Trans. Patt. Anal. and Machine Intel*. Vol. 18. No. 1, pp 23-37. January. 1996.
15. R.M. Karp and J. Pearl. "Searching for an Optimal Path in a Tree with Random Costs". *Artificial Intelligence*. 21. (1,2), pp 99-116. 1983.
16. M. Kass, A. Witkin, and D. Terzopoulos. "Snakes: Active Contour models". In *Proc. 1st Int. Conf. on Computer Vision*. 259-268. 1987.
17. N. Khaneja, M.I. Miller, and U. Grenander. "Dynamic Programming Generation of Geodesics and Sulci on Brain Surfaces". Submitted to *PAMI*. 1997.
18. D.C. Knill and W. Richards. (Eds). **Perception as Bayesian Inference**. Cambridge University Press. 1996.
19. S. Konishi, A.L. Yuille, J.M. Coughlan, and S.C. Zhu. "Fundamental Bounds on Edge Detection: An Information Theoretic Evaluation of Different Edge Cues". To appear in *Proceedings Computer Vision and Pattern Recognition CVPR'99*. Fort Collins. Colorado. 1999.
20. U. Montanari. "On the optimal detection of curves in noisy pictures." *Communications of the ACM*, pages 335–345, 1971.
21. J. Pearl. **Heuristics**. Addison-Wesley. 1984.
22. B.D. Ripley. **Pattern Recognition and Neural Networks**. Cambridge University Press. 1995.
23. S. Russell and P. Norvig. "Artificial Intelligence: A Modern Approach. Prentice-Hall. 1995.

24. B. Selman and S. Kirkpatrick. "Critical Behaviour in the Computational Cost of satisfiability Testing". Artificial Intelligence. 81(1-2); 273-295. 1996.

25. A.L. Yuille and J. Coughlan. "Twenty Questions, Focus of Attention, and A*: A theoretical comparison of optimization strategies." In *Proceedings of the International Workshop on Energy Minimization Methods in Computer Vision and Pattern Recognition.* Ed. M. Pellilo and E. Hancock. Venice, Italy. Springer-Verlag. May. 1997.

26. A.L. Yuille and J.M. Coughlan. "Convergence Rates of Algorithms for Visual Search: Detecting Visual Contours". In *Proceedings NIPS'98.* 1998.

27. A.L. Yuille and J.M. Coughlan. "Visual Search: Fundamental Bounds, Order Parameters, Phase Transitions, and Convergence Rates". Submitted to *Transactions on Pattern Analysis and Machine Intelligence.* 1999.

28. A.L. Yuille and J. Coughlan. "An A* perspective on deterministic optimization for deformable templates". To appear in *Pattern Recognition Letters.* 1999.

29. S.C. Zhu, Y. Wu, and D. Mumford. "Minimax Entropy Principle and Its Application to Texture Modeling". Neural Computation. Vol. 9. no. 8. Nov. 1997.

30. S.C. Zhu and D. Mumford. "Prior Learning and Gibbs Reaction-Diffusion". IEEE Trans. on PAMI vol. 19, no. 11. Nov. 1997.

31. S.C. Zhu and D. Mumford. "GRADE: A framework for pattern synthesis, denoising, image enhancement, and clutter removal." In Proceedings of International Conference on Computer Vision. Bombay. India. 1998.

32. S-C Zhu, Y-N Wu and D. Mumford. FRAME: Filters, Random field And Maximum Entropy: — Towards a Unified Theory for Texture Modeling. Int'l Journal of Computer Vision 27(2) 1-20, March/April. 1998.

33. S.C. Zhu. "Embedding Gestalt Laws in Markov Random Fields". Submitted to *IEEE Computer Society Workshop on Perceptual Organization in Computer Vision.*

A New Algorithm for
Energy Minimization with Discontinuities

Yuri Boykov, Olga Veksler, and Ramin Zabih

Computer Science Department, Cornell University, Ithaca, NY 14853
yura@cs.cornell.edu, olga@cs.cornell.edu, rdz@cs.cornell.edu

Abstract. Many tasks in computer vision involve assigning a label (such as disparity) to every pixel. These tasks can be formulated as energy minimization problems. In this paper, we consider a natural class of energy functions that permits discontinuities. Computing the exact minimum is NP-hard. We have developed a new approximation algorithm based on graph cuts. The solution it generates is guaranteed to be within a factor of 2 of the energy function's global minimum. Our method produces a local minimum with respect to a certain move space. In this move space, a single move is allowed to switch an arbitrary subset of pixels to one common label. If this common label is α then such a move *expands* the domain of α in the image. At each iteration our algorithm efficiently chooses the *expansion* move that gives the largest decrease in the energy. We apply our method to the stereo matching problem, and obtain promising experimental results. Empirically, the new technique outperforms our previous algorithm [6] both in terms of running time and output quality.

1 Energy minimization in early vision

Many early vision problems require estimating some spatially varying quantity (such as intensity, disparity or texture) from noisy measurements. Such quantities tend to be piecewise smooth; they vary smoothly at most points, but change dramatically at object boundaries. Every pixel $p \in \mathcal{P}$ must be assigned a label in some set $\mathcal{L}$; for motion or stereo, the labels are disparities, while for image restoration they represent intensities. The goal is to find a labeling f that assigns each pixel $p \in \mathcal{P}$ a label $f_p \in \mathcal{L}$, where f is both piecewise smooth and consistent with the observed data.

These vision problems can be naturally formulated in terms of energy minimization. In this framework, one seeks the labeling f that minimizes the energy

$$E_{smooth}(f) \quad + \quad E_{data}(f).$$

Here E_{smooth} measures the extent to which f is not piecewise smooth, while E_{data} measures the disagreement between f and the observed data. Many different energy functions have been proposed in the literature, depending upon the exact vision problem. The form of E_{data} is typically $E_{data}(f) = \sum_{p \in \mathcal{P}} D_p(f_p)$, where

D_p measures how appropriate a label is for the pixel p given the observed data. In the image restoration problem, for example, usually $D_p(f_p) = (f_p - i_p)^2$, where i_p is the observed intensity of the pixel p.

The choice of E_{smooth} is a critical issue, and many different functions have been proposed. For example, in regularization-based vision [2, 12, 15], E_{smooth} makes f smooth everywhere. This leads to poor results at object boundaries. Energy functions that do not have this problem are called *discontinuity-preserving*. A large number of discontinuity-preserving energy functions have been proposed (see for example [11, 14, 19]). Geman and Geman's seminal paper [9] gave a Bayesian interpretation of many energy functions, and proposed a discontinuity-preserving energy function based on Markov Random Fields (MRF's).

The major difficulty with energy minimization for early vision lies in the enormous computational costs. Typically these energy functions have many local minima (i.e., they are non-convex). Worse still, the space of possible labelings has dimension $|\mathcal{P}|$, which is many thousands. There have been numerous attempts to design fast algorithms for energy minimization; we will review this area in section 2. However, as a practical matter the computational problem remains unresolved.

In this paper we address a class of discontinuity-preserving energy functions. Let the neighborhood system $\mathcal{N}$ denote the set of pairs of adjacent pixels in $\mathcal{P}$. We consider functions of the form

$$E_P(f) \quad = \quad \sum_{\{p,q\} \in \mathcal{N}} u_{\{p,q\}} \cdot \delta(f_p \neq f_q) \quad + \quad \sum_{p \in \mathcal{P}} D_p(f_p), \qquad (1)$$

where

$$\delta(f_p \neq f_q) \quad = \quad \begin{cases} 1 & \text{if } f_p \neq f_q, \\ 0 & \text{otherwise.} \end{cases}$$

We allow D_p to be an arbitrary function, as long as it is non-negative and finite.[1]

This energy function is in some sense the simplest energy that preserves discontinuities. The smoothness term provides a penalty $u_{\{p,q\}} \geq 0$ for assigning different labels to two adjacent pixels $\{p,q\}$. This penalty does not depend on the labels assigned, as long as they are different. Such energy functions naturally arise from a particular MRF that we call a generalized Potts model; this derivation is given in [6]. We will therefore refer to the energy function E_P given in equation (1) as the *Potts energy*.

In early vision, there are a few energy functions whose global minimum can be rapidly computed [6, 10, 13]. Unfortunately, we have shown in [6] that minimizing the Potts energy is NP-hard, so it very likely requires exponential time. In this paper we introduce a new approximation algorithm for this energy minimization problem, and apply it to several vision problems. The key properties of our algorithm are that it produces a local minimum in a certain move space, and that the resulting labeling is guaranteed to be within a factor of 2 of the global minimum of the Potts energy.

---

[1] Our results do not require that D_p be finite, but this assumption simplifies the presentation considerably.

We begin with a brief survey of energy minimization methods in computer vision. In section 3 we give an overview of our approach to energy minimization. We define *expansion* moves and prove that a local minimum of the Potts energy with respect to such moves is within a factor of two of the global minimum. Section 4 gives the details of a graph cut technique that efficiently computes the expansion move producing the largest decrease in the Potts energy. Intuitively, this gives the direction of the steepest descent from a current solution. By using this technique iteratively we follow the "fastest" way into a local minimum of the Potts energy with respect to expansion move space. In section 5 we provide some experimental results on the stereo matching problem.

Note that in our earlier work [6] we presented a similar greedy descent algorithm for approximate Potts energy minimization based on *swap* moves. Strictly speaking, expansion moves and swap moves are not directly comparable since there are swap moves that are not expansion moves and vice versa. However, we have both theoretical and experimental evidence that the expansion move algorithm is superior. Theoretically, we will show in section 3.3 that a local minimum in terms of expansion moves is within a factor of 2 of the global minimum, while no such result is available for the swap move algorithm. Experimentally, the results in section 5 suggest that the expansion moves algorithm leads to a better and faster optimization of the Potts energy.

2 Related work

Energy minimization is quite popular in computer vision, and a wide variety of methods have been used. An exhaustive survey is beyond the scope of this paper; however, we will briefly describe the energy minimization methods that are most prevalent in vision.

2.1 Global energy minimization

The problem of finding the global minimum of an arbitrary energy function is obviously intractable (it includes the Potts energy minimization problem as a special case). As a consequence, any general-purpose energy minimization algorithm will require exponential time to find the global minimum, unless P=NP.

Simulated annealing was popularized for vision by [9], and is the only general-purpose global energy minimization method in widespread use. With certain annealing schedules, annealing can be guaranteed to find the global minimum. Unfortunately, the schedules that lead to this guarantee are extremely slow. In practice, annealing is inefficient partly because at each step it changes the value of a single pixel.

Graph cuts can be used to find the global minimum of certain energy functions. These algorithms permit D_p to be arbitrary. [10] addressed the case of $|L| = 2$. This result was generalized by [6, 13] to handle label sets of arbitrary size, when the smoothness energy is of the form $\sum_{\{p,q\} \in \mathcal{N}} |f_p - f_q|$. This smoothness energy, unfortunately, leads to oversmoothing at object boundaries. In addition, there must be a natural isomorphism between the label set L and the integers $\{1, 2, \ldots, k\}$. This rules out some significant problems such as motion.

Another alternative is to use methods that have optimality guarantees in certain cases. Continuation methods, such as GNC [4], are the best-known example. These methods involve approximating an intractable (non-convex) energy function by a sequence of energy functions, beginning with a tractable (convex) approximation. At every step in the approximation, a local minimum is found using the solution from the previous step as the starting point. There are circumstances where these methods are known to compute the optimal solution (see [4] for details). Continuation methods can be applied to a large number of energy functions, but except for these special cases nothing is known about the quality of their output.

2.2 Local energy minimization

Due to the inefficiency of computing a global minimum, many authors have opted for a local minimum. One problem with this is that it is difficult to determine the cause of an algorithm's failures. When an algorithm gives unsatisfactory results, it may be due either to a poor choice of energy function, or to the fact the answer is far from the global minimum. There is no obvious way to tell which of these is the problem.[2] Another issue is that local minimization techniques are naturally sensitive to the initial estimate.

There are several ways in which a local minimum can be computed. By phrasing the energy minimization problem in continuous terms, variational methods can be applied. These methods were popularized by Horn [12]. Variational techniques use the Euler equations, which are guaranteed to hold at a local minimum (although they may also hold elsewhere). A number of methods have been proposed to speed up the convergence of the resulting numerical problems, including (for example) multigrid techniques [18]. To apply these algorithms to actual imagery, of course, requires discretization. An alternative is to use discrete relaxation methods; this has been done by many authors, including [7, 16, 17].

It is important to note that a local minimum is defined relative to a set of allowed moves. Most existing minimization algorithms find a local minimum relative to "small" moves, which typically are defined in terms of the L_2 distance. To be precise, they attempt to compute a labeling f such that $f = \arg\min_{|f-f'|<\epsilon} E_P(f')$, for some small ϵ.

In [6] we described an algorithm for approximate minimization of the Potts energy based on *swap* moves. For a fixed pair of labels α, β, this move swaps the labels between a subset of pixels labeled α and another subset labeled β. The

---

[2]In the special cases where the global minimum can be rapidly computed, it is possible to separate these issues. For example, [10] points out that the global minimum of an Ising energy function is not necessarily the desired solution for image restoration. [5, 10] analyze the performance of simulated annealing in cases with a known global minimum.

algorithm in [6] is based on a graph cut technique that efficiently computes the best α, β-swap move from a current solution. By iterating over all distinct pairs α, β this technique enables the steepest descent search of the local minimum of the Potts energy with respect to swap moves. The properties of such a local minimum are based on the strength of swap moves.

In this paper we describe a new algorithm based on *expansion* moves. The structure of the algorithm is similar to [6]. It is still the greediest descent into a local minimum with respect to a certain move space. However, there are several important differences. Most importantly, the new algorithm produces a local minimum that is guaranteed to be within a factor of 2 from the global minimum. Such a bound is not available for the swap move algorithm in [6]. Moreover, the steepest descent in the new move space requires iterating over distinct labels, not pairs of labels. Altogether this suggest that the new algorithm can potentially produce faster and better solutions. The data we present in section 5 supports this conclusion.

3 The expansion move algorithm

Here we describe the algorithm for approximate minimization of the Potts energy E_P based on expansion moves. In this section, we discuss the expansion moves, which are best described in terms of partitions. We sketch the algorithm and list its basic properties. Then we introduce the notion of a graph cut, which is the basis for our method.

3.1 Partitions and move spaces

Any labeling f can be uniquely represented by a partition of image pixels

$$\mathbf{P} = \{\mathcal{P}_l \mid l \in \mathcal{L}\}$$

where $\mathcal{P}_l = \{p \in \mathcal{P} \mid f_p = l\}$ is a subset of pixels assigned label l. Since there is an obvious one to one correspondence between labelings f and partitions $\mathbf{P}$, we can use these notions interchangingly.

Given a label α, a move from a partition $\mathbf{P}$ (labeling f) to a new partition $\mathbf{P}'$ (labeling f') is called an α-*expansion* if $\mathcal{P}_\alpha \subset \mathcal{P}'_\alpha$ and $\mathcal{P}'_l \subset \mathcal{P}_l$ for any label $l \neq \alpha$. In other words, an α-expansion move allows any set of image pixels to change their labels to α.

Note that a move which gives an arbitrary label α to a single pixel is an α-expansion. As a consequence, the standard move space used in annealing is a special case of our move space.

3.2 Algorithm and properties

The structure of the expansion move algorithm is shown in figure 1. We will call a single execution of steps 3.1–3.2 an *iteration*, and an execution of steps 2–4 a *cycle*. In each cycle, the algorithm performs an iteration for every label in a certain order that can be fixed or random. A cycle is successful if a strictly better labeling is found at any iteration. The algorithm stops after the first unsuccessful cycle since no further improvement is possible.

```
1. Start with an arbitrary labeling f
2. Set success := 0
3. For each label α ∈ L
      3.1. Find f̂ = arg min E_P(f') among f' within one α-expansion of f
      3.2. If E_P(f̂) < E_P(f), set f := f̂ and success := 1
4. If success = 1 goto 2
5. Return f
```

Fig. 1: The expansion move algorithm.

The algorithm have a number of important properties.

- Obviously, a cycle takes $|\mathcal{L}|$ iterations. Note that a cycle in the swap move algorithm [6] takes $|\mathcal{L}|^2$ iterations.
- The algorithm is guaranteed to terminate in a finite number of cycles, although there is no bound beyond the trivial one of $|\mathcal{L}|^{|\mathcal{P}|}$. Nevertheless, in the applications we have considered the algorithm stops after a few cycles. Moreover, most of the improvements occur during the first cycle, as we will show in section 5.
- Once the algorithm has terminated, the energy of the resulting labeling is a local minimum with respect to an expansion move.

3.3 Optimality Guarantees

We now show that if f^* is a local minimum in terms of expansion moves, then $E_P(f^*) \leq 2 \cdot E_P(f^o)$, where f^o is the optimal solution minimizing the Potts energy E_P. Let $\mathbf{P}^o = \{\mathcal{P}^o_\alpha \mid \alpha \in \mathcal{L}\}$ be a partition corresponding to f^o so that

$$\mathcal{P}^o_\alpha = \{p \in \mathcal{P} \mid f^o_p = \alpha\}$$

is a set of pixels assigned to α in the optimal solution. We can produce a labeling f^α within one α-expansion move from f^* as follows:

$$f^\alpha_p = \begin{cases} \alpha & \text{if } p \in \mathcal{P}^o_\alpha \\ f^*_p & \text{if } p \notin \mathcal{P}^o_\alpha \end{cases}$$

The key observation is that since f^* is a local minimum in the expansion move space then for any $\alpha \in \mathcal{L}$

$$E_P(f^*) \leq E_P(f^\alpha). \tag{2}$$

For a given label $\alpha \in \mathcal{L}$ we can split the Potts energy of any labeling f into three terms $E_P(f) = E_{in}^{\alpha}(f) + E_{bd}^{\alpha}(f) + E_{ex}^{\alpha}(f)$ where

$$E_{in}^{\alpha}(f) = \sum_{\substack{\{p,q\} \in \mathcal{N} \\ p,q \in \mathcal{P}_{\alpha}^{o}}} u_{\{p,q\}} \cdot \delta(f_p \neq f_q) + \sum_{p \in \mathcal{P}_{\alpha}^{o}} D_p(f_p)$$

$$E_{bd}^{\alpha}(f) = \sum_{\substack{\{p,q\} \in \mathcal{N} \\ p \in \mathcal{P}_{\alpha}^{o}, q \notin \mathcal{P}_{\alpha}^{o}}} u_{\{p,q\}} \cdot \delta(f_p \neq f_q)$$

$$E_{ex}^{\alpha}(f) = \sum_{\substack{\{p,q\} \in \mathcal{N} \\ p,q \notin \mathcal{P}_{\alpha}^{o}}} u_{\{p,q\}} \cdot \delta(f_p \neq f_q) + \sum_{p \notin \mathcal{P}_{\alpha}^{o}} D_p(f_p)$$

correspond to the parts of the Potts energy $E_P(f)$ concentrated at the pixels inside $\mathcal{P}_{\alpha}^{o}$, at the boundary of $\mathcal{P}_{\alpha}^{o}$, and at the pixels outside of $\mathcal{P}_{\alpha}^{o}$, correspondingly.

Since $f_p^* = f_p^{\alpha}$ for any $p \notin \mathcal{P}_{\alpha}^{o}$ then $E_{ex}^{\alpha}(f^*) = E_{ex}^{\alpha}(f^{\alpha})$. Thus, (2) implies that for any $\alpha \in \mathcal{L}$

$$E_{in}^{\alpha}(f^*) + E_{bd}^{\alpha}(f^*) \leq E_{in}^{\alpha}(f^{\alpha}) + E_{bd}^{\alpha}(f^{\alpha}). \tag{3}$$

Since $f_p^{\alpha} = f_p^{o} = \alpha$ for any $p \in \mathcal{P}_{\alpha}^{o}$ then $E_{in}^{\alpha}(f^{\alpha}) = E_{in}^{\alpha}(f^{o})$. Moreover,

$$E_{bd}^{\alpha}(f^{\alpha}) \leq \sum_{\substack{\{p,q\} \in \mathcal{N} \\ p \in \mathcal{P}_{\alpha}^{o}, q \notin \mathcal{P}_{\alpha}^{o}}} u_{\{p,q\}} = E_{bd}^{\alpha}(f^{o}).$$

Therefore, (3) implies that for any $\alpha \in \mathcal{L}$

$$E_{in}^{\alpha}(f^*) + E_{bd}^{\alpha}(f^*) \leq E_{in}^{\alpha}(f^{o}) + E_{bd}^{\alpha}(f^{o}). \tag{4}$$

Summing up inequality (4) over all labels $\alpha \in \mathcal{L}$ we obtain

$$E_P(f^*) + \sum_{\{p,q\} \in B} u_{\{p,q\}} \cdot \delta(f_p^* \neq f_q^*) \leq E_P(f^{o}) + \sum_{\{p,q\} \in B} u_{\{p,q\}} \tag{5}$$

where $B = \{\{p,q\} \in \mathcal{N} \mid f_p^{o} \neq f_q^{o}\}$ is a set of all pairs of neighboring pixels disconnected in the optimal solution f^{o}. Note that the summations on both sides of (5) show up because each pair of pixels in B is encountered twice when summing up the terms in (4) over $\alpha \in \mathcal{L}$. Finally, since $\sum_{\{p,q\} \in B} u_{\{p,q\}} \leq E_P(f^{o})$ then (5) implies that $E_P(f^*) \leq 2E_P(f^{o})$.

3.4 Graph cuts

The key part of the algorithm is step 3.1, where graph cuts are used to efficiently find $\hat{f}$. Let $\mathcal{G} = \langle \mathcal{V}, \mathcal{E} \rangle$ be a weighted graph with two distinguished vertices called the terminals. A *cut* $\mathcal{C} \subset \mathcal{E}$ is a set of edges such that the terminals are separated in the induced graph $\mathcal{G}(\mathcal{C}) = \langle \mathcal{V}, \mathcal{E} - \mathcal{C} \rangle$. In addition, no proper subset of $\mathcal{C}$

separates the terminals in $\mathcal{G}(\mathcal{C})$. The cost of the cut $\mathcal{C}$ is denoted by $|\mathcal{C}|$ and equals the sum of its edge weights.

The minimum cut problem is to find the cut with smallest cost. This problem can be solved very efficiently by computing the maximum flow between the terminals, according to a theorem due to Ford and Fulkerson [8]. There are a large number of fast algorithms for this problem (see [1], for example). The worst case complexity is low-order polynomial; however, in practice the running time is nearly linear.

Step 3.1 uses a single minimum cut on a graph whose size is $O(|\mathcal{P}|)$. The graph is dynamically updated after each iteration. The next section describes the details of our graph cut technique that allows efficient implementation of step 3.1.

4   Finding the optimal expansion move

Given an input labeling f (partitioning $\mathbf{P}$) and a label α, we wish to find a labeling $\hat{f}$ that minimizes E_P over all labelings within one α-expansion of f. This is the critical step in the algorithm given at the bottom of figure 1. Our technique is based on computing a labeling corresponding to a minimum cut on a graph $\mathcal{G}_\alpha = \langle \mathcal{V}_\alpha, \mathcal{E}_\alpha \rangle$. The structure of this graph is determined by the current partitioning $\mathbf{P}$ and by the label α. The graph dynamically changes after each iteration.

This section is organized as follows. First we describe the construction of $\mathcal{G}_\alpha$ for a given f (or $\mathbf{P}$) and α. We show that cuts $\mathcal{C}$ on $\mathcal{G}_\alpha$ correspond in a natural way to labelings $f^\mathcal{C}$ which are within one α-expansion move of f. Then, based on a number of simple properties, we define a class of *elementary* cuts. Theorem 1 shows that elementary cuts are in one to one correspondence with the set of labelings that are within one α-expansion of f, and also that the cost of an elementary cut is $|\mathcal{C}| = E_P(f^\mathcal{C})$. A corollary from this theorem states our main result that the desired labeling $\hat{f}$ equals $f^\mathcal{C}$ where $\mathcal{C}$ is a minimum cut on $\mathcal{G}_\alpha$.

The structure of the graph is illustrated in Figure 2. For legibility, this figure shows the case of 1D image. In fact, the structure of $\mathcal{G}_\alpha$ will be the same for any image. The set of vertices includes the two terminals α and $\bar{\alpha}$, as well as all image pixels $p \in \mathcal{P}$. In addition, for each pair of neighboring pixels $\{p, q\} \in \mathcal{N}$ separated in the current partition (i.e. $f_p \neq f_q$) we create an *auxiliary vertex* $a_{\{p,q\}}$. Auxiliary nodes are introduced at the boundaries between partition sets $\mathcal{P}_l$ for $l \in \mathcal{L}$. Thus, the set of vertices is

$$\mathcal{V}_\alpha = \alpha \cup \bar{\alpha} \cup \mathcal{P} \cup \left(\bigcup_{\substack{\{p,q\} \in \mathcal{N} \\ f_p \neq f_q}} a_{\{p,q\}} \right).$$

Each pixel $p \in \mathcal{P}$ is connected to the terminals α and $\bar{\alpha}$ by edges t_p^α and $t_p^{\bar{\alpha}}$, correspondingly. For brevity, we will refer to these edges as t-links (terminal

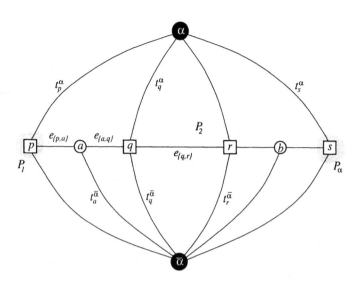

Fig. 2: An example of the graph $\mathcal{G}_\alpha$ for a 1D image. The set of pixels in the image is $\mathcal{P} = \{p, q, r, s\}$ and the current partition is $\mathbf{P} = \{\mathcal{P}_1, \mathcal{P}_2, \mathcal{P}_\alpha\}$ where $\mathcal{P}_1 = \{p\}$, $\mathcal{P}_2 = \{q, r\}$, and $\mathcal{P}_\alpha = \{s\}$. There are two auxiliary nodes $a = a_{\{p,q\}}$, $b = a_{\{r,s\}}$ introduced between neighboring pixels separated in the current partition. Auxiliary nodes are added at the boundary of sets $\mathcal{P}_l$.

links). Each pair of neighboring pixels $\{p, q\} \in \mathcal{N}$ which is not separated by the partition $\mathbf{P}$ (i.e. $f_p = f_q$) is connected by an edge $e_{\{p,q\}}$ which we will call an n-link (neighborhood link). For each pair of neighboring pixels $\{p, q\} \in \mathcal{N}$ such that $f_p \neq f_q$ we create a triplet of edges

$$\mathcal{E}_{\{p,q\}} = \left\{ e_{\{p,a\}}, \ e_{\{a,q\}}, \ t_a^{\bar{\alpha}} \right\}$$

where $a = a_{\{p,q\}}$ is the corresponding auxiliary node. The n-links $e_{\{p,a\}}$ and $e_{\{a,q\}}$ connect pixels p and q to $a_{\{p,q\}}$ and the t-link $t_a^{\bar{\alpha}}$ connects the auxiliary node $a_{\{p,q\}}$ to the terminal $\bar{\alpha}$. Finally, we can write the set of all edges as

$$\mathcal{E}_\alpha = \left(\bigcup_{p \in \mathcal{P}} \{t_p^\alpha, t_p^{\bar{\alpha}}\} \right) \cup \left(\bigcup_{\substack{\{p,q\} \in \mathcal{N} \\ f_p \neq f_q}} \mathcal{E}_{\{p,q\}} \right) \cup \left(\bigcup_{\substack{\{p,q\} \in \mathcal{N} \\ f_p = f_q}} e_{\{p,q\}} \right).$$

The weights assigned to the edges are shown in the table below.

edge	weight	for all
$|t_p^\alpha|$	$D_p(\alpha)$	$p \in \mathcal{P}_\alpha$
$|t_p^{\bar\alpha}|$	∞	
$|t_p^\alpha|$	$D_p(\alpha)$	$p \notin \mathcal{P}_\alpha$
$|t_p^{\bar\alpha}|$	$D_p(f_p)$	
$|e_{\{p,a\}}| = |e_{\{a,q\}}| = |t_a^{\bar\alpha}|$	$u_{\{p,q\}}$	$\{p,q\} \in \mathcal{N},\ f_p \neq f_q$
$|e_{\{p,q\}}|$	$u_{\{p,q\}}$	$\{p,q\} \in \mathcal{N},\ f_p = f_q$

Any cut C on the graph $\mathcal{G}_\alpha$ must sever (include) exactly one t-link for any pixel $p \in \mathcal{P}$: if neither t-link were in C, there would be a path between the terminals; while if both t-links were cut, then a proper subset of C would be a cut. Thus, any cut includes either t_p^α or $t_p^{\bar\alpha}$ for each pixel $p \in \mathcal{P}$. This defines a natural labeling f^C corresponding to a cut C on $\mathcal{G}_\alpha$. Formally,

$$f_p^C = \begin{cases} \alpha & \text{if } t_p^\alpha \in C \\ f_p & \text{if } t_p^{\bar\alpha} \in C \end{cases} \qquad \forall p \in \mathcal{P}. \tag{6}$$

In other words, a pixel p is assigned label α if the cut C severs t-link t_p^α, while p is assigned its old label f_p if C severs $t_p^{\bar\alpha}$. The terminal α stands for the new label and the terminal $\bar\alpha$ stands for the old labels assigned to pixels in the initial labeling f.

Lemma 1. *A cut C on $\mathcal{G}_\alpha$ corresponds to a labeling f^C which is one α-expansion away from the original labeling f.*

Proof. A cut C cannot sever t-links $t_p^{\bar\alpha}$ for any pixel $p \in \mathcal{P}_\alpha$ due to the infinite cost. Thus, $f_p^C = \alpha$ for any $p \in \mathcal{P}_\alpha$. For any pixel $p \notin \mathcal{P}_\alpha$ the value of f_p^C can be either α or f_p. □

It is easy to show that a cut C severs an n-link $e_{\{p,q\}}$ between neighboring pixels $\{p,q\} \in \mathcal{N}$ such that $f_p = f_q$ if and only if C leaves the pixels p and q connected to different terminals. Formally, for any cut C

Property 1. If $t_p^{\bar\alpha}, t_q^{\bar\alpha} \in C$ then $e_{\{p,q\}} \notin C$.
Property 2. If $t_p^\alpha, t_q^\alpha \in C$ then $e_{\{p,q\}} \notin C$.
Property 3.1 If $t_p^{\bar\alpha}, t_q^\alpha \in C$ then $e_{\{p,q\}} \in C$.
Property 3.2 If $t_p^\alpha, t_q^{\bar\alpha} \in C$ then $e_{\{p,q\}} \in C$.

The first two properties follow from the requirement that no proper subset of C should separate the terminals. Properties 3.1 and 3.2 also use the fact that a cut has to separate the terminals.

These properties are illustrated in Figure 3. The following lemma is a consequence of properties 1–3 above and equation 6.

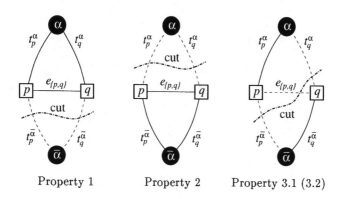

Property 1   Property 2   Property 3.1 (3.2)

Fig. 3: Properties of a cut $\mathcal{C}$ on $\mathcal{G}_\alpha$ for two pixels $p, q \in \mathcal{N}$ such that $f_p = f_q$. Dotted lines show the edges cut by $\mathcal{C}$ and solid lines show the edges remaining in the induced graph $\mathcal{G}_\alpha(\mathcal{C}) = \langle \mathcal{V}_\alpha, \mathcal{E}_\alpha - \mathcal{C} \rangle$.

Lemma 2. *If $\{p, q\} \in \mathcal{N}$ and $f_p = f_q$ then any cut $\mathcal{C}$ on $\mathcal{G}_\alpha$ satisfies*

$$|\mathcal{C} \cap e_{\{p,q\}}| = u_{\{p,q\}} \cdot \delta(f_p^{\mathcal{C}} \neq f_q^{\mathcal{C}}).$$

Consider now the set of edges $\mathcal{E}_{\{p,q\}}$ corresponding to a pair of neighboring pixels $\{p, q\} \in \mathcal{N}$ such that $f_p \neq f_q$. In this case, there are several different ways to cut these edges even when the pair of severed t-links at p and q is fixed. However, a minimum cut $\mathcal{C}$ on $\mathcal{G}_\alpha$ is guaranteed to sever the edges in $\mathcal{E}_{\{p,q\}}$ depending on what t-links are cut at the pixels p and q.

The rule for this case is described in properties 4–6 below. Assume that $a = a_{\{p,q\}}$ is an auxiliary node between the corresponding pair of neighboring pixels. Then a minimum cut $\mathcal{C}$ on $\mathcal{G}_\alpha$ satisfies the following properties.

Property 4. If $t_p^{\bar{\alpha}}, t_q^{\bar{\alpha}} \in \mathcal{C}$ then $\mathcal{C} \cap \mathcal{E}_{\{p,q\}} = t_a^{\bar{\alpha}}$.
Property 5. If $t_p^{\alpha}, t_q^{\alpha} \in \mathcal{C}$ then $\mathcal{C} \cap \mathcal{E}_{\{p,q\}} = \emptyset$.
Property 6.1 If $t_p^{\bar{\alpha}}, t_q^{\alpha} \in \mathcal{C}$ then $\mathcal{C} \cap \mathcal{E}_{\{p,q\}} = e_{\{p,a\}}$.
Property 6.2 If $t_p^{\alpha}, t_q^{\bar{\alpha}} \in \mathcal{C}$ then $\mathcal{C} \cap \mathcal{E}_{\{p,q\}} = e_{\{a,q\}}$.

The first property results from the fact that no subset of $\mathcal{C}$ is a cut. The others follow from the minimality of $|\mathcal{C}|$ and the fact that $e_{\{p,a\}}$, $e_{\{a,q\}}$ and $t_a^{\bar{\alpha}}$ have identical weights. These properties are illustrated in figure 4.

Lemma 3. *If $\{p, q\} \in \mathcal{N}$ and $f_p \neq f_q$ then the minimum cut $\mathcal{C}$ on $\mathcal{G}_\alpha$ satisfies*

$$|\mathcal{C} \cap \mathcal{E}_{\{p,q\}}| = u_{\{p,q\}} \cdot \delta(f_p^{\mathcal{C}} \neq f_q^{\mathcal{C}}).$$

Proof. The equation follows from properties 4-6 above and equation (6). Note that $f_p \neq f_q$ implies that α is the only common label that a cut on $\mathcal{G}_\alpha$ can assign to p and q using our convention in (6). That is, $f_p^{\mathcal{C}} = f_q^{\mathcal{C}}$ if and only if both $f_p^{\mathcal{C}} = \alpha$ and $f_q^{\mathcal{C}} = \alpha$. □

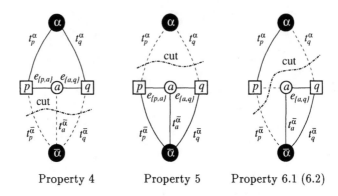

Fig. 4: Properties of a minimum cut C on $\mathcal{G}_\alpha$ for two pixel $p, q \in \mathcal{N}$ such that $f_p \neq f_q$. Dotted lines show the edges cut by C and solid lines show the edges in the induced graph $\mathcal{G}_\alpha(C) = \langle \mathcal{V}_\alpha, \mathcal{E}_\alpha - C \rangle$.

Note that the penalty $u_{\{p,q\}}$ is imposed whenever $f_p^C \neq f_q^C$. This is exactly what the auxiliary pixel construction was designed for. We had to develop a special trick for the case when the original labels for p and q do not agree ($f_p \neq f_q$) in order to get the same effect that lemma 2 establishes for the simpler situation when $f_p = f_q$.

Properties 1–3 hold for any cut, and properties 4–6 hold for a minimum cut. However, there can be other cuts besides the minimum cut that satisfy all six properties. We will define a *elementary* cut on $\mathcal{G}_\alpha$ to be a cut that satisfies properties 1–6.

Theorem 1. *Let the graph $\mathcal{G}_\alpha$ be constructed as above for a given f and α. Then there is a one to one correspondence between the set of all elementary cuts on $\mathcal{G}_\alpha$ and the set of all labelings within one α-expansion of f. Moreover, for any elementary cut C we have $|C| = E_P(f^C)$.*

Proof. We first show that an elementary cut C is uniquely determined by the corresponding labeling f^C. The label f_p^C at the pixel p determines which of the t-links to p is in C. Properties 1-3 show which n-links $e_{\{p,q\}}$ between pairs of neighboring pixels $\{p,q\}$ such that $f_p = f_q$ should be severed. Similarly, properties 4-6 determine which of the links in $\mathcal{E}_{\{p,q\}}$ corresponding to $\{p,q\} \in \mathcal{N}$ such that $f_p \neq f_q$ should be cut.

We now compute the cost of an elementary cut C, which is

$$|C| = \sum_{p \in \mathcal{P}} |C \cap \{t_p^\alpha, t_p^{\bar{\alpha}}\}| + \sum_{\substack{\{p,q\} \in \mathcal{N} \\ f_p = f_q}} |C \cap e_{\{p,q\}}| + \sum_{\substack{\{p,q\} \in \mathcal{N} \\ f_p \neq f_q}} |C \cap \mathcal{E}_{\{p,q\}}|. \quad (7)$$

It is easy to show that for any pixel $p \in \mathcal{P}$ we have $|C \cap \{t_p^\alpha t_p^{\bar{\alpha}}\}| = D_p(f_p^C)$. Lemmas 2 and 3 hold for elementary cuts, since they were based on properties 1-6. These two lemmas give us the second and the third terms in (7). Thus, the total cost of a elementary cut $\mathcal{C}$ is

$$|\mathcal{C}| \;=\; \sum_{p \in \mathcal{P}} D_p(f_p^{\mathcal{C}}) + \sum_{\{p,q\} \in \mathcal{N}} u_{\{p,q\}} \cdot \delta(f_p^{\mathcal{C}} \neq f_q^{\mathcal{C}}) \;=\; E_P(f^{\mathcal{C}}).$$

Therefore, $|\mathcal{C}| = E_P(f^{\mathcal{C}})$. □

Our main result is a simple consequence of this theorem, since the minimum cut is an elementary cut.

Corollary 1. *The optimal α-expansion move from f is $\hat{f} = f^{\mathcal{C}}$ where $\mathcal{C}$ is the minimum cut on $\mathcal{G}_\alpha$.*

5 Experimental results

In this section we apply our method to the stereo matching problem. We compare our method with simulating annealing, using real image pairs, including one with dense ground truth. For D_p we use the method of [3] to reduce the effects of image sampling. We select $u_{\{p,q\}}$ using the information present in a single image, as described in [6].

We experimented with several variants of simulated annealing, including both the standard (Metropolis) sampler and the Gibbs sampler. Our comparative data uses the annealing variant and the choice of cooling schedule that best minimized the energy. Simulated annealing is quite sensitive to the starting point, so we initialized it using the results of normalized correlation. Our methods give very similar answers regardless of the starting point, but we used the same starting point as annealing to make the comparison fair. All running times are given in seconds, on a 200 MHz Pentium Pro.

Figure 5(a) shows the left image of a real stereo pair where the ground truth is known at each pixel. We obtained this image pair from the University of Tsukuba Multiview Image Database. The ground truth is shown in figure 5(b). Our results are shown below, both for the expansion move algorithm presented in this paper and the swap move algorithm introduced in [6]. For comparison, we also show the results from simulated annealing, as well as from normalized correlation (using the window size that minimizes the number of errors). Figure 7 shows the performance of algorithms as a function of time, both in terms of energy and in terms of accuracy with respect to the ground truth.

Figure 6 shows the performance of expansion move algorithm on the CMU meter image, along with the results of simulated annealing. The performance in terms of energy is similar to the results shown in figure 7(a).

Acknowledgements

We thank J. Kleinberg, D. Shmoys and E. Tardos for insightful remarks on the content of this paper. We are also grateful to Y. Ohta and Y. Nakamura for supplying the ground truth imagery from the University of Tsukuba. This research has been supported by DARPA under contract DAAL01-97-K-0104, and by a grant from Microsoft.

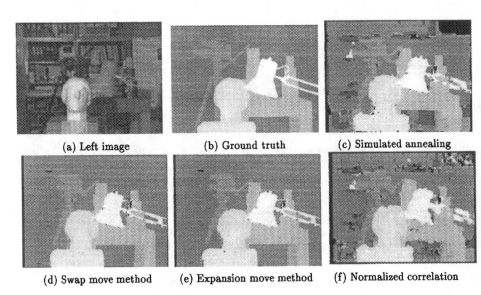

(a) Left image   (b) Ground truth   (c) Simulated annealing

(d) Swap move method   (e) Expansion move method   (f) Normalized correlation

Fig. 5: Performance on real imagery with ground truth

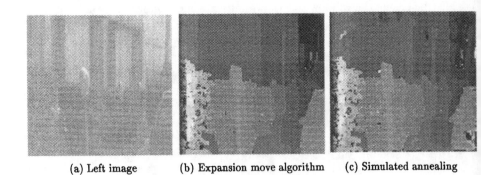

(a) Left image   (b) Expansion move algorithm   (c) Simulated annealing

Fig. 6: CMU meter imagery results

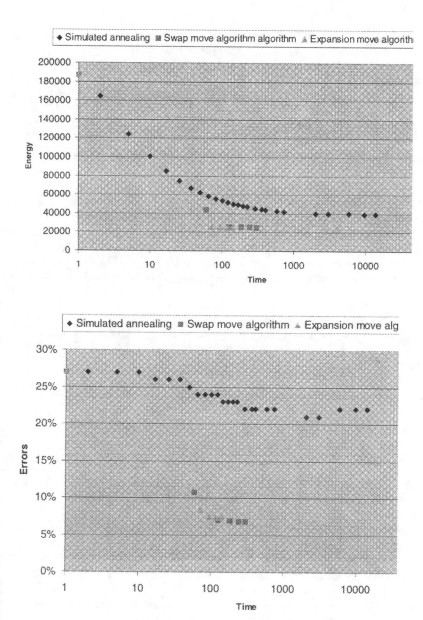

Fig. 7: Performance comparison with simulated annealing, on the imagery shown in figure 5(a). Comparison is done in terms of energy (top) and accuracy with respect to the ground truth (bottom). Each data point for our methods corresponds to a cycle.

References

1. Ravindra K. Ahuja, Thomas L. Magnanti, and James B. Orlin. *Network Flows: Theory, Algorithms, and Applications.* Prentice Hall, 1993.
2. Stephen Barnard. Stochastic stereo matching over scale. *International Journal of Computer Vision*, 3(1):17–32, 1989.
3. Stan Birchfield and Carlo Tomasi. A pixel dissimilarity measure that is insensitive to image sampling. *IEEE Transactions on Pattern Analysis and Machine Intelligence*, 20(4):401–406, April 1998.
4. A. Blake and A. Zisserman. *Visual Reconstruction.* MIT Press, 1987.
5. Andrew Blake. Comparison of the efficiency of deterministic and stochastic algorithms for visual reconstruction. *IEEE Transactions on Pattern Analysis and Machine Intelligence*, 11(1):2–12, January 1989.
6. Yuri Boykov, Olga Veksler, and Ramin Zabih. Energy minimization with discontinuities. In review. Available from http://www.cs.cornell.edu/home/rdz. An earlier version of this paper appeared in *CVPR '98*.
7. P.B. Chou and C.M. Brown. The theory and practice of Bayesian image labeling. *International Journal of Computer Vision*, 4(3):185–210, 1990.
8. L. Ford and D. Fulkerson. *Flows in Networks.* Princeton University Press, 1962.
9. S. Geman and D. Geman. Stochastic relaxation, Gibbs distributions, and the Bayesian restoration of images. *IEEE Transactions on Pattern Analysis and Machine Intelligence*, 6:721–741, 1984.
10. D. Greig, B. Porteous, and A. Seheult. Exact maximum a posteriori estimation for binary images. *Journal of the Royal Statistical Society, Series B*, 51(2):271–279, 1989.
11. W. Eric L. Grimson and Theo Pavlidis. Discontinuity detection for visual surface reconstruction. *Computer Vision, Graphics and Image Processing*, 30:316–330, 1985.
12. B. K. P. Horn and B. Schunk. Determining optical flow. *Artificial Intelligence*, 17:185–203, 1981.
13. H. Ishikawa and D. Geiger. Segmentation by grouping junctions. In *IEEE Conference on Computer Vision and Pattern Recognition*, pages 125–131, 1998.
14. David Lee and Theo Pavlidis. One dimensional regularization with discontinuities. *IEEE Transactions on Pattern Analysis and Machine Intelligence*, 10(6):822–829, November 1988.
15. Tomaso Poggio, Vincent Torre, and Christof Koch. Computational vision and regularization theory. *Nature*, 317:314–319, 1985.
16. A. Rosenfeld, R.A. Hummel, and S.W. Zucker. Scene labeling by relaxation operations. *IEEE Transactions on Systems, Man, and Cybernetics*, 6(6):420–433, June 1976.
17. R.S. Szeliski. Bayesian modeling of uncertainty in low-level vision. *International Journal of Computer Vision*, 5(3):271–302, December 1990.
18. Demetri Terzopoulos. Image analysis using multigrid relaxation methods. *IEEE Transactions on Pattern Analysis and Machine Intelligence*, 8(2):129–139, 1986.
19. Demetri Terzopoulos. Regularization of inverse visual problems involving discontinuities. *IEEE Transactions on Pattern Analysis and Machine Intelligence*, 8(4):413–424, 1986.

Convergence of a Hill Climbing Genetic Algorithm for Graph Matching

Andrew D.J. Cross and Edwin R. Hancock

Department of Computer Science
University of York, York, Y01 5DD, UK.

Abstract. This paper presents a convergence analysis for the problem of consistent labelling using genetic search. The work builds on a recent empirical study of graph matching where we showed that a Bayesian consistency measure could be efficiently optimised using a hybrid genetic search procedure which incorporated a hill-climbing step. In the present study we return to the algorithm and provide some theoretical justification for its observed convergence behaviour. The main conclusion of this study is that the hill-climbing step significantly accelerates convergence, and that the convergence rate is polynomial in the size of the node-set of the graphs being matched.

1 Introduction

Configurational optimisation problems permeate all fields of machine intelligence. Broadly speaking, they are concerned with assigning symbolic or discretely defined variables to sites organised on a regular or irregular network in such a way as to satisfy certain hard constraints governing the structure of the final solution. The problem has been studied for over three decades. Concrete examples include the travelling salesman [7] and N-queens problems [8] together with a variety of network labelling [9, 10] and graph-matching [16] or graph colouring problems. The search for consistent solutions has been addressed using a number of computational techniques. Early examples from the artificial intelligence literature include Mackworth's constraint networks [9, 10], Waltz's use of discrete relaxation to locate consistent interpretations of line drawings [17], together with a host of applications involving the A* algorithm [11, 20]. More recently, the quest for effective search strategies has widened to include algorithms which offer improved global convergence properties. Examples include the use of simulated annealing [5, 7], mean-field annealing [19], tabu-search [14] and most recently genetic search [3].

Despite stimulating a large number of application studies in the machine intelligence literature, the convergence properties of these modern global optimisation methods are generally less well understood than their classical counterparts. For instance, in the case of genetic search, although there has been considerable effort directed at understanding the convergence for infinite populations of linear chromosomes [12, 13, 15], little attention has been directed towards understanding the performance of the algorithm for discrete entities organised on a network structure. In a recent study we have investigated the use of genetic search for graph matching [2]. Here we have used a hill-climbing genetic search procedure to optmise a Bayesian measure of for gauging relational consistency [18]. The main conclusions of our study were threefold. First, the consistent labelling of graphs was only amenable to genetic search if a hill-climbing operator was incorporated. Second, the quality of the final solution was greatly improved if crossover (or genetic recombination) was conducted by exchanging connected subgraphs. Finally, we found the optimisation process to be relatively insensitive to the choice of mutation rate.

Unfortunately, our analysis of the empirical results has hitherto been extremely limited and has been couched only in terms of a rather qualitative model of the pattern-space in which configurational optimisation is performed [2]. This has meant that we have been unable either to predict the convergence behaviour or to account for the three interesting empirical properties listed above. The aim in this paper is to remedy this shortcoming by presenting a detailed analysis of algorithm behaviour. It is important to stress that although there have been several analyses of genetic search, these differ from the study described here in three important ways. First, we are concerned specifically with the graph-matching problem. This means that we present an analysis that is more pertinent to the consistent labelling problem where there is network organisation rather than a linear chromosome. Second, we pose our analysis in terms of discrete assignment variables rather than continuous ones.

2 Relational Graphs

Central to this paper is the aim of matching relational graphs represented in terms of configurations of symbolic labels. We represent such a graph by $G = (V, E)$, where V is the symbolic label-set assigned to the set of nodes and E is the set of edges between the nodes. Formally, we represent the matching of the nodes in the data graph $G_1 = (V_1, E_1)$ against those in the model graph $G_2 = (V_2, E_2)$ by the function $f : V_1 \rightarrow V_2$. In other words, the current state of match is denoted by the set of Cartesian-pairs constituting the function f.

In order to describe local interactions between the nodes at a manageable level, we will represent the graphs in terms of their clique structure. The clique associated with the node indexed j consists of those nodes that are connected by an edge of the graph,i.e. $C_j = j \cup \{i \in V_1 | (i, j) \in E_1\}$. The labelling or mapping of this clique onto the nodes of the graph G_2 is denoted by $\Gamma_j = \{f(i) \in V_2, \forall i \in C_j\}$. Suppose that we have access to a set of patterns that represent feasible relational mappings between the cliques of graph G_1 and those of graph G_2. Typically, these relational mappings would be configurations of consistent clique labellings which we want to recover from an initial inconsistent state of the matched graph G_1. Assume that there are Z_j relational mappings for the clique C_j which we denote by $\Lambda^\mu = \{\lambda_i^\mu \in V_2, \forall i \in C_j\}$ where $\mu \in \{1, 2...Z_j\}$ is a pattern index. According to this notation $\lambda_i^\mu \in V_2$ is the match onto graph G_2 assigned to the node $i \in V_1$ of graph G_1 by the μth relational mapping.

The complete set of legal relational mappings for the clique C_j are stored in a *dictionary* which we denote by $\Theta_j = \{\Lambda^\mu | \mu = 1, 2, ..., Z_j\}$.

The basic measure of consistency underpinning our method is the Hamming distance between the matched supercliques of the data-graph and the consistent relational mappings residing in the dictionary [18]. Using the ingredients outlined above, the Hamming distance between the superclique matching configuration Γ_j and the dictionary item Λ^μ is $H(\Gamma_j, \Lambda^\mu) = \sum_{i \in C_j}(1 - \delta_{f(i),\lambda_i^\mu})$.

According to Wilson and Hancock [18], the probability of match is found by assuming that matching errors occur with a uniform and memoryless error probability P_e. As a result

$$P(\Gamma_j) = \frac{b}{Z_j} \sum_{\mu=1}^{Z_j} \exp[-k_e H(\Gamma_j, \Lambda^\mu)] \tag{1}$$

where $b = (1 - P_e)^{|C_j|}$ and $k_e = \ln \frac{1-P_e}{P_e}$. We use as our global measure of consistency the average of the clique configurational probabilities i.e. $P_G = \frac{1}{|V_1|} \sum_{j \in V_1} P(\Gamma_j)$.

Our recent paper [2], showed how a hill-climbing genetic search procedure could be used to optimise this global consistency measure. In essence the approach relies on generating a population of random initial matching configurations. These undergo cross-over, mutation and selection to locate the optimal configuration of correspondence assignments between the two graphs. The main empirical findings concerning convergence were as follows:

- The method was relatively insensitive to mutation rate. In fact, provided that the mutation probability did not exceed 0.600, then the number of iterations required for convergence was approximately constant.
- The addition of the hill climbing step considerably reduced the number of iterations required for convergence.
- Once the population size exceeded a critical value, then convergence rate was essentially independent of population size.
- The number of iterations required for convergence was approximately polynomial in the number of graph-nodes.

The aim in the remainder of this paper is to provide an analysis which supports these empirical findings.

3 Distribution Analysis

We formulate our investigation of graph-matching as a discrete time process with the states defined over the state-space of all possible correspondences between a pair of graphs. Our analysis of the population is a statistical one, in which we assume that the population is sufficiently large that we can invoke the central limit theorem. For this reason we direct our attention to the modelling of the probability density function for the distribution of solution-vectors.

3.1 Formal Ingredients of the Model

Suppose that the index-set for the population of solutions is $\mathcal{P}$ and that α is the solution index. For the problem of graph matching, each solution vector belonging to the solutions $f^{(\alpha)} : V_1 \to V_2$ represents the labelling of the nodes of a data-graph V_1 with the nodes of a model graph V_2. In order to simplify the analysis, we will assume that the pair of graphs have an identical number of nodes, i.e. $|V| = |V_1| = |V_2|$.

To commence our modelling of the distribution of solution-vectors, we focus our attention on the fraction of mappings that are in agreement with known ground truth. If the configuration of ground-truth correspondence matches is denoted by $\tilde{f}$, then the fraction of correctly assigned matches for the solution-vector indexed α is equal to

$$F_\alpha^{(n)} = \frac{1}{V} \sum_{i \in V} (1 - \delta_{f^{(\alpha)}(i), \tilde{f}(i)}) \tag{2}$$

where α is a population index of the solution-vector, n is the iteration number and δ is the Kronecker delta function. A solution-vector $f^{(\alpha)}$ in which each of the matches is correct would have $F_\alpha^{(n)} = 1$. By contrast a solution vector in which none of the correspondence matches are correct would have $F_\alpha^{(n)} = 0$. In order to analyse how the genetic graph-matching process performs, we wish to evaluate the distribution of $F_\alpha^{(n)}$ over the entire population of candidate solution vectors. At iteration n, we denote the distribution of fractional matching-error by $P_D^{(n)}(F = \gamma)$.

The overall goal of our analysis is to model how the distribution of the fraction of correct matches evolves with iteration number. For reasons of tractability, we largely confine our attention to understanding how the mean fraction of correct matches changes with iteration number. The quantity of interest is

$$\overline{F_\alpha^{(n)}} = \frac{1}{|\mathcal{P}|} \sum_{\alpha \in \mathcal{P}} F_\alpha^{(n)} = \int_\gamma \gamma P_D^{(n)}(F = \gamma) d\gamma \tag{3}$$

We commence our analysis by assuming that the number of correct matches in the initial population follows a binomial distribution. This is justifiable provided that the initial matching errors are governed by a Bernoulli process. If the initial population is chosen in a random fashion, then at the outset the expected fraction of correct matches is given by $F^{(0)} = \frac{1}{|V|}$. Under the binomial assumption, the number of correct matches has mean $F^{(0)}|V|$ and standard deviation $\sqrt{|V|F^{(0)}(1 - F^{(0)})}$. As a result the initial number of correct matches τ is distributed in the following manner

$$P(\tau) = \frac{(|V|)!}{\tau!(|V| - \tau)!} (F^{(0)})^\tau (1 - F^{(0)})^{|V| - \tau} \tag{4}$$

Since we are interested in the fraction of correct matches, we turn our attention to the distribution of the random variable $\gamma = \frac{\tau}{|V|}$. By appealing to the central limit theorem under the assumption that the graphs have large numbers of nodes, we can replace the binomial distribution of the number of correct matches by a Gaussian distribution of the fraction of correct matches. In practice we deal with graphs whose size exceeds 40 nodes, and so this approximation will be a faithful one. The probability distribution for the fraction of correct solutions in the population is therefore

$$P_D^{(0)}(F^{(n)} = \gamma) \approx \frac{1}{\sqrt{2\pi |V| F^{(0)}(1 - F^{(0)})}} \exp\left(-\frac{(|V|\gamma - |V|F^{(0)})^2}{2|V|F^{(0)}(1 - F^{(0)})}\right) \quad (5)$$

Because the distribution is Gaussian, the mode is located at the position $\gamma = |V|\mathcal{F}^{(0)}$.

4 Genetic Operators

In this section we will investigate the role that the three traditional genetic operators, i.e. mutation, crossover and selection, play in the evolution of the average fraction of correct solutions in the genetic population. We will supplement this analysis with a discussion of the hill-climbing process. At this stage, our interest lies not with the prediction of the collective behavior of the operators, but with the effect that each one has in isolation upon the population. Collective behaviour is the topic of Section 5.

4.1 The Mutation Operator

The goal of the mutation operator is to increase population diversity by performing a stochastic state swapping process on the individual node matches residing on each of the different solutions that constitute the current population. This process proceeds independently for both the individual nodes and the individual solutions. This is in contrast with the crossover which serves to exchange information between pairs of solutions in order to form new individuals that are admitted to the population on the basis of their fitness. The uniform probability of any match undergoing a random state swapping process is P_m. For each individual solution vector, there are three possible transitions that can occur in the state of match. First, an individual mutation could increase the number of correct matches by one unit; in this case the increase in the fraction of correct matches is $P_m(1 - F_\alpha^{(n)})\frac{1}{|V|}$. The second possible outcome is a reduction in the number of correctly assigned matches by one unit; in this case the decrease in the fraction of correct matches is $P_m F_\alpha^{(n)} \frac{|V|-1}{|V|}$. Finally, the mutation could leave the number of correct correspondences unchanged; in this case the fraction of correct matches remains unchanged at the value $F_\alpha^{(n)}$. For moderate mutation rates, the most likely change to the fraction of matches is due to the second transition. This corresponds to a disruption of the set of correctly assigned matches.

We are interested in the effect that the mutation operator has upon a solution-vector in which the fraction of correct matches is $F_\alpha^{(n)}$ at iteration n. In particular, we would like to compute the average value of the fraction of correct matches at iteration $n + 1$. Based on the three assignment transitions outlined above, the new average fraction of correct matches is

$$F_\alpha^{(n+1)} = F_\alpha^{(n)} + P_m(1 - F_\alpha^{(n)})\frac{1}{|V|} - P_m F_\alpha^{(n)}\frac{|V| - 1}{|V|} \tag{6}$$

After some straightforward algebra, we can re-write this recursion formula in terms of the fraction of matches correct at the outset, i.e. $F_\alpha^{(0)}$. As a result, the average fraction of correct matches at iteration n is

$$F_\alpha^{(n)} = \frac{1}{|V|} + (F_\alpha^{(0)} - \frac{1}{|V|})\exp(-k_m n) \tag{7}$$

where $k_m = \ln\frac{1}{(1-P_m)}$. There are a number of interesting features of this formula that deserve further comment. First, the equation represents an exponential decay that tends towards a minimum value of $\frac{1}{|V|}$, i.e. the probability of randomly assigning a correct match. The rate of decay is determined by the logarithm of the probability that a mutation operation does not take place, i.e. $1 - P_m$. In qualitative terms, the mutation process represents an exponential drift towards highly unfit solutions. The rate of drift is controlled by two factors. The first of these is the mutation rate P_m. As the mutation probability increases, then so the disruptive effect of the operator become more pronounced. The second factor is the initial fraction of correct matches that were present prior to mutation. As this initial fraction of correct matches increases, then so does the disruptive effect of the mutation operator. The effect of this second drift process is to impose a higher rate of disruption on solutions in the population that are approaching a consistent state. Poor or highly inconsistent solutions, on the other hand, are not significantly affected. This latter drift effect can be viewed as a natural mechanism for escaping from local optima that may be encountered when the global solution is approached in a complex fitness landscape.

Figure 1 illustrates how the peak of the population distribution drifts towards the origin in the fashion predicted. Moreover, the width of the distribution becomes narrower as it approaches the origin. In order to quantify this process we plot the most probable fraction of correct matches in the population against iteration number. This plot is shown in Figure 2. We note that there is a good agreement between our prediction of exponential decay and what is observed experimentally.

4.2   An Analysis of the Selection Operator

In contrast with the mutation operator which is a uniform random process, selection draws upon the fitness function to determine the probability that the different solution vectors survive into the next population generation. Because of this task-specific nature of the fitness function, it is not possible to undertake a

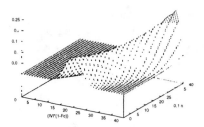

Fig. 1. A numerical simulation of the distribution of correct solutions in the population using only the mutation operator.

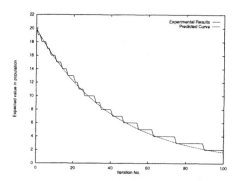

Fig. 2. The comparison of the analytic predictions of the mutation operator with the simulation run results.

general analysis of the selection process. Moreover, in the case of graph-matching the compound exponential structure of our fitness measure further complicates the analysis. To overcome this problem, we present an approximation to our Bayesian consistency measure which allows us to relate the survival probability to the fraction of correct matches. This approximate expression for the survival probability turns out to be polynomial in the fraction of correct matches.

We commence by writing the fitness using the expression for the super-clique matching probability given in equation (1). To make the role of error-probability more explicit, we re-write the matching probability in terms of Kronecker delta functions that express the compatibility between the current matching assignments and the consistent matches demanded by the configuration residing in the dictionary. As a result

$$P(\Gamma_j) = \frac{1}{|\Theta_j|} \sum_{\Lambda^\mu \in \Theta_j} \prod_{i \in C_j} P_e^{(1-\delta_{f(i),\lambda_i^\mu})} (1 - P_e)^{\delta_{f(i),\lambda_i^\mu}} \tag{8}$$

Our aim is to compute the average value of the global consistency measure, P_G^α. Because the consistency function averages the matching probability $P(\Gamma_j)$, the expected-value of the global probability is equal to

$$P_G^\alpha = E\left[\frac{1}{|\Theta_j|} \sum_{\Lambda^\mu \in \Theta_j} \prod_{i \in C_j} P_e^{(1-\delta_{f(i),\lambda_i^\mu})}(1-P_e)^{\delta_{f(i),\lambda_i^\mu}}\right] \tag{9}$$

We now note that the expected value of the exponential function under the product can be re-expressed in terms of the assignment probabilities in the following manner

$$E\left[P_e^{(1-\delta_{f(i),\lambda_i^\mu})}(1-P_e)^{\delta_{f(i),\lambda_i^\mu}}\right] = P_e\, P(f(i) \neq \lambda_i^\mu) + (1-P_e)\, P(f(i) = \lambda_i^\mu) \tag{10}$$

As a result, the expected-value of the global matching probability, i.e. the probability of survival, is equal to

$$P_G^\alpha = \frac{1}{|\Theta_j|} \sum_{\Lambda^\mu \in \Theta_j} \prod_{i \in C_j} \left\{ P_e\, P(f(i) \neq \lambda_i^\mu) + (1-P_e)\, P(f(i) = \lambda_i^\mu) \right\} \tag{11}$$

Unfortunately this expression still contains reference to the dictionary of structure preserving mappings. In order to further simplify matters, we observe that when the configuration of assigned matches becomes consistent then, provided the error probability P_e is small, we would expect the sum of exponentials appearing in equation (1) to be dominated by the single dictionary item that is fully congruent with the ground-truth match. The remaining dictionary items make a negligible contribution. Suppose that $\hat{\mu}$ is the index of the correctly matching dictionary item, then we can write

$$\exp[-k_e H(\Gamma_j, \Lambda^{\hat{\mu}})] \gg \sum_{\Lambda^\mu \in \Theta_j - \Lambda^{\hat{\mu}}} \exp[-k_e H(\Gamma_j, \Lambda^\mu)] \tag{12}$$

We can now approximate the super-clique matching probability by considering only the dominant dictionary item. This allows us to neglect the summation over dictionary items. Finally, we note that the average value of the probability of correspondence match, i.e. $P(f(i) = \lambda_i^\mu)$, is simply equal to the fraction of correct matches $F_\alpha^{(n)}$. By assuming that all super-cliques are of approximately the same average cardinality, denoted by $|C|$, we can approximate the global probability of match in the following manner:

$$P_G^\alpha = \frac{1}{|\Theta_j|}\left[P_e(1 - F_\alpha^{(n)}) + (1-P_e)F_\alpha^{(n)}\right]^{|C|} \tag{13}$$

In other words, our measure of relational consistency is polynomial in the fraction of correct matches. Moreover, the order of the polynomial is equal to the average node connectivity $|C|$. As the average neighbourhood size or node connectivity in the graphs increases, then so the discriminating power of the cost function becomes more pronounced.

At iteration n of the algorithm, we can use the selection probability to compute the distribution of the fraction of correct matches using the relationship

$$P_D^{(n)}(F^{(n+1)} = \gamma) = P_D^{(0)}(\gamma)P_G(\gamma)^n \tag{14}$$

Substituting for the approximate initial distribution given in Equation 5 together with our approximation to the cost function from Equation 13, we find

$$P_D^{(n)}(F^{(n)} = \gamma) \approx \frac{1}{\sqrt{2\pi|V|F^{(0)}(1 - F^{(0)})}} \exp\left(-\frac{(|V|\gamma - |V|F^{(0)})^2}{2|V|F^{(0)}(1 - F^{(0)})}\right)$$
$$\times (P_e(1 - \gamma) + (1 - P_e)\gamma)^{(|C|n)} \tag{15}$$

The required distribution is simply a Gaussian distribution that is modulated by a polynomial of order $|C|n$. This demonstrates that the average fraction of correct matches in the population will tend to increase as the value of n increases. In other words, the iteration process improves the fraction of correct matches. By confining our attention to the solutions that occur most frequently in the population, we can track the iteration dependence of the mode or peak, $F_{max}^{(n)}$, of the distribution of correct matches. To locate the most frequently occurring solution in the population, we proceed as follows. First, we evaluate the derivative of the distribution function in Equation 15 with respect to the fraction of correct mappings, i.e. γ. Next, we set the derivative equal to zero. By solving the resulting saddle-point equation for $F^{(n)}$ and after rejecting the non-physical values of γ that fall outside the interval $[0, 1]$, we find that the maximum value P_D is located at the position

$$F_{max}^{(n)} = \frac{1}{2}\left(F^{(0)} + \frac{P_e}{\kappa_2} + \frac{\sqrt{\kappa_1^2(P_e^2 + 2P_eF^{(0)}\kappa_2 + F^{(0)}\kappa_2^2) + 2\kappa_1\kappa_2^2|C|n}}{\kappa_1\kappa_2}\right) \tag{16}$$

where $\kappa_1 = \frac{(|V|)}{2F^{(0)}(1-F^{(0)})}$ and $\kappa_2 = 1 - 2P_e$.

With this model of the iteration dependance of $F_{max}^{(n)}$ under the selection operator to hand, we are in a position to compute the number of iterations required for algorithm convergence. Our convergence condition is that the modal fraction of correct matches is unity. We identify the value of the iteration index n that satisfies this condition by setting $F_{max}^{(n)} = 1$ in Equation 16. Furthermore, we assume that the initial population is randomly chosen and as a result $F^{(0)} = \frac{1}{|V|}$. Solving for n, we find the number of iterations required for convergence to be equal to

$$n_{converge} = \frac{(|V|)^2(1 - P_e)}{|C|(1 - 2P_e)} \tag{17}$$

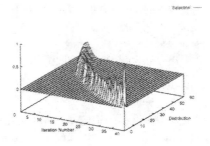

Fig. 3. A numerical simulation of the distribution of correct solutions in a genetic population using only the selection operator.

In other words, commencing from a simple model of the selection process that uses a number of domain specific assumptions concerning our Bayesian consistency measure, we have shown that we would expect the number of iterations required for convergence to be polynomial with respect to the number of nodes in the graphs under match.

In order to provide some justification for our modelling of the population mode, we have investigated how the selection operator modifies the distribution of correct correspondences in the genetic population. Figure 3 shows the distribution $P_D^{(n)}$ as a function of the iteration number. The main point to note is that the width of the distribution remains narrow as the iterations proceed. This is because only the selection operator is used. It is important to stress that there is no diversification process at play.

4.3 An Analysis of the Hill-climbing Operator

Since the hill-climbing operator is only used to make local changes that increase P_G, it is clear that it can only improve the quality of the match. In this section of our analysis, we aim to determine to what extent the hill-climbing operator effects the overall convergence rate of our algorithm.

Modelling the behavior of the gradient ascent step of the algorithm is clearly a difficult problem since it is highly dependent on the local structure of the global landscape of the fitness measure. One way of simplifying the analysis is to adopt a semi-empirical approach. Here we aim to Monte-Carlo the gradient ascent process and extract a parameterisation of the iteration dependance of the required distribution parameters. Our starting point is to generate 1000 random graphs. Commencing from a controlled fraction of initially correct matches, we perform gradient ascent until the configuration of matches stabilises and no more updates can be made. We plot the final fraction of correct matches against the fraction initially correct in Figure 4. The best-fit to the data gives the following iteration dependance

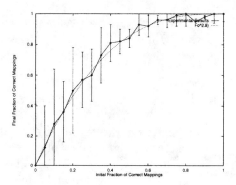

Fig. 4. Empirical results demonstrating how we expect gradient ascent to perform. The dotted curve represents the best fit that was found.

$$F_\alpha^{(n+1)} = 1 - (1 - F_\alpha^{(n)})^{2.8} \tag{18}$$

This result relates the fraction of correct matches at iterations n and $n+1$ resulting from the application of the hill-climbing operator.

By expanding the recursion in iteration number, we can obtain the dependance on the initial fraction of correct matches. At iteration n, the fraction of correct matches is given by

$$F_\alpha^{(n)} = 1 - (1 - F_\alpha^{(0)})^{2.8n} \tag{19}$$

This assumes we can immediately restart the gradient ascent with the solutions from the previous iteration. Since at each iteration gradient ascent is performed until a local maximum is reached, the convergence rate calculated in this section is mainly of theoretical value. Also, when comparing the convergence rate of selection and hill-climbing, the number of function evaluations per iteration clearly should be taken into account as well.

We can use the empirical iteration dependance of the expected fraction of correct solutions to make a number of predictions about the convergence rate of population based hill-climbing. We commence by assuming that the initial set of matches is selected in a random manner. As before, this corresponds to the case $F^{(0)} = \frac{1}{|V|}$. Our condition for convergence is that less than one of the matches per solution is in error, i.e. $F^{(n)} > \frac{|V|-1}{|V|}$. By substituting this condition into equation 19 and solving for n, we find

$$n = 0.36 \frac{\ln(|V|)}{\ln[\frac{|V|}{|V-1|}]} \tag{20}$$

The number of iterations required for convergence increases slowly with graph-size. For modest numbers of nodes the increase is approximately linear.

It is interesting to contrast the dependance on graph-size with that for the selection operator. Whereas hill-climbing has a slow dependance on graph-size, in the case of selection there is a more rapid polynomial dependance. As a figure of merit, for a graph of size 50 nodes, the number of iterations required by selection is a factor of 10 larger than that required by hill-climbing.

5   An Analysis of the Combined Operators

In this section we provide an analysis of the combined effect of the mutation, selection and hill-climbing operators. Since the effect of crossover is simply to blur the fitness distribution, we omit it from our analysis. We provide convergence analysis for both the standard genetic algorithm and a hill-climbing variant.

5.1   Standard Genetic Search

Our aim is to extend the analysis of the individual operators presented in Section 6 by deriving a sufficient condition that ensures a monotonic increase of the expected fraction of correct solutions when composite operators are applied. To embark on this study, we must first consider the order in which the different genetic operators are applied. As the population of candidate solutions enters the new iteration $n + 1$ from the preceding iteration n, we first perform the crossover operation. As we have discussed earlier, this process results in a post-crossover population that is distributed according to a Gaussian distribution. As a result the mode of the distribution is located where the fraction of correct solutions is equal to $F^{(n)}$, while we let the standard deviation of the distribution be equal to $\sigma^{(n)}$. The mutation operator is applied after the crossover process. The main effect is to shift the mode of the Gaussian distribution to a lower fitness value by an amount

$$\Delta F^{mutation} = -\frac{P_m(F^{(n)}|V| - 1)}{|V|} \tag{21}$$

As expected, there is a decrease in the expected fraction of correct solutions. Immediately following the application of the mutation operator, we do not know the exact distribution of the fraction of correct matches. However, as demonstrated earlier, we know that for a large number of mutation operations, the distribution is binomial, which in turn can be approximated well by a Gaussian for large $|V|$. As a result, the required distribution can be approximated in a Gaussian manner. The mode of the distribution is located at the position

$$F^{mutation} = F^{(n)} + \Delta F^{mutation} \tag{22}$$

If the mutation probability P_m is relatively small, as is usually the case, after a single mutation operation, then we can assume that the standard deviation of the Gaussian, i.e. $\sigma^{(n)}$, remains unchanged.

In order to determine how the peak or mode of this distribution is shifted by the selection operator we recall Equation 16. Our interest is now with the change fraction of correct matches that the peak of the distribution undergoes under the combined selection and mutation operators. This quantity is equal to the peak value offset by the shift due to mutation, i.e.

$$\Delta F^{selection} = F_{max}^{selection} - \Delta F^{mutation} \qquad (23)$$

It is important to note that the distribution used as input to the selection operator is the result of the sequential application of the crossover and mutation processes. Computing distribution shift after selection is straightforward, but algebraically tedious. For this reason we will not reproduce the details here. Given that we now have a prediction of how we expect the peak of the population distribution to evolve under the processes of crossover, mutation and subsequent selection, we are in a position to construct a condition for monotonic convergence. Clearly, for the population to converge, the downward shift (i.e. fitness reduction) due to the mutation operator must be smaller than the upward shift (i.e. fitness increase) resulting from selection. In order to investigate this balance of operators, we consider the break-even point between mutation and selection which occurs where

$$\Delta F^{selection} = \Delta F^{mutation} \qquad (24)$$

Substituting from Equations 21 and 23, and solving for P_m, the break-even condition is satisfied when

$$P_m \leq \frac{|V| \cdot |C|(1 - 2P_e)}{(P_e/F^{(n)} + 1)(1 - 2P_e) - F^{(n)}(1 - 2P_e) - |V|P_e} \frac{\sigma^{(n)}}{F^{(n)}} \qquad (25)$$

It is important to note that this condition on the mutation probability is very similar to that derived by Qi and Palmieri [12, 13]. In fact, the maximum mutation rate is proportional to the ratio of the variance of the fraction of correct mappings in the population to the current expected fraction of matches correct. Moreover, the limiting value of the mutation is proportional to the total number of edges in the graphs, i.e. $|V| \cdot |C|$. Finally, we note that as the fraction of correct matches increases, then the mutation rate must be reduced in order to ensure convergence. It is important to emphasize that we have confined our attention to deriving the condition for monotonic convergence of the expected fitness value. This condition does not guarantee that the search procedure will converge to the global optimum. Neither does it make any attempt to capture the possibility of premature convergence to a local optimum. Moreover, the analysis assumes that all members of the population participate in crossover at each iteration, and that enough crossover is performed to arrive at an approximately Gaussian distribution. This may not be realistic since fitness distributions are typically observed to be quite asymmetric.

5.2 Hybrid Hill Climbing

Having derived the monotonic convergence condition for the combined effect of the three standard genetic operators, we will now turn our attention to the hybrid hill-climbing algorithm used in our empirical study of graph-matching

[2]. As before, we compute the change in the fraction of correct matches that we would expect to result from the additional application of the hill-climbing operator. Since this step immediately follows mutation, the population shift is given by

$$\Delta F^{hillclimb} = 1 - (1 - F^{mutation})^{2.8} - F^{mutation} \tag{26}$$

For completeness, our analysis should next focus on the effect of the selection operator. However, as we demonstrated in Section 6.4, the rate of convergence for the selection operator is significantly slower than that of the hill-climbing operator. This observation suggests that we can neglect the effects of selection when investigating the hybrid hill-climbing algorithm.

In order to identify the monotonic convergence criterion for the hybrid hill-climbing algorithm, we focus on the interplay between opposing population shifts caused by mutation and hill climbing. This analysis is entirely analogous to the case presented in the previous subsection where we consider the interplay between mutation and selection for the standard genetic algorithm. In the case of the hybrid hill climbing algorithm, the break-even occurs when

$$\Delta F^{hillclimb} \geq -\Delta F^{mutation} \tag{27}$$

When the size of the graphs is large, i.e. $|V| \gg 1$, the convergence condition on the mutation probability is

$$P_m \leq \frac{\left((1 - F^{(n)})^{\frac{1}{2.8}} + F^{(n)} - 1 \right)}{F^{(n)}} \tag{28}$$

This limiting mutation rate is plotted in Figure 5 as a function of $F^{(n)}$. In practice, we must select the operating value of P_m to fall within the envelope defined by the curve in Figure 5. As the fraction of correct matches approaches unity (i.e. the algorithm is close to convergence), then the mutation rate should be annealed towards zero. More interestingly, we can use the convergence condition to determine the largest value of the mutation rate for which convergence can be obtained. By taking the limit as the fraction of correct matches approaches zero, we find that $\lim_{F^{(n)} \to 0} P_m = 0.6430$. This agrees well with the empirical findings reported in our previous work [2].

Before concluding, we return to the question of population size. The aim here is to use our Monte-Carlo study to assess how the theoretical predictions, made under the central limit assumption for large population size, degrade as the population size becomes relatively small. In Figure 6 we plot $F^{(n)}$ as a function of iteration number for increasing population sizes. It is clear that beyond a population size of about 50 solutions, the convergence curves for the different population sizes become increasingly similar.

6 Conclusions

In this paper our aim has been to understand the role that the different genetic operators play in the convergence of a genetic algorithm. The mutation operator

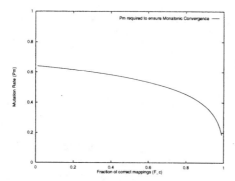

Fig. 5. The maximum mutation rate that may be used to ensure monotonic convergence of a hybrid Genetic Hill-climbing optimisation scheme.

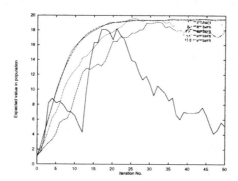

Fig. 6. The expected value for $F^{(n)}$ under the traditional genetic operators for various population sizes.

was found to produce an exponential drift of the population distribution towards incorrect mappings. The drift-rate was found to depend on both the mutation rate and the current fraction of correct correspondences. In other words, there is greater disruption when the population is dominated by a single consistent solution. In the case when the population contains a large number of dissimilar yet poor solutions, there is less disruptive drift. By contrast with the other operators, the role of the crossover operator is to exchange information via recombination. The net effect is to blur the distribution of the fraction of correct solutions in a Gaussian manner. In other words, the mean fraction of correct solutions remains stationary, while the corresponding variance increases.

Based on this operator-by-operator analysis, we have found conditions for convergence. We have obtained two interesting results First, the convergence rate for the standard genetic algorithm was found to be limited by the mutation probability; the limiting value is proportional to the total number of graph-edges. The second result was to show that in the case of the hill-climbing genetic algorithm, the corresponding limiting mutation probability is independent of the structure of the graphs. This result accords well with our previous empirical findings.

References

1. Aarts E. and Korst J., "Simulated Annealing and Boltzmann Machines", *John Wiley and Sons, New York*, 1989.
2. Cross A.D.J., Wilson R.C. and Hancock E.R., "Inexact Graph Matching using Genetic Search", *Pattern Recognition*, **30**, pp. 953–970, 1997.
3. Fogel D.B., "An Introduction to Simulated Evolutionary Optimisation", *IEEE Transactions on Neural Networks*, **5**, pp. 3–14, 1994.
4. Geman S. and D Geman, "Stochastic relaxation, Gibbs distributions and Bayesian restoration of images," *IEEE PAMI*, **PAMI-6**, pp.721–741, 1984.
5. Gidas B., "A Re-normalisation-Group Approach to Image Processing Problems", *IEEE PAMI*, **11**, pp. 164–180, 1989.
6. Goldberg D. " Genetic Algorithms in Search, Optimisation and Learning", *Addison-Wesley*, 1989.
7. Kirkpatrick S., C.D. Gelatt and M.P. Vecchi, "Optimisation by Simulated Annealing", *Science*, **220**, pp. 671–680, 1983.
8. Kumar, V., "Algorithms for Constraint-Satisfaction Problems: A Survey", *AI Magazine*, **13**, pp. 32–44, 1992.
9. Mackworth, A.K. and Freuder, E.C., "The Complexity of Some Polynomial Network Consistency Algorithms for Constraint Satisfaction Problems", Artificial Intelligence, **25**, pp. 65-74, 1985.
10. Mackworth, A.K., "Consistency in a Network of Relations", *Aritificial Intelligence*, **8**, pp. 99-118, 1977.
11. Pearl J., "Heuristics: intelligent search strategies for computer problem solving". *Addison Wesley*, 1984.
12. Qi, X.F. and F. Palmieri, "Theoretical Analysis of Evolutionary Algorithms with an Infinite Population in Continuous Space: Basic Properties of Selection and Mutation" *IEEE Transactions on Neural Networks*, **5**, pp. 102–119, 1994.
13. Qi, X.F. and F. Palmieri, "Theoretical Analysis of Evolutionary Algorithms with an Infinite Population in Continuous Space: Analysis of the Diversification Role of Crossover" *IEEE Transactions on Neural Networks*, **5**, pp. 120–129, 1994.
14. Rolland E., H. Pirkul and F. Glover, "Tabu search for graph partitioning", *Annals of Operations Research*, **63**, pp. 290–232, 1996.
15. Rudolph G., "Convergence Analysis of Canonical Genetic Algorithms", *IEEE Transactions on Neural Networks*, **5**, 96-101, 1994.
16. Sanfeliu A. and Fu K.S., "A Distance Measure Between Attributed Relational Graphs for Pattern Recognition", *IEEE SMC*, **13**, pp 353–362, 1983.
17. Waltz D., "Understanding Line Drawings of Scenes with Shadows", *in "The Psychology of Computer Vision", edited by P.H. Winton*, McGraw-Hill, 1975.
18. Wilson R.C. and Hancock E.R., "Structural Matching by Discrete Relaxation", *IEEE PAMI*, **19**, pp634–648, 1997.
19. Yuille A., "Generalised Deformable Models, Statistical Physics and Matching Problems", *Neural Computation*, **2**, pp. 1-24, 1990.
20. Yuille A.M. and Coughlan J., "Twenty Questions, Focus of Attention and A*: A Theoretical Comparison of Optimisation Strategies", *Energy Minimisation Methods in Computer Vision and Pattern Recognition, Edited by M. Pelillo and E.R. Hancock*, Lecture Notes in Computer Science 1223, pp. 197–212, 1997.

A New Distance Measure for Non-rigid Image Matching

Anand Rangarajan[1], Haili Chui[1], and Eric Mjolsness[2]

[1] Departments of Diagnostic Radiology and Electrical Engineering,
Yale University,
New Haven, CT, USA
[2] Jet Propulsion Laboratory,
California Institute of Technology,
Pasadena, CA, USA

Abstract. We construct probabilistic generative models for the non-rigid matching of point-sets. Our formulation is explicitly Platonist. Beginning with a Platonist super point-set, we derive real-world point-sets through the application of four operations: i) spline-based warping, ii) addition of noise, iii) point removal and iii) amnesia regarding the point-to-point correspondences between the real-world point-sets and the Platonist source. Given this generative model, we are able to derive new non-quadratic distance measures w.r.t. the "forgotten" correspondences by a) eliminating the spline parameters from the generative model and by b) integrating out the Platonist super point-set. The result is a new non-quadratic distance measure which has the interpretation of weighted graph matching. The graphs are related in a straightfoward manner to the spline kernel used for non-rigid warping. Experimentally, we show that the new distance measure outperforms the conventional quadratic assignment distance measure when both distances use the same weighted graphs derived from the spline kernel.

1 Introduction

The need for non-rigid image matching arises in many domains within the field of computer vision. Some form of non-rigid matching is required to match objects that have undergone complex deformations. The extent of the deformation required to achieve non-rigid matching is an important quantity as it provides a convenient measure of distance between the two objects.

Non-rigid matching methods can be broadly divided into two categories; intensity-based and feature-based. Intensity-based methods attempt to calculate the optical flow between the two images. Usually, these methods require fairly strong brightness constancy assumptions between the two images. Feature-based methods attempt to match two sets of sparse features that have been extracted from the underlying image intensities. Usually, these methods require both object and deformation models in order to constrain the set of allowed matches and deformations.

Object models are typically constructed using a hierarchy of features: points, lines, curves, surfaces, etc. If matching is performed using generic, unlabeled point features, then the correspondence problem is acute. On the other hand, if high-level feature representations are used, the correspondence problem is alleviated but the matching is not likely to be robust against missing features. In addition, the constraints on the deformation model become more complex when high-level features are used.

In this paper, we are mainly concerned with deriving new distance measures for non-rigid matching of unlabeled point features. The new distance measure is a function of the unknown point-to-point correspondences and can handle outliers as well. Since we mostly focus on the new distance measure, at this point we are not presenting an algorithm to minimize this distance.

The formulation of our problem is explicitly Platonist. We begin by assuming a Platonist super point-set of unlabeled features. By using a probabilistic thin-plate spline warping model, we are able to generate real-world point feature sets. Outliers are explicitly modeled by forcing each real-world warped point-set to be a strict subset of the Platonist super point-set. The final step in this generative model is the loss of information of the point-to-point correspondences between the real-world point-set and the Platonist super point-set.

After exploiting a Platonist analogy in formulating this model, we then derive a new distance measure between the real-world point-sets. First, we eliminate the thin-plate spline warpings from the model by setting these parameters to their maximum *a posteriori* estimates. Then, in typical Bayesian fashion, we integrate out the hidden Platonist super point-set. The result is a new non-quadratic distance measure between all of the real-world point-sets defined solely in terms of the unknown correspondences.

Having derived the new non-quadratic distance measure, we present comparisons with the more traditional quadratic assignment distance measures. As a by-product of our derivation, we are able to show that the new distance measure is closely related to a weighted graph matching distance measure with the "graphs" determined by the thin-plate spline kernel. Both distance measures (quadratic and non-quadratic) use the same graphs derived from the spline kernels. Finally, we show that our new distance measure significantly outperforms the quadratic distance measure indicating a payoff resulting from our principled derivation.

2 Review

The various approaches to non-rigid image matching can be broadly grouped into two categories—intensity-based and feature-based. Intensity-based methods begin by assuming some form of brightness constancy reminiscent of optical flow methods [2]. Most methods in this class attempt to minimize an energy function that consists of two terms. The first term simply sums over the square of the intensity differences between the two images at each pixel. The second term is an elastic matching term which is typically derived from considerations of smoothness of the displacement field [7]. A free parameter is used to tradeoff between these two terms. The principal difficulty with this entire class of methods is that the brightness constancy assumption is frequently violated. Recently, there has been considerable interest in using entropy and mutual information-based intensity distance measures [30, 16] to overcome these limitations. A second problem with these methods is related to the lack of explicit object modeling. Since no attempt is made to construct object models, these methods cannot enforce *correspondence* constraints on structures that are *a priori* known to match. Recently [5], there has been some effort to overcome this limitation by including region segmentation information into the computation of optical flow. However, it is fair to say that at the present time, intensity-based image matching methods have yet to fully solve the aforementioned (two) problems by incorporating segmentation information into mutual information-based estimation of displacement fields (flow).

Feature-based image matching methods form the second class of methods. In contrast to the optical flow-based intensity matching methods, feature-based methods are more varied. One way of dividing the space of feature-based methods is along the lines of sparse versus dense features. Labeled landmark points are the most popular kind of sparse features since non-rigid matching of landmarks does not require a solution to the point-to-point correspondence problem For instance in [6], thin-plate splines (TPS) [31] are used to characterize the deformation of landmarks. Basically, the non-rigid matching problem is solved by minimizing the bending energy of a thin-plate spline while forcing corresponding landmarks extracted from the two images to perfectly match. Landmark positioning "jitter" can be accounted for in this model by allowing a trade-off between the landmark position *least-squares* matching energy term and the spline bending energy term. This is analogous to the "vanilla" optical flow image matching method mentioned above. The major drawback of this method is the over-reliance on a few landmarks. Extracting labeled and corresponding landmarks from the two images is a difficult problem. Moreover, the method is quite sensitive to the number and choice of landmarks.

Dense feature-based matching methods run the gamut of matching points, lines, curves, surfaces and even volumes [4]. These methods usually begin with an object parameterization. Then, the allowable ways in which the object can deform is specified [18, 23]. The methods that fall into this class differ in object parameterizations and in the specification of the kinds of allowed deformations. In most cases, curves and/or surfaces are first fitted to features extracted from the images and then matched [18, 28, 27, 10]. These methods work well when the surfaces (and curves) to be matched are reasonably smooth. Also, the surface fitting step that precedes matching is predicated on good feature extraction. These methods have not been widely accepted in domains such as brain matching due to the extreme variability of cortical surfaces.

One of the principal reasons for the emphasis on object modeling in non-rigid matching is that it allows us to circumvent the correspondence problem. For example, once a smooth curve is fitted to a set of feature points, the matching can be taken up at the curve level rather than at the point level. Curve correspondence is easier than point correspondence [28] due to the strong constraint imposed by the smooth curve on the space of possible point-to-point correspondences. While the surface case is more complicated, surfaces can be approximately matched when they are smooth and the allowed deformations are not very complex [18, 27]. The downside is the lack of robustness. Sensor noise sometimes makes it difficult to fit smooth curves and surfaces to an underlying set of feature points. In such cases, while point feature locations may still be trustworthy, the fitting of surface normals and other higher-order features becomes problematic. Consequently, these higher-order features cannot be used.

In this paper, we begin with an integrated pose and correspondence formulation using point features. Essentially, we modify the pose parameters to include non-rigid deformations. We now turn to a review of recent approaches that attempt to integrate the search for correspondence in non-rigid matching. While the correspondence problem has a long history in rigid, affine and projective point matching [14], there is relatively a dearth of literature on non-rigid point matching. Recently, there has been some interest in using point-based correspondence strategies in non-rigid matching [23, 10, 29, 32, 21]. The modal matching approach in [23] relies on the point correspondence approach pioneered in [24] and further developed in [25]. The basic idea here is to use a pairing matrix that is built up from the Gaussian of the distances between any point feature in one set and the other. The modes of this matrix are used to obtain the correspondence. In [23], following [8], the deformation modes of the point-sets are obtained from the principal components of the covariance matrix of a pre-specified training set of shapes. The main drawback of this approach is that it does not use the *spatial relationships* between the points in each set to constrain the search for the correspondences and the mapping. In [9], after pointing out this drawback, the inter-relationships between the points is taken into account by building a graph representation from Delaunay triangulations. The search for correspondence is accomplished using inexact graph matching [26]. However, the spatial mappings are restricted to be affine or projective. In [1], decomposable graphs are hand-designed for deformable template matching and minimized with dynamic programming. However, the graphs are not automatically generated and there is no direct relationship between the deformable model and the graphs that are used. In [17], a maximum clique approach [20, 12] is used to match relational sulcal graphs. Again, the graphs are hand designed and not related to spatial deformations.

3 Deriving the distance measure

We first present background material on the thin-plate spline—our choice for the non-rigid spatial mapping. Then, we bring in the unknown correspondences and proceed with the derivation of the new distance measures.

3.1 Thin-plate splines

Our main reason for choosing the thin-plate spline is due its well understood behavior in landmark matching [6]. Essentially, the thin-plate spline produces a smoothly interpolated spatial mapping (with adherence to landmarks handled by the data term). The thin-plate spline (TPS) formulation required here is for 2D and 3D point matching. In both cases, we'll consider smoothness terms comprising of second-order derivatives of the interpolating function. Lack of space does not permit us to present the thin-plate spline in great detail. Instead, we merely present a "bare-bones" derivation. The interested reader is referred to [31] for the general formulation and to [6] for the application to landmark matching. In Figure 1, we depict an example thin-plate spline warping. Note the decomposition into the affine and warping components—a special characteristic of thin-plate spline mappings.

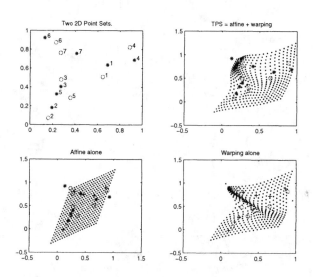

Fig. 1. Top Left: Original and warped point-sets. Top Right: Visualization of the thin-plate mapping. Bottom Left: Affine component of the mapping. Bottom right. Warping component of the mapping.

Assume for the moment that we have N pairs of corresponding points in either 2D or in 3D. Denote the two point-sets by Z and X respectively. The representations of the point-set Z is shown for the cases of 2D and 3D below:

$$
\text{In 2D } Z = \begin{bmatrix} 1 & z_1^1 & z_1^2 \\ 1 & z_2^1 & z_2^2 \\ 1 & z_3^1 & z_3^2 \\ \cdot & \cdot & \cdot \\ \cdot & \cdot & \cdot \\ 1 & z_N^1 & z_N^2 \end{bmatrix}, \text{ and in 3D } Z = \begin{bmatrix} 1 & z_1^1 & z_1^2 & z_1^3 \\ 1 & z_2^1 & z_2^2 & z_2^3 \\ 1 & z_3^1 & z_3^2 & z_3^3 \\ \cdot & \cdot & \cdot \\ \cdot & \cdot & \cdot & \cdot \\ 1 & z_N^1 & z_N^2 & z_N^3 \end{bmatrix} \quad (1)
$$

A similar representation holds for the point-set X as well. The representations in (1) are the so called *homogeneous* coordinates.

We now set up a thin-plate spline mapping from X to Z. Thin-plate splines are asymmetric in the sense that a mapping from X to Z cannot be easily inverted to yield a mapping from Z to X. Minimizing the following energy function gives us a smooth spline interpolant capable of warping points in X arbitrarily close to points in Z. A regularization parameter λ determines the closeness of the fit. In 2D,

$$E_{\text{tps}}(f) = \sum_{i=1}^{N} \|Z_i - f(X_i)\|^2$$

$$+\lambda \int_{-\infty}^{\infty} \int_{-\infty}^{\infty} \left[\left(\frac{\partial^2 f}{\partial (x^1)^2} \right)^2 + 2 \left(\frac{\partial^2 f}{\partial x^1 \partial x^2} \right)^2 + \left(\frac{\partial^2 f}{\partial (x^2)^2} \right)^2 \right] dx^1 dx^2. \tag{2}$$

A similar expression holds in 3D.

Define $t = (x^1, x^2, \ldots, x^D)$ where $D = 2$ for 2D and $D = 3$ for 3D. Then $t_i \overset{\text{def}}{=} (x_i^1, x_i^2, \ldots, x_i^D)$. Also, in 2D, $[\phi_1(t), \phi_2(t), \phi_3(t)] \overset{\text{def}}{=} [1, x^1, x^2]$ with a straightforward extension to 3D. For the thin-plate spline energy function given in (2), it is possible to show that there exists a unique minimizer f_λ given by

$$f_\lambda(t) = \sum_{k=1}^{D+1} d_k \phi_k(t) + \sum_{i=1}^{N} c_i E(t - t_i), \tag{3}$$

where $E(t - t_i)$ is the Green's function for the thin-plate spline: $E(\tau) = \tau^2 \log \tau$ in 2D and $-|\tau|$ in 3D. Here $|t - t_i| = \sqrt{\sum_{k=1}^{D} (x^k - x_i^k)^2}$. The minimizer f of the thin-plate spline energy function given in (3), is specified in terms of two unknowns c and d. Using (3), it is possible to eliminate f from the thin-plate spline energy function. When this is done, we get

$$E_{\text{tps2}}(c, d) = \|Z - Xd - Kc\|^2 + \lambda \text{ trace } c^T K c. \tag{4}$$

In (4), Z and X are the $N \times (D + 1)$ point-sets, d is a $(D + 1) \times (D + 1)$ *affine* transformation consisting of translation, rotation and shear components, c is a $N \times (D + 1)$ matrix of warping parameters (with all entries in the first column set to zero), and K is a $N \times N$ matrix corresponding to the Green's function (which is different for 2D and 3D). The principal difference between $E(t - t_i)$ and K is that the latter is defined only at the landmark points: the matrix entry $K_{ij} = E(t_i - t_j)$.

As it stands, finding least-squares solutions for the pair (c, d) by directly minimizing (4) is awkward. Instead, a QR decomposition is used to separate the affine and warping spaces. For more details, please see [31]:

$$X = [Q_1, Q_2] \begin{pmatrix} R \\ 0 \end{pmatrix} \tag{5}$$

where Q_1 and Q_2 are $N \times (D + 1)$ and $N \times (N - D - 1)$ orthonormal matrices, respectively. R is upper triangular. With this transformation in place, (4) becomes

$$E_{\text{tps-final}}(\gamma, d) = \|Q_2^T Z - Q_2^T K Q_2 \gamma\|^2 + \|Q_1^T Z - Rd - Q_1^T K Q_2 \gamma\|^2$$
$$+ \lambda \gamma^T Q_2^T K Q_2 \gamma, \quad (6)$$

where $c = Q_2 \gamma$ and γ is a $(N-D-1) \times (D+1)$ matrix. Given this definition of c, $X^T c = 0$. The least-squares energy function in (6) can be first minimized w.r.t. γ and then w.r.t. the affine transformation d. The final result after minimization is

$$\hat{\gamma} = (Q_2^T K Q_2 + \lambda I_{(N-D-1)})^{-1} Q_2^T Z, \text{ and } \hat{d} = R^{-1}(Q_1^T X - K Q_2 \gamma). \quad (7)$$

The bending energy of the thin-plate spline after eliminating (c, d) is

$$E_{\text{bending}}(Z) = \text{trace}\left[Z^T Q_2 (Q_2^T K Q_2 + \lambda I_{N-D-1})^{-1} Q_2^T Z \right]. \quad (8)$$

3.2 A Platonist formulation

Having described the thin-plate spline spatial mapping in its two conventional (integral and matrix kernel) forms, we turn to the integrated pose and correspondence formulation.

First, we no longer assume that the correspondence between the point-sets Z and X is known. We introduce a correspondence matrix M which obeys the following constraints.

1. The correspondences are binary: $M_{ai} \in \{0, 1\}$.
2. Every point in X is matched to one point in Z: $\sum_a M_{ai} = 1$.
3. Every point in Z is matched to one point in X or is an *outlier* w.r.t. X: $\sum_i M_{ai} \leq 1$.

In informal terms, M is a matrix with binary entries whose columns sum to one and whose rows may either sum to one or be all zero. The one-to-one correspondence constraint is not sacrosanct and can be modified to a classification (many-to-one) constraint. This is because, in non-rigid mapping, a one-to-one constraint is incorrect when, for example, points are deformed into lines.

The bending energy of the thin-plate spline needs to be modified to take into account the introduction of the correspondence matrix M. Note that M allows us to generalize to the case of unequal point counts between X and Z.

We now write a probabilistic generative model for obtaining Z given X and a set of spline parameters (c, d). Since the correspondence M is unknown, it is included in the generative model as a hidden variable.

$$p(Z, M | X, c, d) = \frac{\exp[-E_1(Z, M, c, d)]}{Z_{\text{part1}}} \quad (9)$$

where

$$E_1(Z, M, c, d) = \sum_{ai} M_{ai} \|Z_a - (Xd)_i - (Kc)_i\|^2. \quad (10)$$

In (10), $(Xd)_i$ and $(Kc)_i$ are the ith elements of the vectors Xd and Kc respectively. The partition function Z_{part1} is a normalization constant. Equation

(10) is reminiscent of a Gaussian mixture model with Z playing the role of the cluster centers and M being the complete data classification matrix [15]. If M is presumed known, the least-squares term in (10) reduces to the thin-plate spline least squares term.

The pure bending energy term is exactly the same as in the thin-plate spline:

$$p(c, d|X, \lambda) = \frac{\exp[-E_2(c)]}{Z_{\text{part2}}}, \text{ where } E_2(c) = \lambda \text{ trace } c^T K c. \tag{11}$$

The two energy terms in (10) and (11) can be combined into one. The resulting probabilistic generative model for Z can be written (after some algebraic manipulation) as

$$p(Z, M, c, d|X, \lambda) = \frac{\exp[-E(Z, M, c, d]}{Z_{\text{part}}}, \tag{12}$$

where

$$E(Z, M, c, d) = \|MZ - Xd - Kc\|^2 + \lambda \text{ trace } c^T K c$$
$$+ \text{trace } Z^T[\text{diag}(\sum_i M_{ai}) - M^T M]Z. \tag{13}$$

The diag operator above takes a vector and rearranges it into a square matrix with the vector entries appearing along the diagonal. The remaining entries are zero. The binary nature of the entries of M makes the last term redundant since it is zero. However, the term should be kept in mind if and when the binary constraint on the entries of M are relaxed; in that event, the last term becomes significant once again.

A few key observations can be made regarding the integrated correspondence-spline energy function in (13). After we define $Z_{\text{perm}} \stackrel{\text{def}}{=} MZ$, note that the form of the energy function is exactly the same as that of the original thin-plate spline bending energy in (4). (The last term does not contain the warping parameters (c, d) and from the perspective of solving for (c, d) the previous statement holds.) Consequently, we can exploit all of the properties of the thin-plate spline that were briefly derived in the previous section to separate out the warping and affine spaces. We can eliminate (c, d) from (13) and this step is quite similar to the work in [33]. The bending energy [after eliminating (c, d)] is

$$E_{\text{corr-bend}}(Z, M) = \text{trace}[Z^T M^T G M Z], \tag{14}$$

where

$$G \stackrel{\text{def}}{=} Q_2(Q_2^T K Q_2 + \lambda I_{N-D-1})^{-1} Q_2^T. \tag{15}$$

In deriving (14), we have dropped the last term in (13).

We are now in a position to extend this formulation to the simultaneous non-rigid matching of several point-sets $X^1, X^2, \ldots, X^K$. We find that the traditional Platonist metaphor suits us admirably. The point-set Z assumes the role of the "light beyond the cave" and each point-set X^k is cast in the role of a "shadow perceived on the cave wall." We model the Platonist super point-set Z as a superset of all the points present in each of the real-world point-sets X^k, $k \in \{1, \ldots, K\}$. We assume the following generative model for obtaining the real-world point-sets from the archetype Z. Each real-world point-set X^k is obtained by (i) warping Z using a thin-plate spline, (ii) removing a subset of points from Z, (iii) adding additive white Gaussian noise (AWGN) to the remaining points and finally, (iv) erasing or forgetting the correspondence information between Z and the newly created point-set X^k.

Since we have already worked out the bending energy expression [in (14)] between Z and a single real-world point-set X, we now extend the formulation to cover the simultaneous matching of Z to *all* of the real-world points-sets X^k, $k \in \{1, \ldots, K\}$. Henceforth, we denote the set comprising all real-world point-sets by $\mathbf{X}$. The sets of all correspondences, warping and affine parameters are denoted by $\mathbf{M}$, $\mathbf{c}$ and $\mathbf{d}$ respectively.

The likelihood model for the Platonist super point-set Z is

$$p(Z, \mathbf{M}, \mathbf{c}, \mathbf{d} | \mathbf{X}) = \frac{\exp\left[-\sum_{k=1}^{K} E(Z, M^k, c^k, d^k)\right]}{Z_{\mathrm{partall}}}$$

$$= \prod_{k=1}^{K} \frac{\exp\left[-E(Z, M^k, c^k, d^k)\right]}{Z_{\mathrm{part}}^k}. \tag{16}$$

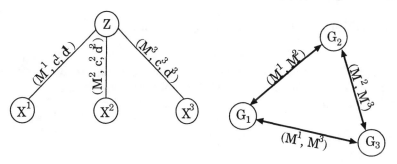

Fig. 2. Left: Platonist Formulation. Right: Real-world reformulation.

An important (and somewhat remarkable) fact about (16) is its separability. The Platonist super point-set Z is the sole bottleneck in the network of connections between the real-world point-sets $\mathbf{X}$. Consequently, with Z fixed, we can easily solve in closed-form for the entire set of thin-plate spline warping parameters $\mathbf{c}$ and $\mathbf{d}$. Note that the set of correspondence matrices $\mathbf{M}$ is also held fixed. This calculation is merely a generalization of the earlier calculation involving Z and X. Here, we have a set of point-sets $\mathbf{X}$ and Z. Our approach schema is depicted in Figure 2. On the left in Figure 2 is the original Platonist formulation with the Platonist super-point set Z acting as a generator for the point-sets X^k. On the right in Figure 2 is the real world reformulation. With $(\mathbf{c}, \mathbf{d})$ eliminated and Z integrated out, we obtain a distance measure between all of the real-world point-sets. Note that the point-sets X^k have been replaced by the corresponding graphs G^k.

3.3 Eliminating the spatial mapping

The spline parameter set $(\mathbf{c}, \mathbf{d})$ is eliminated exactly as before in (14). The only difference is that the elimination is carried out K times—once for each set of parameters $\{c^k, d^k\}$, $k \in \{1, \ldots, K\}$. We will not repeat this derivation. The bending energy after eliminating $(\mathbf{c}, \mathbf{d})$ is now a sum over all K bending energies:

$$E_{\text{corr-bend-total}}(Z, \mathbf{M}) = \sum_{k=1}^{K} \text{trace} \left[Z^T (M^k)^T G^k M^k Z \right], \qquad (17)$$

where

$$G^k \overset{\text{def}}{=} Q_2^k \left[(Q_2^k)^T K^k Q_2^k + \lambda I_{N-D-1} \right]^{-1} (Q_2^k)^T.$$

With the above solution for the spatial mapping parameters $(\mathbf{c}, \mathbf{d})$, we may write the likelihood for Z as

$$p(Z, \mathbf{M}, \hat{\mathbf{c}}, \hat{\mathbf{d}} | \mathbf{X}) = \prod_{k=1}^{K} \frac{\exp \left[-E_{\text{corr-bend-total}}(Z, M^k) \right]}{Z_{\text{part}}^k}. \qquad (18)$$

3.4 Integrating out the Platonist super point-set

Before integrating out Z, we wish to point out the need for this step. In a standard Bayesian formulation [3], integrating out the latent variables is recommended because the probabilistic structure is preserved by integration.

The distance measure between the real-world point-sets $\mathbf{X}$ is defined as

$$D(\mathbf{M}) \overset{\text{def}}{=} -\log \int p(Z, \mathbf{M} | \mathbf{X}) dZ. \qquad (19)$$

Note that the distance measure is a function of the unknown correspondences between each real-world point set X^k and the Platonist super point-set Z.

In (19), we have used $p(Z, \mathbf{M} | \mathbf{X})$ as shorthand for $p(Z, \mathbf{M}, \hat{\mathbf{c}}, \hat{\mathbf{d}} | \mathbf{X})$. The Platonist super point-set Z is now integrated out:

$$D(\mathbf{M}) = -\log \int \exp \left[-Z^T \left(\sum_{k=1}^{K} (M^k)^T G^k M^k \right) Z \right] dZ + \text{ terms indep. of } \mathbf{M}$$

$$= \frac{1}{2} \log \det \left[\sum_{k=1}^{K} (M^k)^T G^k M^k \right] \qquad (20)$$

This is our non-rigid matching distance measure. It is a function of only the set of correspondences **M**. The thin-plate spline warping parameters have been eliminated and the Platonist super point-set Z has been integrated out.

We now specialize to the case of non-rigid matching of two point-sets X and Y. The distance measure between X and Y is

$$D_{\log-\det}(M^X, M^Y) = \tfrac{1}{2} \log \det \left[(M^X)^T G^X M^X + (M^Y)^T G^Y M^Y \right], \quad (21)$$

where the "graph" G^X is defined as

$$G^X = \lambda Q_2^X \left[(Q_2^X)^T K^X Q_2^X + \lambda I \right]^{-1} (Q_2^X)^T, \quad (22)$$

with a similar expression holding for G^Y. The graph G^X has the nice property that it is *symmetric and non-negative definite*. It can easily be made positive definite which aids in the computation of (21).

3.5 Comparison with traditional quadratic assignment distance measures

The new log-det distance measure in (21) can be directly compared with more traditional quadratic assignment (QAP) distance measures. The QAP distance measure is the obvious foil for comparison since it is the basic quadratic distance measure that is popular and widely used. All the comparisons below are based on the non-rigid matching of two point-sets X and Y. The QAP distance measure is a quadratic distance between the two "graphs" G^X and G^Y. Note that the derivation of the "graphs" from thin-plate spline kernels is a new contribution— one which is quasi-independent of the choice of distance between the two graphs. The QAP distance measure is

$$D_{\text{qap}}(M^X, M^Y) = -\text{trace} \, (M^X)^T G^X M^X (M^Y)^T G^Y M^Y. \quad (23)$$

Due to the cyclical property of the trace operator, the QAP distance can be simplified as $D(M^{XY}) = -\text{trace} \, (M^{XY})^T G^X M^{XY} G^Y$ where $M^{XY} \stackrel{\text{def}}{=} M^X (M^Y)^T$.

4 Results

Figures 3 and 4 compare the QAP distance with the new log-det distance. In Figure 3, we've compared the log-det distance measure with the quadratic distance measure. There are no outliers going from the Platonist super point-set to the two real-world point-sets shown at the top left of the figure. All three distance measures perform well with the log-det distance showing the greatest separation. In Figure 4, we've compared the log-det distance measure with the quadratic distance measure. Somewhat surprising is the degree to which the log-det distance measure outperforms the quadratic distance. On the x-axis, we've plotted permutations over a fixed number of points.

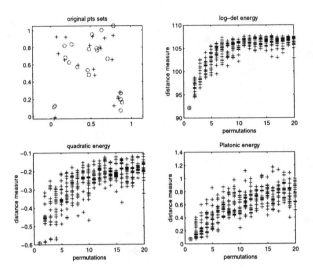

Fig. 3. Top Left: Two warped 2D 20 point-sets originating from a 20 point-set. Top Right: Log-det distance measure. Bottom Left: Quadratic distance measure. Bottom right: Platonic distance measure. The distance measures (ordinate) are plotted against permutations (abscissa). The abscissa value indicates how many points were permuted to obtain the distances. When zero points are permuted, the distance corresponding to the "true" answer is obtained [⊕ is the true distance and + is a distance point for a given permutation] and is plotted at the extreme left on the figure.

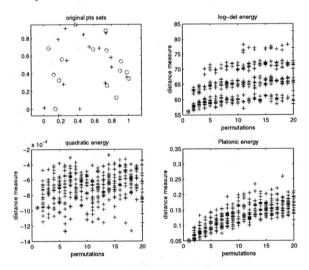

Fig. 4. Top Left: Two warped 2D 15 point-sets originating from a 20 point-set. Top Right: Log-det distance measure. Bottom Left: Quadratic distance measure. Bottom right: Platonic distance measure. The distance measures (ordinate) are plotted against permutations (abscissa). The abscissa value indicates how many points were permuted to obtain the distances. When zero points are permuted, the distance corresponding to the "true" answer is obtained [⊕ is the true distance and + is a distance point for a given permutation] and is plotted at the extreme left on the figure.

As more points are permuted, the distance measure ought to increase. We find that this is the case for the log-det distance measure but not for the quadratic distance. The Platonic distance is the quadratic distance between Z and the real-world point-sets. Since Z contains all the information, this idealized distance performs quite well. Note that there is obviously a question as to what the "true" answer ought to be. However, the bending energy returned by the log-det distance seems to concur with the Platonic bending energy which is reassuring since the latter is the closest you can get to a "gold standard."

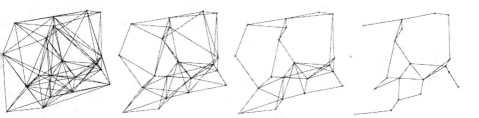

Fig. 5. The topology of the same weighted graph is shown using different thresholds. The node attributes and link weights are absent from this figure. As the threshold is increased (left to right), the topology becomes sparser as expected. The regularization parameter λ is held fixed while the threshold (for displaying a graph weight) is increased. The interplay between the regularization parameter and the graph topology needs to be further investigated.

In Figure 5, we show a point-set with 20 points and associated weighted graphs that have been derived from the spline kernel corresponding to the point-set. We wanted to explore the topology of the graph to see if the spline kernel seemed to use nearest neighbor heuristics in assigning weights. We thresholded the graphs (with increasing thresholds from left to right) after taking teh absolute value of each element. The topology clearly has a nearest neighbor bias which is more evident at larger thresholds. (The threshold for getting a certain number of connections is proportional to the regularization parameter λ.)

In Figure 6, we took a point-set and obtained several point-sets from it by progressively increasing the warping. We depict the topology (using the same threshold for all graphs) for all the warped point-sets. It should be clear from the figure that the topology gets increasingly distorted relative to the origianl topology as you go from smaller to larger warps. However, a family resemblence to the original parent is unmistakeable.

5   Discussion and Conclusion

The two main contributions of this paper are: i) a new definition of weighted graphs based on the thin-plate spline kernel and ii) a new non-quadratic distance measure that significantly outperforms the conventional quadratic assignment

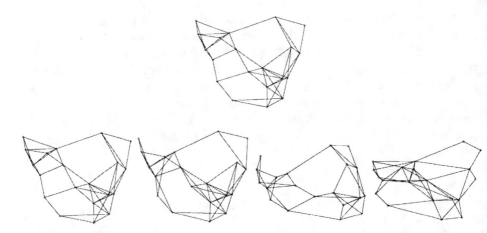

Fig. 6. On the top an original point-set is shown along with its graph depicted for a certain threshold. Below, we show four different thin-plate warped point-sets and their associated graphs. The warping increases as you go from left to right. Note the increased distortion in the topology going from left to right. The same threshold was used for all the graphs.

distance measure. To a certain extent, these two contributions are independent of one another. For instance, it should be possible to take our definition of weighted graphs and use a different distance measure. From the weighted graph standpoint, we have seen that the topology of the thin-plate spline kernel graphs (after thresholding) is somewhat similar to graphs derived from Delaunay triangulations with the important difference being that the spline-based graphs are not planar. The similarity stems from the fact that local connections are favored over more long range ones. We can also use different deformation mappings which should lead to different weighted graph definitions. For example, if we used a radial basis function (RBF) spline for the spatial mapping [33], the weighted graph would have a RBF kernel at its core. From the standpoint of the distance measure, we think that it is very significant that the new log-det distance measure outperforms the quadratic assignment distance. For binary graphs, it has already been shown that non-quadratic distance measures outperform QAP distances [11] and that seems to apply here as well. Enthusiasm must be tempered, however, until fast algorithms can be designed to find good, local minima of the new log-det distance measure.

There are several ways to proceed on the algorithm front. First, it may be possible to extend current deterministic annealing algorithms to the new distance. For instance, we could reduce the difficulty of algorithm design by choosing appropriate Legendre transformations [19] or by using Taylor series approximations of the log-det distance. Another approach would be to take the two topologies (after suitable thresholding) and apply the new maximum clique-based algorithms developed in [22]. After matching the topologies, further refinement using the weights can be performed using the softassign weighted graph matching algorithm [13].

In summary, we have shown that weighted graphs arise naturally in non-rigid point matching problems. The graphs directly depend on the parameterization of the deformations. In addition, we have found that a principled Bayesian Platonist formulation of the problem naturally leads to a new non-quadratic distance measure that outperforms the traditional quadratic assignment distance measure. It remains to be seen if effective algorithms can be designed that can take advantage of the better properties of the new distance measure.

Acknowledgements

A. R. would like to thank Zoubin Ghahramani for a helpful discussion. A. R. and H. C. are supported by a grant from the Whitaker Foundation.

References

1. Y. Amit. Graphical templates for model recognition. *IEEE Trans. Patt. Anal. Mach. Intell.*, 18(4):225–236, 1996.
2. R. Bajcsy and S. Kovacic. Multiresolution elastic matching. *Computer Vision, Graphics and Image Processing*, 46:1–21, 1989.
3. J. Bernardo and A. Smith. *Bayesian Theory*. John Wiley and Sons, New York, NY, 1994.
4. P. J. Besl and N. D. McKay. A method for registration of 3-D shapes. *IEEE Trans. Patt. Anal. Mach. Intell.*, 14(2):239–256, Feb. 1992.
5. M. Black and A. Jepson. Estimating optical flow in segmented images using variable-order parametric models and local deformations. *IEEE Trans. Patt. Anal. Mach. Intell.*, 18(10):972–986, 1996.
6. F. L. Bookstein. Principal warps: Thin-plate splines and the decomposition of deformations. *IEEE Trans. Patt. Anal. Mach. Intell.*, 11(6):567–585, June 1989.
7. G. Christensen, S. Joshi, and M. Miller. Volumetric transformation of brain anatomy. *IEEE Trans. Med. Imag.*, 16(6):864–877, 1997.
8. T. Cootes, C. Taylor, D. Cooper, and J. Graham. Active shape models: Their training and application. *Computer Vision and Image Understanding*, 61(1):38–59, 1995.
9. A. D. J. Cross and E. R. Hancock. Graph matching with a dual-step EM algorithm. *IEEE Trans. Patt. Anal. Mach. Intell.*, 20(11):1236–1253, 1998.
10. J. Feldmar and N. Ayache. Rigid, affine and locally affine registration of free-form surfaces. *Intl. J. Computer Vision*, 18(2):99–119, May 1996.
11. A. M. Finch, R. C. Wilson, and E. R. Hancock. An energy function and continuous edit process for graph matching. *Neural Computation*, 10(7):1873–1894, 1998.
12. M. R. Garey and D. S. Johnson. *Computers and intractability: a guide to the theory of NP-completeness*. W. H. Freeman, San Francisco, CA, 1979.
13. S. Gold and A. Rangarajan. A graduated assignment algorithm for graph matching. *IEEE Trans. Patt. Anal. Mach. Intell.*, 18(4):377–388, 1996.
14. E. Grimson. *Object Recognition by Computer: The Role of Geometric Constraints*. MIT Press, Cambridge, MA, 1990.

15. T. Hofmann and J. M. Buhmann. Pairwise data clustering by deterministic annealing. *IEEE Trans. Patt. Anal. Mach. Intell.*, 19(1):1–14, Jan. 1997.

16. B. Kim, J. L. Boes, K. A. Frey, and C. R. Meyer. Mutual information for automated unwarping of rat brain autoradiographs. *NeuroImage*, 5:31–40, 1997.

17. G. Lohmann and D. von Cramon. Sulcal basin and sulcal strings as new concepts for describing the human cortical topography. In *Workshop on Biomedical Image Analysis*, pages 41–54. IEEE Press, June 1998.

18. D. Metaxas, E. Koh, and N. I. Badler. Multi-level shape representation using global deformations and locally adaptive finite elements. *Intl. J. Computer Vision*, 25(1):49–61, 1997.

19. E. Mjolsness and C. Garrett. Algebraic transformations of objective functions. *Neural Networks*, 3:651–669, 1990.

20. H. Ogawa. Labeled point pattern matching by Delaunay triangulations and maximal cliques. *Pattern Recognition*, 19:35–40, 1986.

21. S. Pappu, S. Gold, and A. Rangarajan. A framework for non-rigid matching and correspondence. In D. S. Touretzky, M. C. Mozer, and M. E. Hasselmo, editors, *Advances in Neural Information Processing Systems 8*, pages 795–801. MIT Press, Cambridge, MA, 1996.

22. M. Pelillo. Replicator equations, maximal cliques and graph isomorphism. *Neural Computation*, 11, 1999. (in press).

23. S. Sclaroff and A. P. Pentland. Modal matching for correspondence and recognition. *IEEE Trans. Patt. Anal. Mach. Intell.*, 17(6):545–561, Jun. 1995.

24. G. Scott and C. Longuet-Higgins. An algorithm for associating the features of two images. *Proc. Royal Society of London*, B244:21–26, 1991.

25. L. Shapiro and J. Brady. Feature-based correspondence: an eigenvector approach. *Image and Vision Computing*, 10:283–288, 1992.

26. L. G. Shapiro and R. M. Haralick. Structural descriptions and inexact matching. *IEEE Trans. Patt. Anal. Mach. Intell.*, 3(9):504–519, Sept. 1981.

27. R. Szeliski and S. Lavallee. Matching 3D anatomical surfaces with non-rigid deformations using octree splines. *Intl. J. Computer Vision*, 18:171–186, 1996.

28. H. Tagare, D. O'Shea, and A. Rangarajan. A geometric criterion for shape based non-rigid correspondence. In *Fifth Intl. Conf. Computer Vision (ICCV)*, pages 434–439, 1995.

29. S. Umeyama. Parameterized point pattern matching and its application to recognition of object families. *IEEE Trans. Patt. Anal. Mach. Intell.*, 15(1):136–144, Jan. 1993.

30. P. Viola and W. M. Wells III. Alignment by maximization of mutual information. In *Fifth Intl. Conf. Computer Vision (ICCV)*, pages 16–23. IEEE Press, 1995.

31. G. Wahba. *Spline models for observational data*. SIAM, Philadelphia, PA, 1990.

32. T. Wakahara. Shape matching using LAT and its application to handwritten character recognition. *IEEE Trans. Patt. Anal. Mach. Intell.*, 16(6):618–629, 1994.

33. A. L. Yuille and N. M. Grzywacz. A mathematical analysis of the motion coherence theory. *Intl. J. Computer Vision*, 3(2):155–175, June 1989.

Continuous-Time Relaxation Labeling Processes

Andrea Torsello and Marcello Pelillo

Dipartimento di Informatica
Università Ca' Foscari di Venezia
Via Torino 155, 30172 Venezia Mestre, Italy

Abstract. We study the dynamical properties of two new relaxation labeling schemes described in terms of differential equations, and hence evolving in continuous time. This contrasts with the customary approach to defining relaxation labeling algorithms which prefers discrete time. Continuous-time dynamical systems are particularly attractive because they can be implemented directly in hardware circuitry, and the study of their dynamical properties is simpler and more elegant. They are also more plausible as models of biological visual computation. We prove that the proposed models enjoy exactly the same dynamical properties as the classical relaxation labeling schemes, and show how they are intimately related to Hummel and Zucker's now classical theory of constraint satisfaction. In particular, we prove that, when a certain symmetry condition is met, the dynamical systems' behavior is governed by a Liapunov function which turns out to be (the negative of) a well-known consistency measure. Moreover, we prove that the fundamental dynamical properties of the systems are retained when the symmetry restriction is relaxed. We also analyze the properties of a simple discretization of the proposed dynamics, which is useful in digital computer implementations. Simulation results are presented which show the practical behavior of the models.

1 Introduction

Relaxation labeling processes are a popular class of parallel, distributed computational models aimed at solving (continuous) constraint satisfaction problems, instances of which arise in a wide variety of computer vision and pattern recognition tasks [1, 9]. Almost invariably, all the relaxation algorithms developed so far evolve in discrete time, i.e., they are modeled as difference rather than as differential equations. The main reason for this widespread practice is that discrete-time dynamical systems are simpler to program and simulate on digital computers. However, continuous-time dynamical systems are more attractive for several reasons. First, they can more easily be implemented in parallel, analog circuitry (see, e.g., [4]). Second, the study of their dynamical properties is simplified thanks to the power of differential calculus, and proofs are more elegant and more easily understood. Finally, from a speculative standpoint, they are more plausible as models of biological computation [7].

Recently, there has been some interest in developing relaxation labeling schemes evolving in continuous time. In particular, we cite the work by Stoddart [16] motivated by the Baum-Eagon inequality [12], and the recent work by Li *et al.* [10] who developed a new relaxation scheme based on augmented Lagrangian multipliers and Hopfield networks. Yu and Tsai [19] also used a continuous-time Hopfield network for solving labeling problems. All these studies, however, are motivated by the assumption that the labeling problem is formulated as an energy-minimization problem, and a connection to standard theories of consistency [8] exists only when the compatibility coefficients are assumed to be symmetric. This is well-known to be a restrictive and unrealistic assumption. When the symmetry condition is relaxed the labeling problem is equivalent to a variational inequality problem, which is indeed a generalization of standard optimization problems [8].

In this paper, we study the dynamical properties of two simple relaxation labeling schemes which evolve in continuous time, each being described in terms of a system of coupled differential equations. The systems have been introduced in the context of evolutionary game theory, to model the evolution of relative frequencies of species in a multi-population setting [18], and one of them has also recently been proposed by Stoddart *et al.* [16], who studied its properties only in the case of symmetric compatibilities. Both schemes are considerably simpler that Hummel and Zucker's continuous-time model [8] which requires a complicated projection operator. Moreover, the first scheme has no normalization phase, and this makes it particularly attractive for practical hardware implementations. Since our models automatically satisfy the constraints imposed by the structure of the labeling problem, they are also much simpler than Yu and Tsai's [19] and Li *et al.*'s [10] schemes, which have to take constraints into account either in the form of penalty functions or Lagrange multipliers.

The principal objective of this study is to analyze the dynamics of these relaxation schemes and to relate them to the classical theory of consistency developed by Hummel and Zucker [8]. We show that all the dynamical properties enjoyed by standard relaxation labeling algorithms do hold for ours. In particular, we prove that, when symmetric compatibility coefficients are employed, the models have a Liapunov function which rules their dynamical behavior, and this turns out to be (the negative of) a well-known consistency measure. Moreover, and most importantly, we prove that the fundamental dynamical properties of the systems are retained when the symmetry restriction is relaxed. We also study the properties of a simple discretization of the proposed models, which is useful in digital computer implementations. Some simulation results are presented which show how the models behave in practice and confirm their validity.

The outline of the paper is as follows. In Section 2, we briefly review Hummel and Zucker's consistency theory, which is instrumental for the subsequent development. In Section 3 we introduce the models and in Section 4 we present the main theoretical results, first for the symmetric and then for the non-symmetric case. Section 5 describes two ways of discretizing the models, and proves some results. In Section 6 we present our simulation results, and Section 7 concludes the paper.

2 Consistency and its properties

The labeling problem involves a set of objects $B = \{b_1, \cdots, b_n\}$ and a set of possible labels $\Lambda = \{1, \cdots, m\}$. The purpose is to label each object of B with one label of Λ. To accomplish this, two sources of information are exploited. The first one relies on *local* measurements which capture the salient features of each object viewed in isolation; classical pattern recognition techniques can be practically employed to carry out this task. The second source of information, instead, accounts for possible interactions among nearby labels and, in fact, incorporates all the contextual knowledge about the problem at hand. This is quantitatively expressed by means of a real-valued four-dimensional matrix of compatibility coefficients $R = \{r_{ij}(\lambda, \mu)\}$. The coefficient $r_{ij}(\lambda, \mu)$ measures the strength of compatibility between the hypotheses "b_i has label λ" and "b_j has label μ:" high values correspond to compatibility and low values correspond to incompatibility. In our discussion, the compatibilities are assumed to be nonnegative, i.e., $r_{ij}(\lambda, \mu) \geq 0$, but this seems not to be a severe limitation because all the interesting concepts involved here exhibit a sort of "linear invariance" property [12]. In this paper, moreover, we will not be concerned with the crucial problem of how to derive the compatibility coefficients. Suffice it to say that they can be either determined on the basis of statistical grounds [11, 15] or, according to a more recent standpoint, adaptively learned over a sample of training data [14, 13].

The initial local measurements are assumed to provide, for each object $b_i \in B$, an m-dimensional vector $\bar{p}_i^0 = (p_i^0(1), \cdots, p_i^0(m))^T$ (where "T" denotes the usual transpose operation), such that $p_i^0(\lambda) \geq 0$, $i = 1 \ldots n$, $\lambda \in \Lambda$, and $\sum_\lambda p_i^0(\lambda) = 1$, $i = 1 \ldots n$. Each $p_i^0(\lambda)$ can be regarded as the initial, non-contextual degree of confidence of the hypothesis "b_i is labeled with label λ." By simply concatenating $\bar{p}_1^0, \bar{p}_2^0, \cdots, \bar{p}_n^0$ we obtain a weighted labeling assignment for the objects of B that will be denoted by $\bar{p}^0 \in \mathbb{R}^{nm}$. A relaxation labeling process takes as input the initial labeling assignment $\bar{p}^0$ and iteratively updates it taking into account the compatibility model R.

At this point, we introduce the space of weighted labeling assignments:

$$\mathbb{K} = \left\{ \bar{p} \in \mathbb{R}^{nm} \ \Big| \ p_i(\lambda) \geq 0, \ i = 1 \ldots n, \ \lambda \in \Lambda \ \text{ and } \ \sum_{\lambda=1}^m p_i(\lambda) = 1, \ i = 1 \ldots n \right\}$$

which is a linear convex set of $\mathbb{R}^{nm}$. Every vertex of $\mathbb{K}$ represents an *unambiguous* labeling assignment, that is one which assigns exactly one label to each object. The set of these labelings will be denoted by $\mathbb{K}^*$:

$$\mathbb{K}^* = \left\{ \bar{p} \in \mathbb{K} \ \big| \ p_i(\lambda) = 0 \text{ or } 1, \ i = 1 \ldots n, \ \lambda \in \Lambda \right\}.$$

Moreover, a labeling $\bar{p}$ in the interior of $\mathbb{K}$ (i.e., $0 < p_i(\lambda) < 1$, for all i and λ) will be called *strictly ambiguous*.

Now, let $\bar{p} \in \mathbb{K}$ be any labeling assignment. To develop a relaxation algorithm that updates $\bar{p}$ in accordance with the compatibility model, we need to define, for each object $b_i \in B$ and each label $\lambda \in \Lambda$, what is called a *support* function. This should quantify the degree of agreement between the hypothesis that b_i is labeled with λ, whose confidence is expressed by $p_i(\lambda)$, and the context. This measure is commonly defined as follows:

$$q_i(\lambda; \bar{p}) = \sum_{j=1}^{n} \sum_{\mu=1}^{m} r_{ij}(\lambda, \mu) p_j(\mu) .$$ (1)

Putting together the instances $q_i(\lambda; \bar{p})$, for all the $p_i(\lambda)$, we obtain an nm-dimensional support vector that will be denoted by $\bar{q}(\bar{p})$.[1]

The following updating rule

$$p_i^{t+1}(\lambda) = \frac{p_i^t(\lambda) q_i^t(\lambda)}{\sum_{\mu} p_i^t(\mu) q_i^t(\mu)}$$ (2)

where $t = 0, 1, \ldots$ denotes (discrete) time, defines the original relaxation labeling operator of Rosenfeld, Hummel, and Zucker [15], whose dynamical properties have recently been clarified [12]. In the following discussion we shall refer to it as the "classical" relaxation scheme.

We now briefly review Hummel and Zucker's theory of constraint satisfaction [8] which commences by providing a general definition of consistency. By analogy with the unambiguous case, which is more easily understood, a weighted labeling assignment $\bar{p} \in \mathbb{K}$ is said to be *consistent* if

$$\sum_{\lambda=1}^{m} p_i(\lambda) q_i(\lambda; \bar{p}) \geq \sum_{\lambda=1}^{m} v_i(\lambda) q_i(\lambda; \bar{p}) , \quad i = 1 \ldots n$$ (3)

for all $\bar{v} \in \mathbb{K}$. Furthermore, if strict inequalities hold in (3), for all $\bar{v} \neq \bar{p}$, then $\bar{p}$ is said to be *strictly consistent*. It can be seen that a necessary condition for $\bar{p}$ to be strictly consistent is that it is an unambiguous one, that is $\bar{p} \in \mathbb{K}^*$.

In [8], Hummel and Zucker introduced the *average local consistency*, defined as

$$A(\bar{p}) = \sum_{i=1}^{n} \sum_{\lambda=1}^{m} p_i(\lambda) q_i(\lambda)$$ (4)

and proved that when the compatibility matrix R is symmetric, i.e., $r_{ij}(\lambda, \mu) = r_{ji}(\mu, \lambda)$ for all i, j, λ, μ, then any local maximum $\bar{p} \in \mathbb{K}$ of A is consistent. Basically, this follows immediately from the fact that, when R is symmetric, we have $\nabla A(\bar{p}) = 2\bar{q}$, $\nabla A(\bar{p})$ being the gradient of A at $\bar{p}$. Note that, in general, the converse need not be true since, to prove this, second-order derivative information would be required. However, by demanding that $\bar{p}$ be strictly consistent, this does happen [12].

---

[1] Henceforth, when it will be clear from context, the dependence on $\bar{p}$ will not be stated.

3 Continuous-time relaxation labeling processes

The two relaxation labeling models studied in this paper are defined by the following systems of coupled differential equations:

$$\frac{d}{dt}p_i(\lambda) = p_i(\lambda)\left(q_i(\lambda) - \sum_\mu p_i(\mu)q_i(\mu)\right) \tag{5}$$

and

$$\frac{d}{dt}p_i(\lambda) = p_i(\lambda)\frac{q_i(\lambda) - \sum_\mu p_i(\mu)q_i(\mu)}{\sum_\mu p_i(\mu)q_i(\mu)} . \tag{6}$$

For the purpose of the present discussion, $q_i(\lambda)$ denotes the linear support as defined in equation (1). As a matter of fact, many of the results proved below do not depend on this particular choice. More generally, the only requirements are that the support function be nonnegative and, to be able to grant the existence and uniqueness of the solution of the differential equations, that it be of class C^1 [6].

In the first model we note that, although there is no explicit normalization process in the updating rule, the assignment space $\mathbb{K}$ is invariant under dynamics (5). This means that any trajectory starting in $\mathbb{K}$ will remain in $\mathbb{K}$. To see this, simply note that:

$$\sum_\lambda \frac{d}{dt}p_i(\lambda) = \sum_\lambda p_i(\lambda)\left(q_i(\lambda) - \sum_\mu p_i(\mu)q_i(\mu)\right) = 0$$

which means that the interior of $\mathbb{K}$ is invariant. The additional observation that the boundary too is invariant completes the proof. The same result can be proven for the other model as well, following basically the same steps. The lack of normalization makes the first model, which we call the *standard* model, more attractive than Hummel and Zucker's projection-based scheme [8], since it makes it more amenable to hardware implementations and more acceptable biologically. The interest in the other model, called the *normalized* model and also studied by Stoddart [16], derives from the fact that, in a way, it is the continuous-time translation of the classical Rosenfeld-Hummel-Zucker relaxation scheme [15]. This will be clearer when we show the discretizations of the models. We note that, using a linear support function (1), the dynamics of the models is invariant under a rescaling of the compatibility coefficients $r_{ij}(\lambda, \mu)$. That is, if we define a set of new compatibility coefficients $r_{ij}^*(\lambda, \mu) = \alpha r_{ij}(\lambda, \mu) + \beta$, with $\alpha > 0$ and $\beta \geq 0$, the orbit followed by the model remains the same, while the speed at which the dynamics evolve changes by a factor α.

As stated in the Introduction, one attractive feature of continuous-time systems is that they are readily mapped onto hardware circuitry. In [17] we show a circuit implementation for the standard and the normalized models, respectively. As expected, the standard model leads to a more economic implementation.

The fixed (or equilibrium) points of our dynamical systems are characterized by $\frac{d}{dt}\bar{p} = 0$ or, more explicitly, by $p_i(\lambda)\left[q_i(\lambda) - \sum_\mu p_i(\mu)q_i(\mu)\right] = 0$ for all $i = 1\ldots n$, $\lambda \in \Lambda$. This leads us to the condition

$$p_i(\lambda) > 0 \Rightarrow q_i(\lambda) = \sum_\mu p_i(\mu)q_i(\mu) \tag{7}$$

which is the same condition we have for the Rosenfeld-Hummel-Zucker and Hummel-Zucker models.

The next result follows immediately from a characterization of consistent labelings proved in [12, Theorem 3.1].

Proposition 1. *Let $\bar{p} \in \mathbb{K}$ be consistent. Then $\bar{p}$ is an equilibrium point for the relaxation dynamics (5) and (6). Moreover, if $\bar{p}$ is strictly ambiguous the converse also holds.*

This establishes a first connection between our continuous-time relaxation labeling processes and Hummel and Zucker's theory of consistency.

4 The dynamical properties of the models

In this section we study the dynamical properties of the proposed dynamical systems. Specifically, we show how our continuous-time relaxation schemes are intimately related to Hummel and Zucker's theory of consistency, and enjoy all the dynamical properties which hold for the classical discrete-time scheme (2), and Hummel and Zucker's projection-based model.

Before going into the technical details, we briefly review some instrumental concepts in dynamical systems theory; see [6] for details. Given a dynamical system, an equilibrium point $\bar{x}$ is said to be *stable* if, whenever started sufficiently close to $\bar{x}$, the system will remain near to $\bar{x}$ for all future times. A stronger property, which is even more desirable, is that the equilibrium point $\bar{x}$ be *asymptotically stable*, meaning that $\bar{x}$ is stable and in addition is a *local attractor*, i.e., when initiated close to $\bar{x}$, the system tends towards $\bar{x}$ as time increases. One of the most fundamental tools for establishing the stability of a given equilibrium point is known as the Liapunov's direct method. It involves seeking a so-called *Liapunov* function, i.e., a continuous real-valued function defined in state space which is non-increasing along a trajectory.

4.1 Symmetric compatibilities

We present here some results which hold when the compatibility matrix R is symmetric, i.e., $r_{ij}(\lambda, \mu) = r_{ji}(\mu, \lambda)$, for all $i, j = 1\ldots n$ and $\lambda, \mu \in \Lambda$. The following instrumental lemma, however, holds for the more general case of asymmetric matrices.

Lemma 1. *For all $\bar{p} \in \mathbb{K}$ we have*

$$\bar{q}(\bar{p}) \cdot \frac{d}{dt}\bar{p} \geq 0 \; ,$$

where "·" represents the inner product operator, for both the standard and normalized relaxation schemes (5) and (6).

Proof: Let $\bar{p}$ be an arbitrary labeling assignment in $\mathbb{K}$. For the standard model we have:

$$\bar{q}(\bar{p}) \cdot \frac{d}{dt}\bar{p} = \sum_{i,\lambda} q_i(\lambda)p_i(\lambda)\left(q_i(\lambda) - \sum_{\mu} p_i(\mu)q_i(\mu)\right)$$

$$= \sum_{i}\left[\sum_{\lambda} p_i(\lambda)q_i^2(\lambda) - \left(\sum_{\lambda} p_i(\lambda)q_i(\lambda)\right)^2\right]$$

Using the Cauchy-Schwartz inequality we obtain, for all $i = 1 \ldots n$,

$$\left(\sum_{\lambda} p_i(\lambda)q_i(\lambda)\right)^2 = \left(\sum_{\lambda} \sqrt{p_i(\lambda)} \cdot \sqrt{p_i(\lambda)q_i^2(\lambda)}\right)^2$$

$$\leq \sum_{\lambda} p_i(\lambda) \cdot \sum_{\lambda} p_i(\lambda)q_i^2(\lambda) = \sum_{\lambda} p_i(\lambda)q_i^2(\lambda)$$

Hence, since $\sum_{\lambda} p_i(\lambda)q_i^2(\lambda) \geq (\sum_{\lambda} p_i(\lambda)q_i(\lambda))^2$, we have $\bar{q}(\bar{p}) \cdot \frac{d}{dt}\bar{p} \geq 0$.
The proof for the normalized model is identical; we just observe that:

$$\bar{q}(\bar{p}) \cdot \frac{d}{dt}\bar{p} = \sum_{i} \frac{\sum_{\lambda} q_i(\lambda)p_i(\lambda)\left(q_i(\lambda) - \sum_{\mu} p_i(\mu)q_i(\mu)\right)}{\sum_{\mu} p_i(\mu)q_i(\mu)}$$

$\square$

A straightforward consequence of the previous lemma is the following important result, which states that, in the symmetric case, the average local consistency is always non-decreasing along the trajectories of our dynamical systems.

Theorem 1. *If the compatibility matrix R is symmetric, we have*

$$\frac{d}{dt}A(\bar{p}) \geq 0$$

for all $\bar{p} \in \mathbb{K}$. In other words, $-A$ is a Liapunov function for the relaxation models (5) and (6).

Proof: Assuming $r_{ij}(\lambda, \mu) = r_{ji}(\mu, \lambda)$, we have:

$$\frac{d}{dt} A(\bar{p}) = \sum_{i\lambda} \sum_{j\mu} r_{ij}(\lambda, \mu) p_j(\mu) \frac{d}{dt} p_i(\lambda)$$

$$= \bar{q}(\bar{p}) \cdot \frac{d}{dt} \bar{p} \geq 0$$

$\square$

As far as the normalized scheme is concerned, this result has been proven by Stoddart [16]. By combining the previous result with the fact that strictly consistent labelings are local maxima of the average local consistency (see [12, Proposition, 3.4]) we readily obtain the following proposition.

Theorem 2. *Let $\bar{p}$ be a strictly consistent labeling and suppose that the compatibility matrix R is symmetric. Then $\bar{p}$ is an asymptotically stable stationary point for the relaxation labeling processes (5) and (6) and, consequently, is a local attractor.*

Therefore, in the symmetric case our continuous-time processes have exactly the same dynamical properties as the classical Rosenfeld-Hummel-Zucker model [12] and the Hummel-Zucker projection-based scheme [8].

4.2 Arbitrary compatibilities

In the preceding subsection we have restricted ourselves to the case of symmetric compatibility coefficients and have shown how, under this circumstance, the proposed continuous-time relaxation schemes are closely related to the theory of consistency of Hummel and Zucker. However, although symmetric compatibilities can easily be derived and asymmetric matrices can always be made symmetrical (i.e., by considering $R + R^T$), it would be desirable for a relaxation process to work also when no restriction on the compatibility matrix is imposed [8]. This is especially true when the relaxation algorithm is viewed as a plausible model of how biological systems perform visual computation [20].

We now show that the proposed relaxation dynamical systems still perform useful computations in this case, and their connection with the theory of consistency continues to hold. The main result is the following:

Theorem 3. *Let $\bar{p} \in \mathbb{K}$ be a strictly consistent labeling. Then $\bar{p}$ is an asymptotically stable equilibrium point for the continuous-time relaxation labeling schemes defined in equations (5) and (6).*

Proof: The first step in proving the theorem is to rewrite the models in the following way:

$$\frac{d}{dt} \bar{p} = F(\bar{p})$$

where, for all $i = 1 \ldots n$ and $\lambda \in \Lambda$,

$$F_i(\lambda)(\bar{p}) = p_i(\lambda) \left(q_i(\lambda) - \sum_\mu p_i(\mu) q_i(\mu) \right)$$

for the standard model, and

$$F_i(\lambda)(\bar{p}) = p_i(\lambda) \left(\frac{q_i(\lambda)}{\sum_\mu p_i(\mu) q_i(\mu)} - 1 \right)$$

for the normalized model.

Let $DF(\bar{p})$ be the differential of F in $\bar{p}$. We will show that if $\bar{p}$ is strictly consistent all eigenvalues of $DF(\bar{p})$ are real and negative. This means that $\bar{p}$ is a *sink* for the dynamical system and therefore an asymptotically stable point [6].

We begin by recalling that a strictly consistent labeling is necessarily non-ambiguous. Denoting by $\lambda(i)$ the unique label assigned to object b_i, we have:

$$p_i(\lambda) = \begin{cases} 0 & \text{if } \lambda \neq \lambda(i) \\ 1 & \text{if } \lambda = \lambda(i) \end{cases} = \delta_{\lambda\lambda(i)}$$

where δ is the Kronecker delta, i.e., $\delta_{xy} = 1$ if $x = y$, and $\delta_{xy} = 0$ otherwise. Furthermore, we have $q_i(\lambda(i)) > q_i(\lambda)$ for all $\lambda \neq \lambda(i)$.

We first prove the theorem for the standard model. Deriving F with respect to $p_j(\rho)$, we have:

$$\frac{\partial F_i(\lambda)}{\partial p_j(\rho)}(\bar{p}) = \delta_{ij}\delta_{\lambda\rho} \left(q_i(\lambda) - \sum_\mu p_i(\mu) q_i(\mu) \right) +$$

$$p_i(\lambda) \left(\frac{\partial q_i(\lambda)}{\partial p_j(\rho)} - \delta_{ij} q_i(\rho) - \sum_\mu p_i(\mu) \frac{\partial q_i(\mu)}{\partial p_j(\rho)} \right) \qquad (8)$$

If we arrange the assignment vector in the following way:

$$p = (p_1(\lambda_1), \cdots, p_1(\lambda_m), \cdots, p_n(\lambda_1), \cdots, p_n(\lambda_m))^T$$

and define the matrices $C_{ij} = (C_{ij}(\lambda, \mu))_{\lambda,\mu}$ as $C_{ij}(\lambda, \mu) = \frac{\partial F_i(\lambda)}{\partial p_j(\mu)}$, the differential takes the form:

$$DF = \begin{pmatrix} C_{11} & C_{12} & \ldots & C_{1n} \\ C_{21} & C_{22} & \ldots & C_{2n} \\ \vdots & \vdots & \ddots & \vdots \\ C_{n1} & C_{n2} & \ldots & C_{nn} \end{pmatrix}$$

We can show that, if $\bar{p}$ is strictly consistent, $C_{ij} = 0$ if $i \neq j$. In fact, we have:

$$\frac{\partial F_i(\lambda)}{\partial p_j(\rho)}(\bar{p}) = p_i(\lambda)\left(\frac{\partial q_i(\lambda)}{\partial p_j(\rho)} - \sum_\mu p_i(\mu)\frac{\partial q_i(\mu)}{\partial p_j(\rho)}\right)$$

$$= \delta_{\lambda\lambda(i)}\left(\frac{\partial q_i(\lambda)}{\partial p_j(\rho)} - \frac{\partial q_i(\lambda(i))}{\partial p_j(\rho)}\right) = 0$$

In this case the differential takes the form:

$$DF = \begin{pmatrix} C_{11} & & 0 \\ & \ddots & \\ 0 & & C_{nn} \end{pmatrix}$$

Analyzing the matrices C_{ii} we can see that these too take a particular form on strictly consistent assignments. In fact we have:

$$\frac{\partial F_i(\lambda)}{\partial p_i(\rho)}(\bar{p}) = \delta_{\lambda\rho}\left(q_i(\lambda) - \sum_\mu p_i(\mu)q_i(\mu)\right) +$$

$$p_i(\lambda)\left(\frac{\partial q_i(\lambda)}{\partial p_i(\rho)} - q_i(\rho) - \sum_\mu p_i(\mu)\frac{\partial q_i(\mu)}{\partial p_i(\rho)}\right)$$

$$= \delta_{\lambda\rho}\left(q_i(\lambda) - q_i(\lambda(i))\right) + \delta_{\lambda\lambda(i)}\left(\frac{\partial q_i(\lambda)}{\partial p_i(\rho)} - q_i(\rho) - \frac{\partial q_i(\lambda(i))}{\partial p_i(\rho)}\right)$$

$$= \delta_{\lambda\rho}\left(q_i(\lambda) - q_i(\lambda(i))\right) - \delta_{\lambda\lambda(i)}q_i(\rho)$$

As we can notice, the non-zero values of C_{ii} are on the main diagonal and on the row $C_{ii}(\lambda\rho)$ with $\lambda = \lambda(i)$. Thus the eigenvalues of C_{ii} are the elements on the main diagonal. These are:

$$\begin{cases} q_i(\lambda) - q_i(\lambda(i)) & \text{for } \lambda \neq \lambda(i), \\ -q_i(\lambda(i)) & \text{otherwise.} \end{cases} \tag{9}$$

Since $\bar{p}$ is strictly consistent, $q_i(\lambda) < q_i(\lambda(i))$ so all the eigenvalues are real and negative and not lower than $-q_i(\lambda(i))$. This tells us that the assignment is a sink, and hence an asymptotically stable point for the dynamical system.

We now prove the theorem for the normalized model. The fundamental steps to follow are the same as for the standard model; we mainly have to derive the new values for the partial derivatives:

$$\frac{\partial F_i(\lambda)}{\partial p_j(\rho)}(\bar{p}) = \frac{\delta_{ij}\delta_{\lambda\rho}q_i(\lambda) + p_i(\lambda)\frac{\partial q_i(\lambda)}{\partial p_j(\rho)}}{\sum_\mu p_i(\mu)q_i(\mu)}$$

$$- \frac{p_i(\lambda)q_i(\lambda)\left(\delta_{ij}q_i(\rho) + \sum_\mu p_i(\mu)\frac{\partial q_i(\mu)}{\partial p_j(\rho)}\right)}{\left(\sum_\mu p_i(\mu)q_i(\mu)\right)^2} - \delta_{ij}\delta_{\lambda\rho}$$

$$= \delta_{ij}\delta_{\lambda\rho}\frac{q_i(\lambda)}{q_i(\lambda(i))} + \delta_{\lambda\lambda(i)}\frac{\frac{\partial q_i(\lambda)}{\partial p_j(\rho)}}{q_i(\lambda(i))} - \delta_{ij}\delta_{\lambda\lambda(i)}\frac{q_i(\lambda)q_i(\rho)}{q_i(\lambda(i))^2}$$

$$- \delta_{\lambda\lambda(i)}\frac{q_i(\lambda)\frac{\partial q_i(\lambda(i))}{\partial p_j(\rho)}}{q_i(\lambda(i))^2} - \delta_{ij}\delta_{\lambda\rho}$$

$$= \delta_{ij}\delta_{\lambda\rho}\frac{q_i(\lambda)}{q_i(\lambda(i))} - \delta_{ij}\delta_{\lambda\lambda(i)}\frac{q_i(\rho)}{q_i(\lambda(i))} - \delta_{ij}\delta_{\lambda\rho}$$

As the standard model, we have $C_{ij} = 0$ for $i \neq j$, and the matrices C_{ii} are non-zero only on the main diagonal and on the row related to the assignment $\lambda(i)$. Once more, then, the eigenvalues are equal to the elements on the main diagonal. These are:

$$\begin{cases} \frac{q_i(\lambda) - q_i(\lambda(i))}{q_i(\lambda(i))} & \text{for } \lambda \neq \lambda(i), \\ -1 & \text{otherwise.} \end{cases} \tag{10}$$

Thus the eigenvalues are all real and negative and not lower than -1, i.e., strictly consistent assignments are sinks for system (6). □

The previous theorem is the analog to the fundamental local convergence result of Hummel and Zucker [8, Theorem 9.1], which is also valid for the classical relaxation scheme (2) [12, Theorem 6.4]. Note that, unlike Theorem 2, no restriction on the structure of the compatibility matrix is imposed here.

5 Discretizing the models

In order to simulate the behavior of the models on a digital computer, we need to make them evolve in discrete rather than continuous time steps. Two well known techniques to approximate differential equations are the Euler method and the Runge-Kutta method. With the Euler method we have:

$$p_i^{t+h}(\lambda) = p_i^t(\lambda) + hF_i^t(\lambda)(\bar{p}) \tag{11}$$

where h is the step size. This equation is advantageous since it can be computed in a very efficient way, so it is the ideal candidate for our simulations. We will prove that, given a certain integration step h, this model enjoys all the dynamical properties shown for the continuous models it approximates.

In order to determine the difference in global behavior between the continuous models and the discrete approximations, we also use a finer discretization model: the IV grade Runge-Kutta method. This has been done on the assumption that this model would have a global dynamic behavior very similar to that of the continuous models. We have chosen the following Runge-Kutta scheme:

$$p_i^{t+h}(\lambda) = p_i^t(\lambda) + \frac{1}{6}k_1(i,\lambda) + \frac{2}{6}k_2(i,\lambda) + \frac{2}{6}k_3(i,\lambda) + \frac{1}{6}k_4(i,\lambda)$$

where the coefficients k_1, k_2, k_3, k_4 represent:

$$\begin{cases} k_1(i,\lambda) = hF_i(\lambda)(\bar{p}) \\ k_2(i,\lambda) = hF_i(\lambda)\left(\bar{p} + \frac{1}{2}\bar{k}_1\right) \\ k_3(i,\lambda) = hF_i(\lambda)\left(\bar{p} + \frac{1}{2}\bar{k}_2\right) \\ k_4(i,\lambda) = hF_i(\lambda)(\bar{p} + \bar{k}_3) \end{cases}$$

We will prove that the models discretized with Euler's method are well defined, that is, they map points in the assignment space $\mathbb{K}$ onto $\mathbb{K}$. Euler's scheme applied to our standard relaxation model (5) gives:

$$p_i^{t+h}(\lambda) = p_i^t(\lambda) + hp_i^t(\lambda)\left(q_i^t(\lambda) - \sum_\mu p_i^t(\mu)q_i^t(\mu)\right)$$

We note that when h equals 1 the process is identical to the one recently proposed by Chen and Luh [2,3]. Their model imposes strict constraints on the compatibility coefficients to insure that $\mathbb{K}$ be invariant with respect to iterations of the process. However, it can be proven that, if an appropriate integration step h is chosen, it is not necessary to impose such constraints.

It is easy to prove that $\sum_\lambda p_i(\lambda)$ always equals 1:

$$\sum_\lambda p_i^{t+h}(\lambda) = 1 + h\left(\sum_\lambda p_i^t(\lambda)q_i^t(\lambda) - \sum_\lambda p_i^t(\lambda)\sum_\mu p_i^t(\mu)q_i^t(\mu)\right) = 1$$

But we have to prove that the iteration of the process never leads to negative assignments.

Proposition 2. *Let $h \leq 1/q_i(\lambda; \bar{p})$ for all $i, \lambda, \bar{p}$. Denoting by E the function generated applying Euler's scheme to the model (5), then for all $\bar{p} \in \mathbb{K}$, we have $E_i(\lambda)(\bar{p}) \geq 0$.*

Proof: We have:

$$p_i^{t+h}(\lambda) \geq p_i^t(\lambda) + hp_i(\lambda)\left(q_i^t(\lambda) - \sum_\mu p_i^t(\mu)\frac{1}{h}\right)$$

$$= p_i^t(\lambda) + hp_i^t(\lambda)\left(q_i^t(\lambda) - \frac{1}{h}\right) \geq p_i^t(\lambda) - hp_i^t(\lambda)\frac{1}{h} = 0$$

which proves the proposition. □

If we use the linear support function (1), the integration step can be:

$$h \leq \frac{1}{\max_{i\lambda} \left\{ \sum_j \max_\mu r_{ij}(\lambda, \mu) \right\}}$$

It can readily be seen that this model also corrects deviation from the assignment space, provided that $p_i^t(\lambda) \geq 0$. In fact, given $\sum_\lambda p_i^t(\lambda) = 1 + \varepsilon$ we have:

$$\sum_\lambda p_i^{t+h}(\lambda) = \sum_\lambda p_i^t(\lambda) + h \sum_\lambda p_i^t(\lambda) \left(q_i(\lambda) - \sum_\mu p_i^t(\mu) q_i^t(\mu) \right)$$

$$= (1 + \varepsilon) + h \left[\left(1 - \sum_\lambda p_i^t(\lambda) \right) \sum_\mu p_i^t(\mu) q_i^t(\mu) \right]$$

$$= 1 + \varepsilon - \varepsilon h \sum_\mu p_i^t(\mu) q_i^t(\mu)$$

As far as the normalized model is concerned, Euler's scheme yields:

$$p_i^{t+h}(\lambda) = (1 - h)p_i^t(\lambda) + h \frac{p_i^t(\lambda) q_i^t(\lambda)}{\sum_\mu p_i^t(\mu) q_i^t(\mu)}$$

As can easily be seen, with $h = 1$, this is the same equation that defines the classical model. Thus for $h = 1$ the model is well defined.

With an h lower than 1 the resulting assignment is a convex linear combination of $\bar{p}$ and the assignment resulting from applying one iteration of the classical method to $\bar{p}$. Since the assignment space $\mathbb{K}$ is convex, the resulting assignment will also be in $\mathbb{K}$.

We can see that this model is also numerically stable. In fact, with $h = 1$, if we have $p_i(\lambda) \geq 0$, the model corrects any deviation from $\mathbb{K}$ in one step. On the other hand, with $h < 1$, if $\sum_\lambda p_i^t(\lambda) = 1 + \varepsilon$, we have:

$$\sum_\lambda p_i^{t+h}(\lambda) = \sum_\lambda (1 - h)p_i^t(\lambda) + \sum_\lambda h \frac{p_i^t(\lambda) q_i^t(\lambda)}{\sum_\mu p_i^t(\mu) q_i^t(\mu)} = (1 - h)(1 + \varepsilon) + h$$

$$= (1 - h\epsilon)$$

That is, the iteration of the model reduces the deviation from $\mathbb{K}$ at every step.

It is easy to prove that strictly consistent assignments are local attractors for these discrete models. In order to do this we must note that the differential of E is $I + hDF$; so, given an eigenvalue a of DF, there is an eigenvalue of DE equal to $1 + ha$. Furthermore, this property defines all eigenvalues of DE. As we have seen in (9), the eigenvalues of DF calculated for the standard model are all not lower than $\max_i \left\{ -q_i \big(\lambda(i) \big) \right\}$ and all strictly lower than 0; so, for any integration step lower than $1/q_i(\lambda)$ for all i and all λ, we have , for any eigenvalue b of DE, $b = 1 + ha \geq 1 - h\frac{1}{h} = 0$ and $b = 1 + ha < 1$. Thus, strictly consistent assignments are hyperbolic attractors for the system [5]. The eigenvalues of DF calculated for the normalized model are all not lower than -1 and all strictly lower than 0 (10); so, for $h \leq 1$ the eigenvalues of DE are all not lower than 0 and all strictly lower than 1. Thus, in this case as well, strictly consistent assignments are hyperbolic attractors for the system.

6   Experimental results

In order to evaluate the practical behavior of the proposed models we conducted two series of experiments. Our goal was to verify that the models exhibit the same dynamical behavior as the classical relaxation scheme (2). The experiments were conducted using both the Euler and the Runge-Kutta discretizations described in the previous section.

The first set of simulations were conducted over the classical "triangle" problem introduced as a toy example in the seminal paper by Rosenfeld, Hummel and Zucker [15]. The problem is to label the edges of a triangle as convex, concave, right- or left-occluding. Here, only eight possible labelings are possible (see [15] for details). The compatibility coefficients used were the same as those given in [15]. As a first control we verified whether the models' behaviors differ, starting from the eight initial assignments given in [15]. From these starting points all the models gave the same sets of classifications. After this preliminary test, we generated 100 random assignments and used them as starting points for each model. The iterations were stopped when the sum of Kullback's I-directed divergence between two successive assignments was lower than 10^{-7}. All the models converged to a non-ambiguous assignment. Moreover the Euler discretizations of our dynamics gave the same results as the classical model for all initial assignments, while the Runge-Kutta discretizations gave a different result only for one initial assignment. This single assignment was reached with the highest number of iterations of all the assignments generated. This is probably due to the symmetry of the problem: a similar problem can be seen with a uniform probability distribution among assignments. The iteration of each model should converge to the *a priori* probability of each classification, that is 3/8 for each occluding edge and 1/8 for convex or concave edges. What really happens is that the assignments start by heading towards the *a priori* distribution, but, after a few iterations, they head towards a non-ambiguous assignment. This happens because the *a priori* probability is not a hyperbolic attractor for the system. It is possible that a similar problem affected the only initial assignment that gave different results: the models headed toward different non-ambiguous solutions from a unique non-hyperbolic equilibrium that separates the orbits. The average number of iterations that the models needed to reach the stopping criterion is shown in Table 1.

Model	Iterations
classical (eq. (2))	79.1
standard, discretized with Euler's scheme	118.8
normalized discretized with Runge-Kutta scheme	81.4
standard, discretized with Runge-Kutta scheme	87.2

Table 1. Average number of iterations for the triangle labeling problem.

The second set of simulations was carried out by generating random sets of (asymmetric) compatibility matrices. This is the set of tests which most effectively point out differences in the dynamic behavior of the models. Since there was no underlying scheme on the pattern of compatibility coefficients, we do not expect the models to converge to a non-ambiguous assignment each time. In fact, in our experiments, the classical model converged to a non-ambiguous assignment only 22% of the time. The aim of this set of tests was to verify whether, when the classical model converges to a non-ambiguous assignment, the other models converge to the same assignment. Ten random coefficients matrices were generated for this experiment and for each matrix the various models were started from ten random assignments. Hence we made a hundred tests for each model. The assignment space dimension was five objects ($n = 5$)and three labels ($m = 3$). The stopping criterion was the same as the previous set of experiments. Here, the classical model (2) converged to a non-ambiguous assignment 22 times. The Runge-Kutta discretization of both models converged to the same assignments 20 times, while Euler discretization of the standard model reported the same assignments 17 times. Table 2 reports the average number of iterations needed to reach the stopping criterion.

Model	Iterations
classical (eq. (2))	294.0
standard, discretized with Euler's scheme	643.1
normalized discretized with Runge-Kutta scheme	260.1
standard, discretized with Runge-Kutta scheme	324.5

Table 2. Average number of iterations for the random compatibility experiments.

7 Conclusions

In this paper we have presented and analyzed two relaxation labeling processes. In contrast with the standard approach, these models evolve through continuous-time rather than discrete-time dynamics. This fact permits the design of analog hardware implementations, makes the study of its properties simpler and more elegant, and makes the model more plausible biologically. We have analyzed the dynamical behavior of the models and shown how it is intimately related to Hummel and Zucker's classical theory of consistency. We have proven that the models enjoy exactly the same dynamical properties which have already been proven for the classical processes. The dynamics of the models discretized through Euler's scheme has also been studied. Experimental results confirm the validity of the proposed models.

References

1. D. H. Ballard and C. M. Brown, *Computer Vision*. Prentice-Hall, Englewood Cliffs, NJ, 1982.
2. Q. Chen and J. Y. S. Luh. Ambiguity reduction by relaxation labeling. *Pattern Recognition*, 27(1):165–180, 1994.
3. Q. Chen and J. Y. S. Luh. Relaxation labeling algorithm for information integration and its convergence. *Pattern Recognition*, 28(11):1705–1722, 1995.
4. A. Cichoki and R. Unbehauen. *Neural Networks for Optimization and Signal Processing*. Wiley, 1993.
5. R. L. Devaney. *An Introduction to Chaotic Dynamical Systems*. Addison Wesley, 1989.
6. M. W. Hirsch and S. Smale. *Differential Equations, Dynamical Systems, and Linear Algebra*. Academic Press, New York, 1974.
7. J. J. Hopfield. Neurons with graded response have collective computational properties like those of two-state neurons. *Proc. Natl. Acad. Sci. USA*, 81:3088–3092, 1984.
8. R. A. Hummel and S. W. Zucker. On the foundations of relaxation labeling processes. *IEEE Trans. Pattern Anal. Machine Intell.*, 5:267–287, 1983.
9. J. Kittler and J. Illingworth. Relaxation labeling algorithms–A review. *Image Vision Comput.*, 3:206–216, 1985.
10. S. Z. Li, W. Y. C. Soh, and E. K. Teoh. Relaxation labeling using augmented Lagrange-Hopfield method. *Pattern Recognition*, 31(1):73–81, 1998.
11. S. Peleg and A. Rosenfeld. Determining compatibility coefficients for curve enhancement relaxation processes. *IEEE Trans. Syst. Man Cybern.*, 8:548–555, 1978.
12. M. Pelillo. The dynamics of nonlinear relaxation labeling processes. *J. Math. Imaging Vision*, 7(4):309–323, 1997.
13. M. Pelillo and A. M. Fanelli. Autoassociative learning in relaxation labeling networks. *Pattern Recognition Lett.*, 18(1):3–12, 1997.
14. M. Pelillo and M. Refice. Learning compatibility coefficients for relaxation labeling processes. *IEEE Trans. Pattern Anal. Machine Intell.*, 16(9):933–945, 1994.
15. A. Rosenfeld, R. A. Hummel, and S. W. Zucker. Scene labeling by relaxation operations. *IEEE Trans. Syst. Man Cybern.*, 6(6):420–433, 1976.
16. A. J. Stoddart, M. Petrou, and J. Kittler. On the foundations of probabilistic relaxation with product support. *J. Math. Imaging Vision*, 9:29–48, 1998.
17. A. Torsello and M. Pelillo. Continuous-time relaxation labeling processes. *Pattern Recognition*, submitted.
18. J. W. Weibull, *Evolutionary Game Theory*. MIT Press, Cambridge, MA, 1995.
19. S.-S. Yu and W.-H. Tsai. Relaxation by the Hopfield neural network. *Pattern Recognition*, 25(2):197–209, 1992.
20. S. W. Zucker, A. Dobbins, and L. Iverson. Two stages of curve detection suggest two styles of visual computation. *Neural Computat.*, 1:68–81, 1989.

Realistic Animation Using Extended Adaptive Mesh for Model Based Coding

Lijun Yin and Anup Basu[**]

Department of Computing Science, University of Alberta,
Edmonton, AB, T6G 2H1, Canada
{lijun, anup}@cs.ualberta.ca

Abstract. Accurate localization and tracking of facial features are crucial for developing high quality model-based coding (MPEG-4) systems. For teleconferencing applications at very low bit rates, it is necessary to track eye and lip movements accurately over time. These movements can be coded and transmitted to a remote site, where animation techniques can be used to synthesize facial movements on a model of a face. In this paper we describe the integration of simple heuristics which are effective in improving the results of well-known facial feature detection with robust techniques for adapting a dynamic mesh for animation. A new method of generating a self-adaptive mesh using an extended dynamic mesh (EDM) is proposed to overcome the convergence problem of the dynamic-motion-equation method (DMM). The new method consisting of two-step mesh adaptation (called coarse-to-fine adaptation) can enhance the stability of the DMM and improve the performance of the adaptive process. The accuracy of the proposed approach is demonstrated by experiments on eye model animation. In this paper, we focus our discussion only on the detection, tracking, modeling and animation of eye movements.

1 Introduction

From the image analysis point of view, images can be considered as having structural features or objects such as contours and regions. These image features or objects have been exploited to encode images at very low bit rates. Research on this approach, known as model-based coding, which is related to both image analysis and computer graphics, has recently intensified. Up to now, most of the contributions to 3D model-based coding have focused on human facial images. Although a number of schemes for model-based coding have been proposed [4, 13], automatic facial feature detection and tracking along with facial expression analysis and synthesis still poses a big challenge to the problem of finding accurate features and their motion.

A variety of approaches have been proposed for detection of facial features. These

---

[**] This work is supported in part by the Canadian Natural Sciences and Engineering Research Council.

include deformable template matching [8, 24, 16], hough transforms [5], and color image processing [3, 21]. Matching deformable templates requires a fairly accurate initial localization of the template because the energy minimization process only finds a local minimum. Other problems are caused by using several energy terms and weighting factors during the different epochs of matching. Because of the definition of the energy terms in [24] the template also inclines to shrink. In this paper we overcome some of these difficulties by improving the initial localization process. We show that simple processing on color images coupled with Hough transform and deformable template matching can produce very accurate results.

Another important component in model-based coding is synthesizing facial movements and expression at a remote site using the motion parameters detected on an actual face image and animation on a model of this face. To represent a facial expression, several approaches have been proposed relying on feature detection [2, 24], facial motion and expression analysis [4, 22, 9], and facial expression synthesis [19, 4]. However, little work has been done specifically on accurate eye expression synthesis. Because the eyes are one of the most significant organs contributing to a vivid face expression, subtle changes in eye movements can result in a different expression. Therefore accurate eye expression analysis and synthesis are necessary. Recently, face animation methods have been proposed in [10, 14], however, the techniques need to correspond dots on faces to resolve the feature detection issue.

In this paper, we present an approach to synthesize eye movements by us-

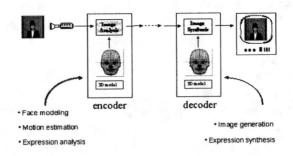

Fig. 1. An example of a MPEG-4 (SNHC) implementation

ing the extracted eye features and the extended dynamic mesh (called extended adaptive mesh) to compute the deformation of the eyes in a 3D model. After creating an individualized 3D face model [23], we map this model to the first frame of the face sequence. Based on the extracted eye features we apply the deformation parameters to the 3D model to synthesize eye movements in successive frames. The work described here is related to the emerging MPEG-4 guidelines for developing very low bit rate coding systems [20]. The Synthetic Natural Hybrid Coding (SNHC) component of MPEG-4 proposes that features and movement of features be detected and coded at a server site; these codes be transmitted following certain standards; and finally the codes be received at a client site and used for animation on a model to graphically emulate the real scene. Figure 1 shows an example of an MPEG-4 (SNHC) multimedia system. It should be noted that MPEG-4 does not specify the techniques to be used for feature detection to realize an actual implementation; this allows researchers (such as our group) to investigate alternative techniques as outlined in this paper. The MPEG-4 animation guideline outlines animation of feature points and suggests interpolation for non-feature points. We demonstrate that an extended adaptive mesh produces much more accurate and realistic animation compared to existing methods ([16]-[20], [13]).

The remainder of this paper is organized as follows: In Section 2, we describe the approach proposed for accurate eye feature detection and tracking. In Section 3, an extended adaptive mesh technique for eye model adaptation and animation is described. Experimental results are shown in Section 4. Finally, concluding remarks are given in Section 5.

2 Robust eye feature detection

Our approach to detecting the eyes is similar to [5] in that it uses Hough transform and deformable template matching, however, our approach also exploits color information to extract the eyes accurately. The algorithm can be outlined as follows:

- Determine two coarse regions of interest for the eyes.
- Search the iris of the eyes using a gradient based Hough transform.
- Determine a fine region of interest for extracting the boundaries of the eyes.
- Using color information get an initial approximation for the eye lids.
- Localize the eye lids using deformable templates.

After detecting the face region two coarse regions of interest in the upper left and upper right half of an image can be defined to detect the eyes. Also a coarse range for the size of the eyes can be derived [1]. Since the iris is the most significant feature of the eye and has a simple circular shape, it is detected first by using a gradient-based Hough transform for circles [6, 12].

Information on the position, magnitude and direction of significant edges is extracted by convolving the intensity image with a Sobel kernel. Figure 2(a) shows an example of the robustness of the Hough transform which extracts the iris circle. After extracting the circles the deformable templates for the eye lids have to be initialized. The template along with all the parameters are shown in Figure 3 with the parameters being set as follows:

$$h_1 = 1.5r_{iris} \qquad h_2 = 0.5r_{iris} \qquad w = 2.2r_{iris} \tag{1}$$

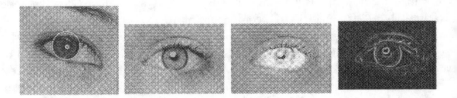

Fig. 2. (a)-(d)(left-right): (a)Extracted circle using gradient based hough transform; Image fields used for computing the potential energy: (b) image, (c) saturation, (d) edge.

where r_{iris} is the radius of the extracted circle. The orientation α is determined by the center points of the two circles. The initialized deformable template is also shown in Figure 3. For the matching of the deformable template two differ-

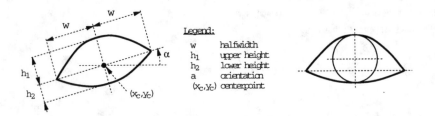

Legend:
w   halfwidth
h_1 upper height
h_2 lower height
a   orientation
(x_c, y_c) centerpoint

Fig. 3. Deformable Template (model, initialization).

ent types of image information are used to create potential fields, one in each epoch: *i.e.*, saturation information and edge information. Examples of the image information extracted from a typical eye is shown in Figure 2(c)(d). The first step in localizing the eye lids is to approximate the position of the deformable template relative to the iris. This is done by minimizing the following energy (E_{sat}) which is similar to the valley energy in [24]:

$$E_{sat} = -\frac{1}{|A_w|} \int_{A_w} \Phi_{sat}(x)dA \qquad (2)$$

A_w is the area inside the parabolas but not inside the circle of the iris and $\Phi_{sat}(x)$ is the inverted saturation value of the color image. Since only the location (not the size) is changed this method does not have the shrinking effect. The next step is to estimate the parameters h_1 and h_2. Two regions of interest on both sides of the iris are defined (see Figure 4). The parameters of these regions are set as follows:

$$w_{ROI} = 5 \qquad h_{ROI} = 3r_{iris} \qquad d_{ROI} = r_{iris} + 5 \qquad (3)$$

Depending on the position of the iris inside the deformable template only the left or the right region of interest is used for further computation. By using the horizontal integral projection ([2]), and by detecting the two most significant opposite gradients in the projection, the position of a point on the upper and lower eye lid can be detected (see Figure 4). The parameters h_1 and h_2 of the template are updated as follows:

$$h_1 = h_1 \frac{|y_{up} - y_c|}{h_1^*} \qquad h_2 = h_2 \frac{|y_{low} - y_c|}{h_2^*} \tag{4}$$

y_{up} and y_{low} are the y-coordinates of the detected points inside the region of interest of the upper and lower eye lids. h_1^* and h_2^* are the heights of the actual parabolas inside the region of interest.

The last step is to match the deformable template accurately to the eye lids by

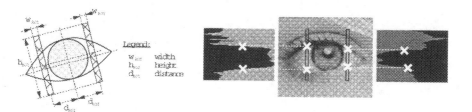

Fig. 4. Integral projections beside iris (template and example).

minimizing the following energy (E_{edge}):

$$E_{edge} = -\frac{1}{|B_w|} \int\limits_{B_w} \Phi_{edge}(\boldsymbol{x}) ds \tag{5}$$

B_w is the boundary of the parabolas and $\Phi_{edge}(\boldsymbol{x})$ is the edge magnitude. During this minimization every parameter of the deformable template (location, orientation, height, width) can be changed.

The tracking of eye features is similar to detection of eye features with the following differences:

- The region of interest as well as the possible size and therefore the Hough space for the extraction of the iris can be restricted by using the position and size of the eye extracted in the previous frame.
- Instead of using the deformable template, the matched template of the previous frame is used for the initialization of the new template.

3 Eye model adaptation and animation with extended adaptive mesh

The eye detection and tracking algorithms extract the contours of the iris and eye lids in each frame of the image sequence. These eye features are used to synthesize the real motions of the eye on a 3D facial model. To make the eye animation realistic, we apply an extended adaptive mesh approach (*i.e.*, extended dynamic mesh (EDM)), instead of the interpolation method [1], to animate the eye movement. Adaptive mesh (*i.e.*, dynamic mesh (DM)) is a well known approach for adaptive sampling of images [17,18] and physically based modeling of non-rigid objects [19,11,16]. The results shown by the previous works [17–19, 11] demonstrated that this technique has become the basis for many powerful approaches in computer vision and computer graphics. The adaptive mesh can be assembled from nodal points connected by adjustable springs. The fundamental equation [17] is a non-linear second order differential equation, which can be written as:

$$m_i \frac{d^2\mathbf{x}_i}{dt^2} + \gamma_i \frac{d\mathbf{x}_i}{dt} + \mathbf{g}_i = \mathbf{f}_i; i = 1, \ldots, N \tag{6}$$

where $\mathbf{x}_i$ is the position of node i, m_i is a point mass of node i, γ_i is the damping coefficient dissipating kinetic energy in the mesh through friction, $\mathbf{f}_i$ is the external force acting on node i, $\mathbf{g}_i$ is the internal force on node i due to the springs connected to neighboring nodes j.

To simulate the dynamics of the deformable mesh, the equations of motion are numerically integrated forward through time until the mesh is *nearly* stabilized. Although a number of numerical methods to solve this equation have been used (*e.g.*, Euler method, Runge-Kutta method [15,19]), the stability is still the main concern in achieving a satisfactory solution. For example, when a node moves across an image feature boundary associated with an abrupt change in its image intensity, the stiffness of those springs connecting with the node changes rapidly and results in a possible reversal of the nodal force, which may lead to perpetual oscillation of the node. In this type of situations a new equilibrium state cannot be reached. To make the mesh converge to a stable state, m_i and γ_i must be carefully chosen. The overdamped behavior (*i.e.*, large values of m_i and γ_i) will contribute to enhance the stability of the numerical simulation, however it is at the expense of accuracy of the solution. To make the solution more stable and accurate, we extend the conventional dynamic mesh method ([17]) by introducing a so called "energy-oriented mesh" (EOM) to refine adaptive meshes. The major differences between conventional dynamic mesh (DM) and the EOM are: (1) EOM makes the mesh movement in the direction of mesh energy decrease instead of decreasing the node velocities and accelerations; (2) EOM checks the node energy in each motion step without considering the velocity, it is independent of the DM and can be the supplemental step to DM for stabilizing mesh movements. Therefore, our model adaptation procedure consists of two major steps: (1) coarse adaptation, which applies DM method to make the large movement converge quickly to the region of an object; (2) fine adaptation, which applies EOM method to finely adjust the mesh obtained after the first step and make the adaptation more "tight" and "accurate".

3.1 Principle of Energy Minimization in Energy Oriented Mesh

According to the principle of minimum potential energy, of all possible kinematically admissible displacement configurations that an elastic body can take up, the configuration which satisfies equilibrium makes the total potential energy assume a minimum value [7]. The potential energy stated in the principle of minimum potential energy includes the strain energy and the potential energy formed by external forces. In our model, there is no external force and all the strain energy is stored as the elastic energy in springs. To reach the equilibrium state, the elastic energy in the springs have to be minimized by displacing nodes. If we let node i move under a nodal force while all the neighboring nodes are fixed, the node will move in the direction of the nodal force because the gradient of total spring energy on node i (E_i) is in the same direction as the nodal force (g_i). This implies that for meshes associated with the image observations, if we let nodes move by successive steps based on the principle of minimum potential energy and reduce strain energy at each step, finally we should obtain a fine adaptation on this image.

When a spring mesh is not in equilibrium state, those nodes with non-zero forces acting on them tend to move in the direction of the resultant nodal forces. The movements of nodes will reduce the energy caused by strain. When a final equilibrium state is reached, no further movements will occur. In order to prevent a node from being over-displaced at each step, energy change for each step must be checked to ensure that the step has reduced the energy in a non-increasing way along the direction of movement.

3.2 Detailed Algorithm

Since eye movements can result in subtle expressions on the human face, the accurate and detailed movement of the adaptive model is highly desirable. Therefore, we use a detailed eye model (120 vertices for each eye) in order to achieve more realistic animation. Following is the main procedure for eye animation:

1. Based on the 3D eye model which is part of our existing 3D facial model, 11 feature vertices are defined for each eye as shown in Figure 5. Once the eye lid contour and the iris are detected in the image sequence, the corresponding feature points on the template can be also determined simply by computing the points on the boundary of the parabola and by using the center of the circle.

2. To adapt the remaining vertices (*i.e.*, non-feature vertices) onto the eye image, two steps are applied: (1) coarse adaptation (DM method), (2) fine adaptation (EOM method).

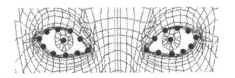

Fig. 5. Defined eye feature vertices.

(a) **Coarse adaptation:**
Solve the dynamic motion equation (6) using conventional explicit Euler time-integration procedure [15] until the motion parameters (velocity $\mathbf{v}_i$ and acceleration $\mathbf{a}_i$) are less than a certain threshold.

$$\mathbf{a}_i^t = \frac{B_i}{m_i}(\mathbf{f}_i^t - \gamma_i \mathbf{v}_i^t - \mathbf{g}_i) \tag{7}$$

$$\mathbf{v}_i^{t+\Delta t} = \mathbf{v}_i^t + \Delta t \mathbf{a}_i^t \tag{8}$$

$$\mathbf{x}_i^{t+\Delta t} = \mathbf{x}_i^t + \Delta t \mathbf{v}_i^{t+\Delta t} \tag{9}$$

where B_i denotes an operator whose role is to enforce boundary conditions or constraints. Equations (7), (8), and (9) are evaluated for all nodes, i.e., $i = 1, \ldots, N$, and consecutive time steps, i.e., $t = 0, \Delta t, 2\Delta t, \ldots$, until $\mathbf{v}_i$ and $\mathbf{a}_i$ are less than a certain threshold.

In our implementation of Equation (6) no external force is involved. The boundary vertices are fixed, which include the *feature vertices* defined on the eye model and the *vertices on the border* of Figure 5. Let node i be connected to M_i other nodes, i.e., node i is attached to M_i springs. The total internal force acting on node i due to these springs movement is:

$$\mathbf{g}_i = \sum_{j \in M_i} C_{ij}(\|\mathbf{r}_{ij}\| - l_{ij}) \frac{\mathbf{r}_{ij}}{\|\mathbf{r}_{ij}\|} \tag{10}$$

where $\mathbf{r}_{ij} = \mathbf{x}_j - \mathbf{x}_i$, $\mathbf{x}_i$ and $\mathbf{x}_j$ are the positions of nodes i and j; l_{ij} is the natural length of the spring connected from nodes i to j; $\|\mathbf{r}_{ij}\|$ is its actual length; and C_{ij} is the stiffness of the spring ij.

Based on the nodal value (*i.e.*, intensity in the nodal position), the springs automatically adjust their stiffness so as to distribute meshes in accordance with the local complexity of the image. Before calculating the stiffness, we apply a Sobel operator to obtain a gradient image, then normalize the intensity values of the gradient image within the range of $[0, 1]$. Suppose the stiffness of a spring changes linearly along with the nodal values on the normalized gradient image, the calculation is then as follows:

$$C_{ij} = -(k_2 I + k_1) \tag{11}$$

where k_1 is the pre-defined minimum stiffness of springs in the mesh; $k_1 +$ k_2 is the maximum stiffness of springs; and I is derived from the nodal values on the normalized gradient image. Unlike the stiffness calculation in [17], which takes the average of two nodal values on a spring, we apply a weighted sum of nodal values as shown in Equation (12). This implies that the node closer to the feature vertices will contribute more to the stiffness.

$$I = \frac{d_j}{d_i + d_j} S_i + \frac{d_i}{d_i + d_j} S_j \tag{12}$$

where S_i and S_j are the nodal values on the normalized gradient image at nodes i and j respectively. d_i (or d_j) is the minimum distance from node i (or j) to the nearest vertex in the set of extracted feature vertices.

To obtain the nodal values (S_i, S_j), we use the conventional finite element concept to calculate the *sub-pixel* values in-between the neighboring pixels. As shown in Figure 6, we split the pixel rectangle into two triangular elements so that sub-pixels within a certain triangle (a plane) are linearly distributed with the same property. The purpose of splitting into two triangular elements is to prevent the node movement over-displace in the next fine adaptation process (*e.g.*, jump across an edge boundary in a motion step). Let A denote a pixel at position (x_A, y_A) having value I_A. Four neighboring pixels A, B, C, D are split into two triangular elements $\triangle ABC$ and $\triangle BCD$. The value of sub-pixel p within $\triangle ABC$ can be obtained from Equation (13) (see [7]):

$$I_p = a_1 x_p + a_2 y_p + a_3 \tag{13}$$

where

$$a_1 = (x_C - x_B)I_A + (x_A - x_C)I_B + (x_B - x_A)I_C \tag{14}$$

$$a_2 = (y_B - y_C)I_A + (y_C - y_A)I_B + (y_A - y_B)I_C \tag{15}$$

$$a_3 = (x_B y_C - x_C y_B)I_A + (x_C y_A - x_A y_C)I_B + (x_A y_B - x_B y_A)I_C \tag{16}$$

Similar equations can be used if the point p is within $\triangle BCD$.

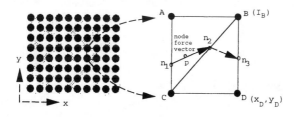

Fig. 6. Node displacement within a triangle area in each step: n_1: node start position; n_2: node end position; n_3: node end position in next step.

(b) **Fine adaptation:**

After a mesh stabilizes, fine adjustments can be done by EOM method for making meshes converge to the image "tightly". Assuming that the image intensity changes continuously over the spatial domain, the grey values in between pixels can be obtained from Equation (13). The criteria of the fine movement of nodes is that *only movements that decrease the node energy stored in the connected springs are allowed.* The nodes energy calculation and the nodes motion rules are described below:

- **obtain node force.** Of all the nodes on the mesh except the boundary vertices, find a node with the largest value of the nodal force using Equation (10), *e.g.*, g_i. Within a sub-pixel domain (triangle area), search sub-pixels along the direction of the g_i vector in order to find one having minimum node energy.

- **obtain node energy.** The strain energy E_{ij} stored in a spring ij is calculated as follows:

$$E_{ij} = C_{ij}(\|\mathbf{r}_{ij}\| - l_{ij})^2 \tag{17}$$

Node energy E_i is defined as the summation of the energy stored in all the springs connected to node i, *i.e.*,

$$E_i = \sum_{j \in M_i} E_{ij} \tag{18}$$

- **Displacement of a node.** Displacement of a node at a step is along the direction of the resultant nodal force. Theoretically, a node should move to a new position within the triangle domain where the node energy is a local minimum. To simplify the computation, in the current implementation, we calculate the node energy in three positions (*i.e.*, node start position (*e.g.*, n_1), middle position (*e.g.*, p), and node end position (*e.g.*, n_2)). The position with the minimum node energy is the new position that the node is allowed to move to. So a displacement at a step is only within a triangular area (including the boundary lines AC and BC, for example in Figure 6 from the node start position to the node end position). The maximum displacement of a node in one step will not exceed the distance between two adjacent pixels.

The rules for moving a node to a new position follows two conditions:

- Check the node energy at the start position and the end position. The energy must decrease; this ensures that the spring system has reached a state with less energy.
- To prevent the reversal of nodal force in the new position, the inner product of the force vector (F_L) at the node position with the force vector (F_M) at the new position must be greater than zero; this prevents the oscillating movement of nodes.

If the two conditions above are satisfied, the node is allowed to move to the new position. Otherwise, the node stays in its current place, and the procedure checks the node with second largest force, repeats the above procedure and continues until no node satisfies the above two conditions or the largest nodal force in the mesh is small enough (less than a certain threshold). Figure 7 shows a flowchart of the extended adaptive mesh algorithm in its current implementation.

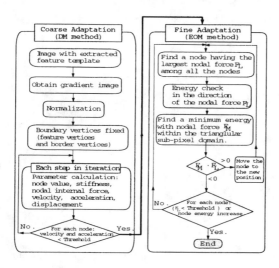

Fig. 7. Schematic diagram of the extended adaptive mesh method.

4 Experimental Results

To evaluate the algorithms developed, experiments with various color images of different eyes and eye sequences were made. The preliminary experiments have shown that the gradient-based Hough transform, which is the first and therefore one of the most important steps, is very robust against noise as well as edges which are not produced by the contour of the iris. The deformable template using color information makes the eye contour detection more robust and accurate. The results on tracking and animating of the iris and the eye lids in one image sequence are shown in Figure 8. The animated eye sequence

Fig. 8. Extracted eye templates in an actual sequence of eye images (frame 1, 24, 68, 169).

(for Figure 8 images) after the first step (DM) and the second step (EOM) are shown in Figure 9, in which a plane mesh is used for testing our extended adaptive mesh algorithm. The overlapped results of coarse and fine adaptations are shown in Figure 10. Figure 11 shows the texture-mapped results using the first frame texture of the original sequence. The improvement of the adaptation accuracy from coarse adaptation (DM) to fine adaptation (EOM) can be clearly seen comparing the top sequence with the bottom sequence in Figure 9 and Figure 10. Figure 11 (bottom) shows that the synthesized eye movements using the extended adaptive mesh are very close to representing the original sequence (Figure 8).

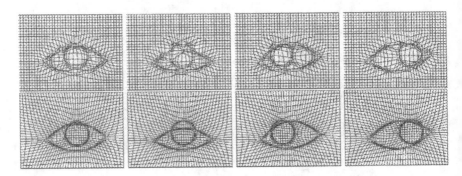

Fig. 9. Eye movement. [top]: coarse adaptation using DM method ($m_i = 1.2, \gamma_i = 1.2, k_1 = 1.0, k_2 = 9.0$, threshold of v_i and a_i are 60.0 at the time of stopping adaptation). [bottom]: fine adaptation using EOM method (the largest nodal force is 0.05 when adaptation is stopped).

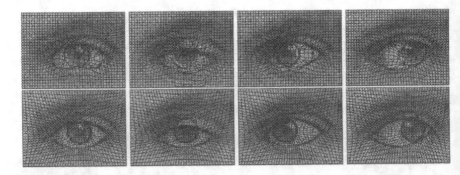

Fig. 10. Overlapped eye mesh: coarse adaptation (top); fine adaptation (bottom).

To apply the extended adaptive mesh method to a real face model, we use a detailed wireframe model with 2954 vertices and 3118 patches, in which there are

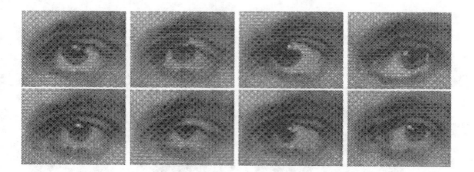

Fig. 11. Synthesized images using the first frame texture: texture-mapped results of coarse adaptation mesh (top) and the fine adaptation mesh (bottom).

120 vertices for each eye. Eye movements are synthesized as follows: First, eye movements are modeled as deformations of the individualized wireframe model (Figure 12) by using our existing face modeling tool [23]. Then, the adapted wireframe models in successive frames are texture-mapped by using the first frame of the sequence. Figure 13 shows a set of synthesized images showing eyes' animation with texture mapped on the facial model.

Observe that the main advantage of our method is the improvement in quality of the animated images compared to ones described in the existing literatures ([16]-[20], [13]). One can clearly see different expressions that a salesman may have depending on his interaction with customers.

Note that even though the DM method is faster the mesh is not as accurate as our EOM improvement. The EOM improvement is essential to create the realistic animation. From the preliminary experimental results, we can see that the algorithms proposed here behaves well for synthesizing eye movements in facial animation. The results are evaluated subjectively (see Figures 8, 11, 12, 13).

Applying the same extended adaptive mesh method, Figure 14 shows lip animation on the same set of faces. The theory and implementation behind tracking and animation of lip movements is similar to eye tracking and animation, however, it is not described in this paper because of limitations on the length of this document.

5 Conclusions and Future Work

In this paper we discussed robust methods for detecting and tracking eye movements, a strategy for eye movement animation, and resulting applications in model-based low bit-rate coding. An energy oriented method is described as an extension to dynamic meshes consisting of a network of springs. The proposed method refines the model adjustment process so as to improve the accuracy of

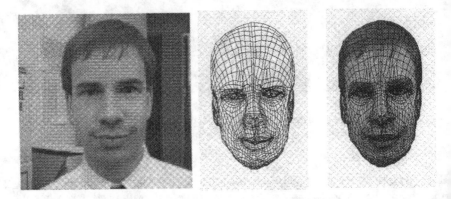

Fig. 12. Original facial image sequence (first frame), the individualized model and the adaptation result (salesman from SGI company).

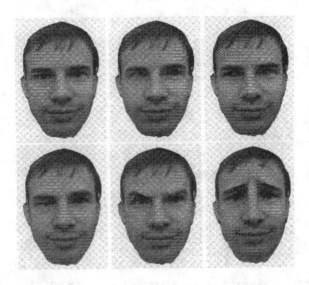

Fig. 13. Synthesized eyes with texture-mapped model (from left to right, top to bottom: frame 1, 36, 81, 109, 125, 160).

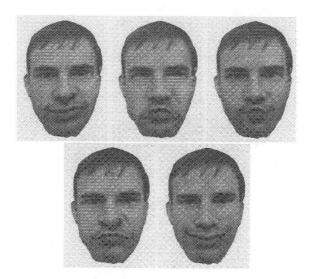

Fig. 14. Synthesized eyes and lips with texture-mapped model (from left to right, top to bottom: frame 200, 253, 304, 375, 410).

model adaptation and tracking. It has also overcome the convergence problem that is commonly encountered in numerical solutions to dynamic-motion equations. Experimental results show that realistic animations with subtle expressions can be achieved by our system. We expect that the algorithms proposed in this paper will contribute towards future modification to MPEG-4(SNHC).

In our current implementation, the maximum force that satisfies the displacement criteria is searched from all the nodes after each move of a node, which takes significant computation time. One alternative is to move a node until the node cannot be moved any further while keeping the neighboring nodes fixed. This change is expected to reduce the search time and improve the efficiency of the EOM adaptation process. Another aspect that should be noted is that the accuracy of the coarse-to-fine adaptation is at the cost of a higher bitrate, we will investigate developing an efficient parameter compression method for non-feature vertices to address this issue.

References

1. S. Bernoegger, L. Yin, A. Basu, and A. Pinz. Eye tracking and animation for MPEG-4 coding. In *Proceedings of ICPR'98, Australia*, pages 1281–1284, August 1998.
2. R. Brunelli and T. Poggio. Face recognition: features versus templates. *IEEE Trans. on PAMI*, 15(10):1042–1052, October 1993.
3. T.C. Chang and T.S. Huang. Facial feature extraction from color images. In *Proceedings of ICPR'94, Jerusalem, Israel*, pages 39–43, October 9-13 1994.

4. C. Choi, K. Aizawa, H. Harashima, and T. Takebe. Analysis and synthesis of facial image sequences in model-based image coding. *IEEE Trans. on Circuit and System for Video Technology*, 4(3):257–275, June 1994.

5. G. Chow and X. Li. Towards a system for automatic facial feature detection. *Pattern Recognition*, 26(12):1739–1755, December 1993.

6. E.R. Davies. A modified hough scheme for general circle location. *Pattern Recognition*, 7:37–43, January 1988.

7. D.J. Dawe. *Matrix and Finite Element Displacement Analysis of Structures.* Clarendon Press, Oxford, UK., 1984.

8. J.Y. Deng and F. Lai. Region-based template deformation and masking for eye-feature extraction and description. *Pattern Recognition*, 30(3):403–419, March 1997.

9. I. A. Essa and A. P. Pentland. Coding, analysis, interpretation, and recognition of facial expressions. *IEEE Trans. on PAMI*, 19(7):757–763, July 1997.

10. B. Guenter, C. Grimm, D. Wood, H. Malvar, and F. Pighin. Making faces. In *SIGGRAPH'98 Proceedings, Orlando,FL.*, pages 55–66, July 1998.

11. W. Huang and D. Goldgof. Adaptive-size meshes for rigid and non-rigid shape analysis and synthesis. *IEEE Trans. on PAMI*, 15(6):611–616, June 1993.

12. P. Kierkegaard. A method for detection of circular arcs based on the hough transform. *Machine Vision and Applications*, 5:249–263, 1992.

13. D.E. Pearson. Developments in model-based video coding. *Proceedings of the IEEE*, 83(6):892–906, June 1995.

14. F. Pighin, J. Hecker, D. Linchinski, R. Szeliski, and D. Salesin. Synthesizing realistic facial expressions from photographs. In *SIGGRAPH'98 Proceedings, Orlando,FL.*, p75-84, July 1998.

15. W. Press, S. Teukolsky, W. Vetterling, and B. Flannery. *Numerical Recipes in C; The Art of Scientific Computing.* Cambridge University Press, 1988.

16. M. Reinders. Model adaptation for image coding. In *Ph.D Thesis, Delft Univ. of Tech.*, 1995.

17. D. Terzopoulos and M. Vasilescu. Sampling and reconstruction with adaptive meshes. In *Proceedings of IEEE CVPR'91*, pages 70–75, 1991.

18. D. Terzopoulos and M. Vasilescu. Adaptive mesh and shells: irregular triangulation, discontinuities and hierarchical subdivision. In *Proceedings of IEEE CVPR'92*, pages 829–832, 1992.

19. D. Terzopoulos and K. Waters. Analysis and synthesis of facial image sequences using physical and anatomical models. *IEEE Trans. on PAMI*, 15(6):569–579, June 1993.

20. MPEG Video&SNHC. Final text for FCD 14496-2:visual. *Doc. ISO/MPEG N2202, Tokyo MPEG Meeting*, March 1998.

21. H. Wu, T. Yokoyama, D. Pramadihanto, and M. Yachida. Face and facial feature extraction from color image. In *Proceedings 2nd IEEE International Conference on Automatic Face and Gesture Recognition, Killington, USA*, p345-350, Oct. 1996.

22. Y. Yacoob and L. S. Davis. Recognizing human facial expressions from long image sequences using optical flow. *IEEE Trans. on PAMI*, 18(6):636–642, June 1996.

23. L. Yin and A. Basu. MPEG4 face modeling using fiducial points. In *Proceedings of IEEE International Conference on Image Processing, Santa Barbara, CA*, pages 109–112, Oct. 1997.

24. A.L. Yuille, P.W. Hallinan, and D.S. Cohen. Feature extraction from faces using deformable templates. *International Journal of Computer Vision*, 8(2):99–111, 1992.

Maximum Likelihood Inference of 3D Structure from Image Sequences

Pedro M. Q. Aguiar* and José M. F. Moura

Carnegie Mellon University
{aguiar,moura}@ece.cmu.edu

Abstract. The paper presents a new approach to recovering the 3D rigid shape of rigid objects from a 2D image sequence. The method has two distinguishing features: it exploits the rigidity of the object over the sequence of images, rather than over a pair of images; and, it estimates the 3D structure directly from the image intensity values, avoiding the common intermediate step of first estimating the motion induced on the image plane. The approach constructs the maximum likelihood (ML) estimate of all the shape and motion unknowns. We do not attempt the minimization of the ML energy function with respect to the entire set of unknown parameters. Rather, we start by computing the 3D motion parameters by using a robust factorization appraoch. Then, we refine the estimate of the object shape along the image sequence, by minimizing the ML-based energy function by a continuation-type method. Experimental results illustrate the performance of the method.

1 Introduction

The recovery of three-dimensional (3D) structure (3D shape and 3D motion) from a two-dimensional (2D) video sequence has been widely considered by the computer vision community. Methods that infer 3D shape from a single frame are based on cues such as shading and defocus. These methods fail to give reliable 3D shape estimates for unconstrained real-world scenes.

If no prior knowledge about the scene is available, the cue to estimating the 3D structure is the 2D motion of the brightness pattern in the image plane. For this reason, the problem is generally referred to as *structure from motion*. The two major steps in *structure from motion* are usually the following: compute the 2D motion in the image plane; and estimate the 3D shape and the 3D motion from the computed 2D motion.

Structure from motion Early approaches to *structure from motion* processed a single pair of consecutive frames and provided existence and uniqueness results to the problem of estimating 3D motion and absolute depth from the 2D motion in the camera plane between two frames, see for example [10]. Two-frame based algorithms are highly sensitive to image noise, and, when the object is far from

---

* The first author is also affiliated with Instituto Superior Técnico, Instituto de Sistemas e Robótica, Lisboa, Portugal. His work was partially supported by INVOTAN.

the camera, i.e., at a large distance when compared to the object depth, they fail even at low level image noise. More recent research has been oriented towards the use of longer image sequences. For example, in [8], the authors use a Kalman filter to integrate along time a set of two-frame depth estimates, while reference [4] uses nonlinear optimization to solve for the rigid 3D motion and the set of 3D positions of feature points tracked along a set of frames. References [9] and [2] estimate the 3D shape and motion by factorizing a measurement matrix whose entries are the set of trajectories of the feature point projections.

The approaches of the references above rely on the matching of a set of features along the image sequence. This task can be very difficult when processing noisy videos. In general, only distinguished points, as brightness corners, can be used as "trackable" feature points. As a consequence, those approaches do not provide dense depth estimates. In [1], we extended the factorization approach of [9] to recover 3D structure from a sequence of optical flow parameters. Instead of tracking pointwise features, we track regions where the optical flow is described by a single set of parameters. The approach of [1] is well suited to the analysis of scenes that can be well approximated with polyhedral surfaces. In this paper we seek dense depth estimates for general shaped surfaces.

To overcome the difficulties in estimating 3D structure through the 2D motion induced onto the image plane, some researchers have used techniques that infer 3D structure directly from the image intensity values. For example [6] estimates directly the 3D structure parameters by using the *brightness change constraint* between two consecutive frames. Reference [5] builds on this work by using a Kalman filter to update the estimates over time.

Proposed approach To formulate the problem of inferring the 3D structure from a video sequence, we use the analogy between the visual perception mechanism and a classical communication system. This analogy has been used to deal with perception tasks involving a single image, such as texture segmentation, and the recovering of shape from texture, see for example [7]. In a communication system, the transmitter receives a message S to be sent to the receiver. The transmitter codes the message and sends the resulting signal I^* through the channel, to the receiver. The receiver gets the signal I, a noisy version of the signal I^*. The receiver decodes I obtaining the estimate $\widehat{S}$ of the message S. In statistical communications theory, we describe statistically the channel distortion and design the receiver according to a statistically optimal criteria. For example, we can estimate $\widehat{S}$ as the message S that maximizes the probability of receiving the signal I, conditioned on the message S sent. This is the *Maximum Likelihood* (ML) estimate.

The communication system is a good metaphor for the problem of recovering 3D structure from video. The message source is the 3D environment. The transmitter is the geometric projection mechanism that transforms the real world S into an ideal image I^*. The channel is the camera that captures the image I, a noisy version of I^*. The receiver is the video analysis system. The task of this system is to recover the real world that has originated the image sequence captured.

According to the analogy above, we recover the 3D structure from the video sequence by computing the ML estimate of all the unknowns: the parameters describing the 3D motion, the object shape, and the object texture. A distinguishing feature of our work is the formulation of the estimate from a set of images, rather than a single pair. This provides accurate estimates for the 3D structure, due to the 3D rigidity of the scene. The formulation of the ML estimate from a set of frames leads to the minimization of a complex energy function. To minimize the ML energy function, we solve for the object texture in terms of the 3D shape and the 3D motion parameters. By replacing the texture estimate, we are left with the minimization of the ML energy function with respect to the 3D shape and 3D motion. We do not attempt the minimization of the ML energy function with respect to the entire set of unknown parameters by using generic optimization methods. Rather, we exploit the specific characteristics of the problem to develop a computationally feasible approximation to the ML solution. We compute the 3D motion by using the factorization method detailed in [2]. In fact, experiments with real videos show that the 3D rigid motion can be computed with accuracy through the optical flow computed across a set of frames for a small number of distinguished points or regions. After estimating the 3D motion, we are left with the minimization of the ML energy function with respect to the 3D shape. We propose a computationally simple continuation method to solve this non-linear minimization. Our algorithm starts by estimating coarse approximations to the 3D shape. Then, it refines the estimate as more images are being taken into account. The computational simplicity of our algorithm comes from the fact that each refinement stage, although non-linear, is solved by a simple Gauss-Newton method that requires no more than one or two iterations.

Our approach provides an efficient way to cope with the ill-posedness of estimating the motion in the image plane. In fact, the local *brightness change constraint* leads to a single restriction, which is insufficient to determine the two components of the local image motion (the so called *aperture problem*). Our method of estimating directly the 3D shape overcomes the *aperture problem* because we are left with the local depth as a single unknown, after computing the 3D motion in a first step.

In this paper we model the image formation process by assuming orthogonal projections. Orthogonal projections have been used as a good approximation to the perspective projection when the object is far from the camera [9, 1, 2]. With this type of scenes, two-frame based methods fail to estimate the absolute depth. Although formulated assuming orthogonal projections, which leads to estimates of the relative depth, our method can be easily extended to cope with perspective projections, which then leads to estimates of the absolute depth.

Paper organization Section 2 formulates the problem. Section 3 discusses the ML estimate. Section 4 summarizes the factorization method used to estimate the 3D motion. Section 5 describes the continuation method used to minimize the ML energy function. Experiments are in section 6. Section 7 concludes the paper.

2 Problem Formulation

We consider a rigid object $\mathcal{O}$ moving in front of a camera. We define the 3D motion of the object by specifying the position of the object coordinate system relative to the camera coordinate system. The position and orientation of $\mathcal{O}$ at time instant f is represented by $m_f = \{t_{uf}, t_{vf}, t_{wf}, \theta_f, \phi_f, \psi_f\}$ where (t_{uf}, t_{vf}, t_{wf}) are the coordinates of the origin of the object coordinate system with respect to the camera coordinate system (3D translation), and $(\theta_f, \phi_f, \psi_f)$ are the Euler angles that determine the orientation of the object coordinate system relative to the camera coordinate system (3D rotation).

Observation model The frame I_f captured at time f, $1 \leq f \leq F$, is modeled as a noisy observation of the projection of the object

$$I_f = \mathcal{P}(\mathcal{O}, m_f) + W_f. \tag{1}$$

We assume that $\mathcal{P}$ is the orthogonal projection operator. For simplicity, the observation noise W_f is zero mean, white, and Gaussian.

The object $\mathcal{O}$ is described by its 3D shape $\mathcal{S}$ and texture $\mathcal{T}$. The texture $\mathcal{T}$ represents the light received by the camera after reflecting on the object surface, i.e., the texture $\mathcal{T}$ is the object brightness as perceived by the camera. The texture depends on the object surface photometric properties, as well as on the environment illumination conditions. We assume that the texture does not change with time.

The operator $\mathcal{P}$ returns the texture $\mathcal{T}$ as a real valued function defined over the image plane. This function is a nonlinear mapping that depends on the object shape $\mathcal{S}$ and the object position m_f. The intensity level of the projection of the object at pixel u on the image plane is

$$\mathcal{P}(\mathcal{O}, m_f)(u) = \mathcal{T}(s_f(\mathcal{S}, m_f; u)), \tag{2}$$

where $s_f(\mathcal{S}, m_f; u)$ is the nonlinear mapping that lifts the point u on the image I_f to the corresponding point on the 3D object surface. This mapping $s_f(\mathcal{S}, m_f; u)$ is determined by the object shape $\mathcal{S}$, and the position m_f. To simplify the notation, we will usually write explicitly only the dependence on f, i.e., $s_f(u)$. Figure 1 illustrates the lifting mapping $s_f(u)$ and the direct mapping $u_f(s)$ for the orthogonal projection of a two-dimensional object. The inverse mapping $u_f(s)$ also depends on $\mathcal{S}$ and m_f, but we will, again, usually show only explicitly the dependence on f. On the left of figure 1, the point s on the surface of the object projects onto $u_f(s)$ on the image plane. On the right, pixel u on the image plane is lifted to $s_f(u)$ on the object surface. We assume that the object does not occlude itself, i.e., we have $u_f(s_f(u)) = u$ and $s_f(u_f(s)) = s$. The mapping $u_f(s)$, seen as a function of the frame index f, for a particular surface point s, is the trajectory of the projection of that point in the image plane, i.e., it is the motion induced in the image plane, usually referred to as *optical flow*.

The observation model (1) is rewritten by using (2) as

$$I_f(u) = \mathcal{T}(s_f(u)) + W_f(u). \tag{3}$$

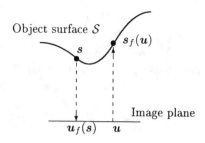

Fig. 1. Mappings $u_f(s)$ and $s_f(u)$.

We consider the estimation of the 3D shape $\mathcal{S}$ and the 3D motion $\{m_f, 1 \leq f \leq F\}$ of the object $\mathcal{O}$ given the video sequence $\{I_f, 1 \leq f \leq F\}$ of F frames.

Maximum Likelihood estimate formulation Given the observation model, the 3D shape and the 3D motion of the object $\mathcal{O}$ are recovered from the video sequence $\{I_f, 1 \leq f \leq F\}$ by estimating all the unknowns: the 3D shape $\mathcal{S}$; the texture $\mathcal{T}$; and the set of 3D positions of the object $\{m_f, 1 \leq f \leq F\}$ with respect to the camera. We formulate the ML solution. When the noise sequence $\{W_f(u)\}$ is zero mean, spatially and temporally white, and Gaussian, the ML estimate minimizes the sum over all the frames of the integral over the image plane of the squared errors between the observations and the model[1],

$$C_{\mathrm{ML}}\left(\mathcal{S}, \mathcal{T}, \{m_f\}\right) = \sum_{f=1}^{F} \int \left[I_f(u) - \mathcal{T}\left(s_f(u)\right)\right]^2 \, du, \qquad (4)$$

$$\left\{\hat{\mathcal{S}}, \hat{\mathcal{T}}, \{\hat{m}_f\}\right\} = \arg\min_{\mathcal{S}, \mathcal{T}, \{m_f\}} C_{\mathrm{ML}}\left(\mathcal{S}, \mathcal{T}, \{m_f\}\right). \qquad (5)$$

In (4), we make explicit the dependence of the cost function C_{ML} on the object texture $\mathcal{T}$. Note that C_{ML} depends on the object shape $\mathcal{S}$ and the object positions $\{m_f\}$ through the mappings $\{s_f(u)\}$.

3 Maximum Likelihood Estimation

We address the minimization of $C_{\mathrm{ML}}\left(\mathcal{S}, \mathcal{T}, \{m_f\}\right)$ by first solving for the texture estimate $\hat{\mathcal{T}}$ in terms of the 3D object shape $\mathcal{S}$ and the object positions $\{m_f\}$. **Texture estimate** We rewrite the cost function C_{ML} given by (4) by changing the integration variable from the image plane coordinate u to the object surface coordinate s. We obtain

$$C_{\mathrm{ML}}\left(\mathcal{S}, \mathcal{T}, \{m_f\}\right) = \sum_{f=1}^{F} \int \left[I_f(u_f(s)) - \mathcal{T}(s)\right]^2 J_f(s) \, ds, \qquad (6)$$

---

[1] We use a continuous spatial dependence for commodity. The variables u and s are continuous while f is discrete.

where $u_f(s)$ is the mapping that projects the point s on the object surface onto the image plane at instant f, see figure 1. The function $J_f(s)$ is the Jacobian of the mapping $u_f(s)$, $J_f(s) = |\nabla u_f(s)|$.

Expression (6) shows that the cost function C_{ML} is quadratic in each intensity value $\mathcal{T}(s)$ of the object texture. The ML estimate $\widehat{\mathcal{T}}(s)$ is

$$\widehat{\mathcal{T}}(s) = \frac{\sum_{f=1}^{F} I_f(u_f(s)) J_f(s)}{\sum_{f=1}^{F} J_f(s)} \tag{7}$$

(see appendix A for the proof). Expression (7) states that the estimate of the texture of the object at the surface point s is a weighted average of the measures of the intensity level corresponding to that surface point. A given region around s on the object surface projects at frame I_f to a region around $u_f(s)$. The size of this projected region changes with time because of the object motion. The more parallel to the image plane is the tangent to the object surface at point s, the larger is the size of the projected region. Expression (7) shows that the larger the Jacobian $J_f(s)$ is, i.e., the larger the region around s is magnified at frame I_f, the larger is the weight given to that frame when estimating the texture $\mathcal{T}(s)$.

Structure from motion as an approximation to ML By inserting the texture estimate $\widehat{\mathcal{T}}$ given by (7) in (6), we can express the cost function C_{ML} in terms of the mappings $\{u_f(s)\}$. After manipulations (see appendix B) we get

$$C_{\mathrm{ML}}(\mathcal{S}, \{m_f\}) = \sum_{f=2}^{F} \sum_{g=1}^{f-1} \int [I_f(u_f(s)) - I_g(u_g(s))]^2 \frac{J_f(s) J_g(s)}{\sum_{h=1}^{F} J_h(s)} \, ds. \tag{8}$$

The cost function C_{ML} in (8) is a weighted sum of the squared differences between all pairs of frames. At each surface point s, the frame pair $\{I_f, I_g\}$ is weighted by $\frac{J_f(s) J_g(s)}{\sum_{h=1}^{F} J_h(s)}$. The larger this weight is, i.e., the larger a region around s is magnified in frames I_f and I_g, the more the square difference between I_f and I_g affects C_{ML}.

Expression (8) also makes clear why the problem we are addressing is referred to as *structure from motion*: having eliminated the dependence on the texture, we are left with a cost function that depends on the *structure* (3D shape $\mathcal{S}$ and 3D motion $\{m_f\}$) only through the *motion* induced in the image plane, i.e., through the mappings $\{u_f(s)\}$. Recall the comment on section 2 that $u_f(\mathcal{S}, m_f; s)$ depends on the shape $\mathcal{S}$ and the motion m_f. The usual approach to the minimization of the functional (8) is in two steps. The first step estimates the motion in the image plane $u_f(s)$ by minimizing an approximation of (8) (in general, only two frames are taken into account). The second step estimates the shape $\mathcal{S}$ and the motion m_f from $\{m_f\}$. Since the motion in the image plane can not be reliably computed in the entire image, these methods cannot provide a reliable dense shape estimate.

Our approach combines the good performance of the factorization method in estimating the 3D motion with the robustness of minimizing the ML energy function with respect to the object shape.

4   Rank 1 Factorization

This section summarizes the factorization method used to estimate the 3D motion. For a detailed description, see [2]. The factorization approach is robust due to the modelization of the rigidity of the moving object along time. This method is also computationally simple because it uses a fast algorithm to factorize a measurement matrix that is rank 1 in a noiseless situation.

A set of N feature points are tracked along an image sequence of F frames. Under orthography, the projection of feature n in frame f, $[u_{fn}, v_{fn}]^T$, is

$$\begin{bmatrix} u_{fn} \\ v_{fn} \end{bmatrix} = \begin{bmatrix} i_{xf} & i_{yf} & i_{zf} \\ j_{xf} & j_{yf} & j_{zf} \end{bmatrix} \begin{bmatrix} x_n \\ y_n \\ z_n \end{bmatrix} + \begin{bmatrix} t_{uf} \\ t_{vf} \end{bmatrix} \tag{9}$$

where $i_{xf}, i_{yf}, i_{zf}, j_{xf}, j_{yf}$, and j_{zf} are entries of the well known 3D rotation matrix, uniquely determined by the Euler angles θ_f, ϕ_f, and ψ_f, see [3], and t_{uf} and t_{vf} are the components of the object translation along the camera plane. We make the object coordinate system and camera coordinate system coincide in the first frame, so we have $u_{1n} = x_n$ and $v_{1n} = y_n$. Thus, the coordinates of the feature points along the camera plane $\{x_n, y_n\}$ are given by their projections in the first frame. The goal of the factorization method is to solve the overconstrained equation system (9) with respect to the following set of unknowns: the 3D positions of the object for $2 \leq f \leq F$, and the relative depths $\{z_n, 1 \leq n \leq N\}$.

By choosing the origin of the object coordinate system to coincide with the centroid of the set of feature points, we get the estimate for the translation as the centroid of the feature point projections. Replacing the translation estimates in the system of equations (9), and defining

$$\begin{bmatrix} \tilde{u}_{fn} \\ \tilde{v}_{fn} \end{bmatrix} = \begin{bmatrix} u_{fn} \\ v_{fn} \end{bmatrix} - \frac{1}{N} \sum_{m=1}^{N} \begin{bmatrix} u_{fm} \\ v_{fm} \end{bmatrix}, \qquad R = \begin{bmatrix} \tilde{u}_{21} & \cdots & \tilde{u}_{2N} \\ \cdots & \cdots & \cdots \\ \tilde{u}_{F1} & \cdots & \tilde{u}_{FN} \\ \tilde{v}_{21} & \cdots & \tilde{v}_{2N} \\ \cdots & \cdots & \cdots \\ \tilde{v}_{F1} & \cdots & \tilde{v}_{FN} \end{bmatrix}, \tag{10}$$

$$M = \begin{bmatrix} i_{x2} \cdots i_{xF} & j_{x2} \cdots j_{xF} \\ i_{y2} \cdots i_{yF} & j_{y2} \cdots j_{yF} \\ i_{z2} \cdots i_{zF} & j_{z2} \cdots j_{zF} \end{bmatrix}^T, \qquad \text{and} \qquad S^T = \begin{bmatrix} x_1 & x_2 & \cdots & x_N \\ y_1 & y_2 & \cdots & y_N \\ z_1 & z_2 & \cdots & z_N \end{bmatrix}, \tag{11}$$

we rewrite (9) in matrix format as

$$R = MS^T. \tag{12}$$

Matrix R is $2(F-1) \times N$ but it is rank deficient. In a noiseless situation, R is rank 3 reflecting the high redundancy in the data, due to the 3D rigidity of the object.

The factorization approach finds a suboptimal solution to the bilinear LS problem of equation (12) where the solution space is constrained by the orthonormality of the rows of the matrix M (11). This nonlinear minimization is solved in two stages. The first stage, *decomposition stage*, solves the unconstrained bilinear problem $R = MS^T$. The second stage, *normalization stage*, computes a set of normalizing parameters by approximating the constraints imposed by the structure of the matrix M.

Decomposition stage Define $M = [M_0, m_3]$ and $S = [S_0, z]$. M_0 and S_0 contain the first two columns of M and S, respectively, m_3 is the third column of M, and z is the third column of S. We decompose the relative depth vector z into the component that belongs to the space spanned by the columns of S_0 and the component orthogonal to this space as $z = S_0 b + a$, with $a^T S_0 = [0\ 0]$. We rewrite R in (12) as $R = M_0 S_0^T + m_3 b^T S_0^T + m_3 a^T$.

The decomposition stage is formulated as

$$\min_{M_0, m_3, b, a} \left\| R - M_0 S_0^T - m_3 b^T S_0^T - m_3 a^T \right\|_F \tag{13}$$

where $\|.\|_F$ denotes the Frobenius norm. The solution for M_0 is given by $\widehat{M}_0 = R S_0 \left(S_0^T S_0 \right)^{-1} - m_3 b^T$. By replacing $\widehat{M}_0$ in (13), we get

$$\min_{m_3, a} \left\| \tilde{R} - m_3 a^T \right\|_F, \quad \text{where} \quad \tilde{R} = R \left[I - S_0 \left(S_0^T S_0 \right)^{-1} S_0^T \right]. \tag{14}$$

We see that the decomposition stage does not determine the vector b. This is because the component of z that lives in the space spanned by the columns of S_0 does not affect the space spanned by the columns of the entire matrix S and the decomposition stage restricts only this last space.

The solution for m_3 and a is given by the rank 1 matrix that best approximates $\tilde{R}$. In a noiseless situation, $\tilde{R}$ is rank 1, see [2] for the details. By computing the largest singular value of $\tilde{R}$ and the associated singular vectors, we get

$$\tilde{R} \simeq u\sigma v^T, \quad \widehat{m}_3 = \alpha u, \quad \hat{a}^T = \frac{\sigma}{\alpha} v^T \tag{15}$$

where α is a normalizing scalar different from 0. To compute u, σ, and v we could perform an SVD, but the rank deficiency of $\tilde{R}$ enables the use of less expensive algorithms to compute u, σ, and v, as detailed in [2].

Normalization stage In this stage, we compute α and b by imposing the constraints that come from the structure of M. We express $\widehat{M}$ in terms of α and b as

$$\widehat{M} = \left[\widehat{M}_0\ \widehat{m}_3 \right] = N \begin{bmatrix} I_{2\times2} & 0_{2\times1} \\ -\alpha b^T & \alpha \end{bmatrix}, \quad N = \left[R S_0 \left(S_0^T S_0 \right)^{-1} u \right]. \tag{16}$$

The constraints imposed by the structure of M are the unit norm of each row and the orthogonality between row j and row $j + F - 1$, where F is the number of frames in the sequence. In terms of N, α, and b, the constraints are

$$n_i^T \begin{bmatrix} I_{2\times 2} & -\alpha b \\ -\alpha b^T & \alpha^2(1 + b^T b) \end{bmatrix} n_i = 1, \quad n_j^T \begin{bmatrix} I_{2\times 2} & -\alpha b \\ -\alpha b^T & \alpha^2(1 + b^T b) \end{bmatrix} n_{j+F-1} = 0 \quad (17)$$

where n_i^T denotes the row i of the matrix N. We compute α and b from the linear LS solution of the system above in a similar way to the one described in [9].

5 Minimization Procedure

After recovering the 3D motion m_f as described is section 4, we insert the 3D motion estimates into the energy function (8) and minimize with respect to the unknown shape S.

We first make explicit the relation between the image trajectories $u_f(s)$ and the 3D shape S and the 3D motion m_f. Choose the coordinate s of the generic point in the object surface to coincide with the coordinates $[x, y]^T$ of the object coordinate system. Under orthography, a point with coordinate s in the object surface is projected on coordinate $u = [x, y]^T = s$ in the first frame, so that $u_1(s) = s$ (remember that we have chosen the object coordinate system so that it coincides with the camera coordinate system in the first frame). At instant f, that point is projected to

$$u_f(s) = u_f\left(\begin{bmatrix} x \\ y \end{bmatrix}\right) = \begin{bmatrix} i_{xf} & i_{yf} & i_{zf} \\ j_{xf} & j_{yf} & j_{zf} \end{bmatrix} \begin{bmatrix} x \\ y \\ z \end{bmatrix} + \begin{bmatrix} t_{uf} \\ t_{vf} \end{bmatrix} = \begin{bmatrix} N_f & n_f \end{bmatrix} \begin{bmatrix} s \\ z \end{bmatrix} + t_f$$

$$= N_f s + n_f z + t_f, \quad (18)$$

where $i_{xf}, i_{yf}, i_{zf}, j_{xf}, j_{yf}$, and j_{zf} are entries of the 3D rotation matrix [3]. The 3D shape is represented by the unknown relative depth z.

Modified image sequence for known motion The 3D shape and the 3D motion are observed in a coupled way through the 2D motion on the image plane, see expression (18). When the 3D motion is known, the problem of inferring the 3D shape from the image sequence is simplified. In fact, the local *brightness change constraint* leads to a single restriction, which is insufficient to determine the two components of the local image motion (this is the so called *aperture problem*). Our method of estimating directly the 3D shape overcomes the *aperture problem* because we are left with the local depth as a single unknown, after computing the 3D motion in the first step. To better illustrate why the problem becomes much simpler when the 3D motion is known, we introduce a modified image sequence $\{\tilde{I}_f, 1 \leq f \leq F\}$, obtained from the original sequence $\{I_f, 1 \leq f \leq F\}$ and the 3D motion. We show that the 2D motion of the brightness pattern on image sequence $\tilde{I}_f$ depends on the 3D shape in a very particular way. This motivates the algorithm we use to minimize (8).

Consider the image $\tilde{I}_f$ related to I_f by the following affine mapping that depends only on the 3D position at instant f,

$$\tilde{I}_f(s) = I_f(N_f s + t_f). \quad (19)$$

From this definition it follows that a point s, that projects to $u_f(s)$ in image I_f, is mapped to $\widetilde{u}_f(s) = N_f^{-1}[u_f(s) - t_f]$ in image $\widetilde{I}_f$. Replacing $u_f(s)$ by expression (18), we obtain for the image motion of the modified sequence $\{\widetilde{I}_f\}$,

$$\widetilde{u}_f(s) = s + N_f^{-1} n_f z. \tag{20}$$

Expression (20) shows that the trajectory of a point s in image sequence $\{\widetilde{I}_f\}$ depends on the relative depth of that point in a very particular way. In fact, the trajectory has the same shape for every point. The shape of the trajectories is given by the evolution of $N_f^{-1} n_f$ across the frame index f. Thus, the shape of the trajectories depends uniquely on the rotational component of the 3D motion. The relative depth z affects only the magnitude of the trajectory. A point with relative depth $z = 0$ is stationary in $\{\widetilde{I}_f\}$, since we get $\widetilde{u}_f(s) = s$ from (20) for arbitrary 3D motion of the object.

Continuation method By minimizing (8) with respect to the relative depth of each point s, we are in fact estimating the magnitude of the trajectory of the point to where the point s maps in image sequence $\{\widetilde{I}_f\}$. The shape of the trajectory is known, since it depends only on the 3D motion. Our algorithm is based on this characteristic of the ML energy function. We use a continuation-type method to estimate the relative depth of each point. The algorithm refines the estimate of the relative depth as more frames are being taken into account. When only a few frames are taken into account, the magnitude of the trajectories on image sequence $\{\widetilde{I}_f\}$ can be only roughly estimated because the length of the trajectories is short and their shape may be quite simple. When enough frames are considered, the trajectories on image sequence $\{\widetilde{I}_f\}$ are long enough, their magnitude is unambiguous, and the relative depth estimates are accurate. Our algorithm does not compute $\{\widetilde{I}_f\}$, it rather uses the corresponding intensity values of $\{I_f\}$.

The advantage of the continuation-type method is that is provides a computationally simple way to estimate the relative depth because each stage of the algorithm updates the estimate by using a Gauss-Newton method, i.e., by solving a linear problem. We consider the relative depth z to be constant in a region $\mathcal{R}$. We estimate z by minimizing the energy resultant from neglecting the weighting factor $\frac{J_f(s)J_g(s)}{\sum_{h=1}^{F} J_h(s)}$ in the ML energy function (8). Thus, we get

$$\hat{z} = \arg\min_z E(z), \qquad E(z) = \sum_{f=2}^{F} \sum_{g=1}^{f-1} \int_{\mathcal{R}} e^2(z)\, ds, \tag{21}$$

where

$$e(z) = I_f(N_f s + n_f z + t_f) - I_g(N_g s + n_g z + t_g). \tag{22}$$

We compute $\hat{z}$ by refining a previous estimate z_0, as

$$\hat{z} = z_0 + \widehat{\delta_z}, \qquad \widehat{\delta_z} = \arg\min_{\delta_z} E(z_0 + \delta_z). \tag{23}$$

The Gauss-Newton method neglects the second and higher order terms of the Taylor series expansion of $e(z_0 + \delta_z)$. By making this approximation, we get

$$\widehat{\delta_z} = -\frac{\sum_{f=2}^{F} \sum_{g=1}^{f-1} \int_{\mathcal{R}} e(z_0)e'(z_0)}{\sum_{f=2}^{F} \sum_{g=1}^{f-1} \int_{\mathcal{R}} [e'(z_0)]^2}, \tag{24}$$

where e' is the derivative of e with respect to z. By differentiating (22), we get

$$
\begin{aligned}
e'(z) = \;& \boldsymbol{I}_{f_x}(\boldsymbol{N}_f \boldsymbol{s} + \boldsymbol{n}_f z + \boldsymbol{t}_f)i_{z_f} + \boldsymbol{I}_{f_y}(\boldsymbol{N}_f \boldsymbol{s} + \boldsymbol{n}_f z + \boldsymbol{t}_f)j_{z_f} \\
& -\boldsymbol{I}_{g_x}(\boldsymbol{N}_g \boldsymbol{s} + \boldsymbol{n}_g z + \boldsymbol{t}_g)i_{z_g} - \boldsymbol{I}_{g_y}(\boldsymbol{N}_g \boldsymbol{s} + \boldsymbol{n}_g z + \boldsymbol{t}_g)j_{z_g},
\end{aligned} \tag{25}
$$

where $\boldsymbol{I}_{f_x}$ and $\boldsymbol{I}_{f_y}$ denote the components of the spatial gradient of image $\boldsymbol{I}_f$.

At the beginning, we start with the initial guess $z_0 = 0$ for any region $\mathcal{R}$. We use square regions where z is estimated as being constant. The size of the regions determines the resolution of the relative depth estimate. We use large regions when processing the first frames and decrease the size of regions as the continuation method takes more frames into account.

6 Experimens

We describe two experiments that illustrate our approach. The first experiment uses a synthetic sequence for which we compare the estimates obtained with the ground truth. The second experiment uses a real video sequence.

Synthetic sequence We consider that the world is 2D and that the images are 1D orthogonal projections of the world. This scenario reflects all the basic properties and difficulties of the *structure from motion* paradigm and corresponds to the real 3D world if we consider only one epipolar plane and assume that the motion occurs on that plane. In figure 2 we show a computer generated sequence of 25 1D images. Time increases from top to bottom. The time evolution of the translational and rotational components of the motion are shown respectively in the left and middle plots of figure 3. The object shape is shown on the right plot of figure 3. The object texture is an intensity function defined over the object contour. We obtained the image sequence in figure 2 by projecting the object texture on the image plane and by adding noise.

In figure 4 we represent the modified image sequence, computed from the original sequence in figure 2, as described in section 5 for the 3D scenario, see expression (19). The motion of the brightness pattern in figure 4 is simpler than the motion in figure 2. In fact, the horizontal positions of the brightness patterns in figure 4 have a time evolution that is equal for the entire image (see, from figure 4 and the left plot of figure 3 that the shape of the trajectories of the brightness patterns is related to the rotational component of the motion). Only the amplitude of the time evolution of the horizontal positions of the brightness patterns in figure 4 is different from an object region to another object region. The amplitude for a given region is proportional to the relative depth of that region. Note that the brightness pattern is almost stationary for regions with

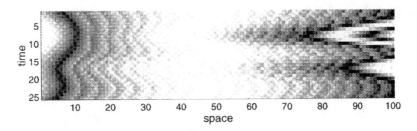

Fig. 2. Sequence of 25 1D images: each horizontal slice is one image.

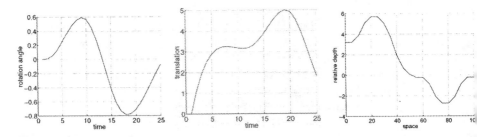

Fig. 3. True motion and true shape. Left: rotational motion; middle: translational motion; right: object shape.

relative depth close to zero (see the regions around pixels 55 and 95 on the right plot of figure 3 and on figure 4). This agrees with the discussion in section 5.

We estimated the relative depth of the object by using the continuation method introduced in section 5. The evolution of the relative depth estimate is represented in the plots of figure 5 for several time instants. The size of the estimation region $\mathcal{R}$ was 10 pixels when processing the first 5 frames, 5 pixels when processing frames 6 to 10, and 3 pixels when processing frames 11 to 25. The true depth shape is shown by the dashed line in the bottom right plot of figure 5. The top left plot was obtained with the first three frames and shows a very coarse estimate of the shape. The bottom right plot was obtained after all 25 frames of the image sequence have been processed. In this plot we made

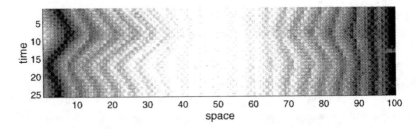

Fig. 4. Modified image sequence for known motion.

a linear interpolation between the central points of consecutive estimation regions. This plot superposes the true and the estimated depths showing a very good agreement between them. The intermediate plots show progressively better estimates of the depth shape.

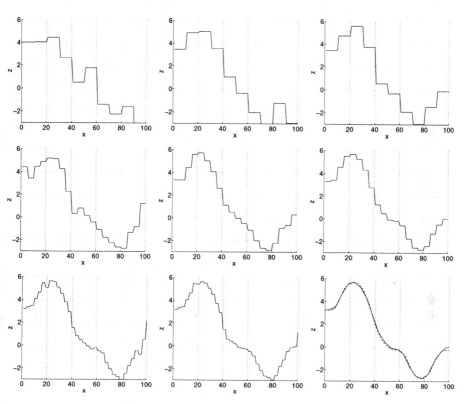

Fig. 5. Continuation method: evolution of the shape estimate. Left to right, top to bottom, after processing F frames where F is successively: $3, 4, 5, 6, 8, 10, 15, 20, 25$. The true shape is shown as the dashed line in the bottom right plot.

Real video We used a sequence of 10 frames from a real video sequence showing a toy clown. Figure 6 shows frames 1 and 5. Each frame has 384×288 pixels. Superimposed on frame 1, we marked with white squares 20 features used in the factorization method. The method used to select the features is reported elsewhere. We tracked the feature points by matching the intensity pattern of each feature along the sequence. Using the factorization approach summarized in section 4 we recovered the 3D motion from the feature trajectories.

We estimated the relative depth of the 3D object by using the continuation method described in section 5. The evolution of the estimate of the relative depth is illustrated by figure 7. The grey level images in this figure code the relative depth estimates. The brighter a pixel is, the closer to the camera it is in the first

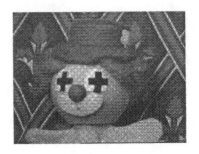

Fig. 6. Clown sequence: frames 1 and 5.

frame. The size of the estimation region $\mathcal{R}$ was 30×30 pixels when processing the first 3 frames, 20×20 pixels when processing frames 4 to 6, and 10×10 pixels when processing frames 7 to 10. The left image was obtained with the first three frames and shows a very coarse estimate of the shape. The right image was obtained after all 10 frames of the image sequence have been processed.

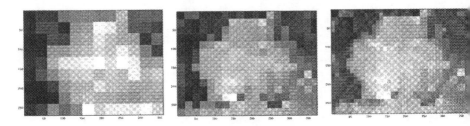

Fig. 7. Relative depth estimate after processing 3, 6, and 10 frames.

7 Conclusion

Final remarks We presented a new approach to the recovery of 3D rigid structure from a 2D video sequence. The problem is formulated as the ML estimation of all the unknowns directly from the intensity values of the set of images in the sequence. We estimate the 3D motion by using a factorization method. We develop a continuation-type algorithm to minimize the ML energy function with respect to the object shape. The experimental results obtained so far are promising and illustrate the good performance of the algorithm.

Future extensions A number of possible extensions of this work are foreseen. First, our methodology can be modified to achieve the estimation of the absolute depth by using the perspective projection model. Other possible extensions include the investigation of different shape models. In this paper we use a dense depth map. Parametric models enable more compact and robust shape representation. The formulation of the problem directly from the image intensity values provides a robust way of dealing with such issues as segmentation (for piece-wise models) and model complexity, by using classic tools such as the Bayesian inference or information-theoretic criteria.

A Texture Estimation

To prove that the ML estimate $\widehat{\mathcal{T}}(s)$ of the texture $\mathcal{T}(s)$ is given by expression (7), we show that it leads to the minimum of the cost function C_{ML}, given by expression (6), over all texture functions $\mathcal{T}(s)$. Consider the candidate $\mathcal{T}(s) = \widehat{\mathcal{T}}(s) + \mathcal{U}(s)$. The functional C_{ML} for texture function $\mathcal{T}(s)$ is

$$
\begin{aligned}
C_{\mathrm{ML}}(\mathcal{T}) = \sum_{f=1}^{F} \int \left[\boldsymbol{I}_f(\boldsymbol{u}_f(s)) - \widehat{\mathcal{T}}(s) - \mathcal{U}(s) \right]^2 J_f(s)\, ds \\
= \sum_{f=1}^{F} \int \left[\boldsymbol{I}_f(\boldsymbol{u}_f(s)) - \widehat{\mathcal{T}}(s) \right]^2 J_f(s)\, ds + \sum_{f=1}^{F} \int \mathcal{U}^2(s) J_f(s)\, ds \\
-2 \sum_{f=1}^{F} \int \left[\boldsymbol{I}_f(\boldsymbol{u}_f(s)) - \widehat{\mathcal{T}}(s) \right] \mathcal{U}(s) J_f(s)\, ds.
\end{aligned}
\tag{26}
$$

The first term of the expression above is $C_{\mathrm{ML}}(\widehat{\mathcal{T}})$. The third term is 0, as comes immediately by replacing $\widehat{\mathcal{T}}(s)$ by expression (7). We have

$$
C_{\mathrm{ML}}(\mathcal{T}) = C_{\mathrm{ML}}(\widehat{\mathcal{T}}) + \sum_{f=1}^{F} \int \mathcal{U}^2(s) J_f(s)\, ds \geq C_{\mathrm{ML}}(\widehat{\mathcal{T}}),
\tag{27}
$$

which concludes the proof. The inequality comes from the fact that we can always choose the texture coordinates s in such a way that the mappings $\boldsymbol{u}_f(s)$ are such that the determinants $J_f(s) = |\nabla \boldsymbol{u}_f(s)|$ are positive. For example, make the texture coordinate s equal to the image plane coordinate $\boldsymbol{u}$ in the first frame $\boldsymbol{I}_1$. The mapping $\boldsymbol{u}_1(s)$ is the identity mapping $\boldsymbol{u}_1(s) = s$ and we have a positive Jacobian $J_1(s) = 1$. Now, draw an oriented closed contour on the surface $\mathcal{S}$, in the neighborhood of s, and containing s in its interior. This contour, which we call $\mathcal{C}_s$ is projected in image $\boldsymbol{I}_1$ in an oriented closed planar contour $\mathcal{C}_{\boldsymbol{u}1}$. It is geometrically evident that the same contour $\mathcal{C}_s$ projects in image $\boldsymbol{I}_f$, in a contour $\mathcal{C}_{\boldsymbol{u}f}$ that has, in general, different shape but the same orientation that the contour $\mathcal{C}_{\boldsymbol{u}1}$ (remember that we are assuming the object does not occlude itself). For this reason, the Jacobian $J_f(s)$ of the function that maps from s to $\boldsymbol{u}_f(s)$ for $2 \leq f \leq F$, has the same signal as the Jacobian $J_1(s)$ of the function that maps from s to $\boldsymbol{u}_1(s)$, so we get $J_f(s) > 0$ for $1 \leq f \leq F$.

B C_{ML} in terms of $\{\boldsymbol{u}_f(s)\}$

We show that the ML-based cost function is expressed in terms of the motion in the image plane as in expression (8). Replace the texture estimate $\widehat{\mathcal{T}}(s)$, given by (7), into the ML-based cost function, given by (6). After simple algebraic manipulations, we get

$$C_{\text{ML}} = \int \sum_{f=1}^{F} \left[\frac{\sum_{g=1}^{F} [I_f(u_f(s)) - I_g(u_g(s))] J_g(s)}{\sum_{h=1}^{F} J_h(s)} \right]^2 J_f(s) \, ds. \quad (28)$$

Expressing the square above in terms of a sum of products and carrying out the products, after algebraic manipulations, we get

$$C_{\text{ML}} = \int \frac{\sum_{f=1}^{F} \sum_{g=1}^{F} \sum_{h=1}^{F} [I_f^2(s) - I_f(s)I_g(s)] J_f(s)J_g(s)J_h(s)}{\left[\sum_{h=1}^{F} J_h(s) \right]^2} \, ds. \quad (29)$$

Now we divide by $\sum_{h=1}^{F} J_h(s)$ both the numerator and the denominator of the integrand function. By using the equality

$$\sum_{f=1}^{F} \sum_{g=1}^{F} [I_f^2(s) - I_f(s)I_g(s)] J_f(s)J_g(s) = \sum_{f=2}^{F} \sum_{g=1}^{f-1} [I_f(s) - I_g(s)]^2 J_f(s)J_g(s), \quad (30)$$

we get

$$C_{\text{ML}} = \int \frac{\sum_{f=2}^{F} \sum_{g=1}^{f-1} [I_f(s) - I_g(s)]^2 J_f(s)J_g(s)}{\sum_{h=1}^{F} J_h(s)} \, ds, \quad (31)$$

and conclude the derivation. Note that by interchanging the integral and the sum in (31), we get the ML-based cost function C_{ML} as in expression (8).

References

[1] P. M. Q. Aguiar and J. M. F. Moura. Video representation via 3D shaped mosaics. In *IEEE International Conference on Image Processing*, Chicago, USA, 1998.

[2] P. M. Q. Aguiar and J. M. F. Moura. Factorization as a rank 1 problem. In *IEEE Int. Conf. on Computer Vision and Pattern Recognition*, June 1999.

[3] N. Ayache. *Artificial Vision for Mobile Robots*. The MIT Press, MA, USA, 1991.

[4] T. Broida and R. Chellappa. Estimating the kinematics and structure of a rigid object from a sequence of monocular images. *IEEE Trans. on PAMI*, 13(6), 1991.

[5] J. Heel. Direct estimation of structure and motion from multiple frames. A. I. Memo 1190, MIT AI Laboratory, MA, USA, 1990.

[6] B. Horn and E. Weldon. Direct methods for recovering motion. *International Journal of Computer Vision*, 2(1), 1988.

[7] D. C. Knill, D. Kersten, and A. Yuille. A Bayesian formulation of visual perception. In *Perception as Bayesian Inference*. Cambridge University Press, 1996.

[8] S. Soatto, P. Perona, R. Frezza, and G. Picci. Recursive motion and structure estimations with complete error characterization. In *IEEE CVPR*, June 1993.

[9] C. Tomasi and T. Kanade. Shape and motion from image streams under orthography: a factorization method. *Int. Journal of Computer Vision*, 9(2), 1992.

[10] R. Tsai and T. Huang. Uniqueness and estimation of three-dimensional motion parameters of rigid objects with curved surfaces. *IEEE Trans. PAMI*, 6(1), 1984.

Magnetic Resonance Imaging Based Correction and Reconstruction of Positron Emission Tomography Images

Jonathan Oakley[1,2], John Missimer[2], and Gábor Székely[1]

[1] Swiss Federal Institute of Technology,
Communication Technology Laboratory,
CH-8092 Zurich, Switzerland
Jonathan.Oakley@vision.ee.ethz.ch
Phone: +41 1 632 5281. Fax: +41 1 632 1199
[2] Paul Scherrer Institut,
CH-5232 Villigen - PSI, Switzerland

Abstract. Due to the inherently limited resolution of Positron Emission Tomography (PET) scanners, quantitative measurements taken from PET images suffer from the partial volume averaging of activity across regions of interest. A correction for this effect in PET activity distributions is therefore essential to distinguish differences due to changes in tracer concentration from those due to changes in the volumes of the active brain tissue. Various consequent image post-processing techniques have been developed to address this problem (see, for example, [1–6]). These operate using associated high resolution anatomical images such as Magnetic Resonance Imaging (MRI), but as well as being highly susceptible to errors in the requisite registration and segmentation procedures, the methods are reliant on unrealistic simplifying assumptions regarding activity distributions in the brain. This work instead couples the correction of PET data to the reconstruction process itself, presenting a two-step scheme using associated MRI data to achieve this. The first step estimates the prior activity distributions from a segmented and intensity transformed MRI image. This is then used in the second step to constrain the Bayesian PET reconstruction with varying degrees of stringency. The prior, or initial correction process, is applied in the form of an energy term, adapted in accordance to an entropy measure taken on the MRI segmentation; i.e., where there is anatomical variation, we assume there also to be activity variation.

1 Introduction

Positron Emission Tomography (PET) is a non-invasive functional imaging technique, unique in its ability to image pathological condition. A radioactive tracer is introduced into the blood stream of the patient to be imaged. Its distribution is then indicative of metabolic activity or blood flow. The isotope itself decays quite quickly, emitting positrons which, within a very short distance, collide with electrons. The annihilation that takes place causes two gamma emissions to occur in almost exactly opposite directions. These emissions are recorded in the various detector crystals that surround the patient, and the projected counts are stored in *sinograms*. The aim of PET reconstruction is to estimate the spatial distribution of the isotope concentration on the basis of the sinogram data.

Because of the inherently limited resolution of PET scanners and the subsequent poor ability to resolve detail in reconstructed images, quantitative measurements taken from PET images suffer from the partial volume averaging of activity across regions of interest. A correction for this partial volume effect (PVE) in PET activity distributions is therefore essential to distinguish differences due to changes in tracer concentration from those due to changes in the volumes of the active brain tissue. Otherwise, as the relative percentage of cerebro-spinal fluid (CSF), grey matter (GM) and white matter (WM) varies - particularly as a result of aging or disease - any given change in the apparent tracer concentration may instead reflect a change in morphology, and without the benefits of an associated anatomical image, we are at loss in deciphering this.

Various image post-reconstruction techniques have been developed to address such issues (see, for example, [1–6]). Operating in accordance to constraints derived from associated high resolution anatomical images, these attempt to restore or redistribute the PET data of the reconstructed images. But, as well as being highly susceptible to errors in the requisite registration and segmentation procedures, such *localisation* methods are dependent on unrealistic simplifying assumptions made on the activity distributions in the brain.

Alternative methods begin with the sinogram data, and must therefore address the problem of tomographic image reconstruction. This task that is inherently *ill-posed*, and as such it is best tackled using statistical methods. Yet artifacts remain, products mainly of the ill-conditioned nature of the system of linear equations to be solved. Methods of *regularisation*, typically found in the form of either a penalising term [7], or of Bayesian priors [8,9], have been used to confront this. When high resolution anatomical information is available, however, stricter, more meaningful constraints can be imposed on the reconstruction solution. That is, the aforementioned image-space correction methods would now be better formulated as part of the reconstruction step. This has been clearly demonstrated in the literature with the incorporation of an accurate projection model [10], and also with the more accurate model of the activity source [11–13].

A new approach is demonstrated in this paper, where the contribution is to firstly derive a more appropriate prior model of the activity distributions, and to secondly present an algorithm capable of its exploitation. The prior model is designed to avoid over-simplifying assumptions regarding the activity distributions, and a Bayesian reconstruction procedure is implemented to accommodate it.

2 Problem Definition

The digitisation of an image grid within the PET scanner's field of view, allows us to assume the following:

- λ is a 2-D image of J pixels, where λ_j denotes the *expected* number of annihilation events occurring at the jth pixel. It is the PET activity distribution to be reconstructed.

- y denotes the I-D measurement vector, where y_i denotes the coincidences counted by the ith of the I detector pairs. This is the sinogram data recorded by the scanner.
- a_{ij} is the probability that an emission originating at the jth pixel is detected in the ith detector pair, where $\sum_j a_{ij} = 1, \forall i$. This forms the stochastic model of the acquisition process, more commonly referred to as the *system matrix*. For the purposes of the algorithm developed in this paper, this model must include the Point Spread Function (PSF) of the scanning device; it should describe, stochastically, how the underlying tracer distribution came to be observed as the PET signal.
- $\bar{y}_i$ denotes the mean, or expected number of coincidences detected by the ith detector pair, such that $\bar{y}_i = \sum_{j=0}^{J-1} a_{ij} \lambda_j$.

That is, we express our PET reconstruction problem on the basis of the following matrix form,

$$\bar{y} = \mathbf{A}\lambda, \tag{1}$$

where $\mathbf{A} \in \mathbb{R}^{I \times J}$ is the above defined system matrix that contains the weight factors between each of the image pixels and each of the projections recorded in the sinogram. Given we are dealing with radioactive decay, the detections are modelled using a Poisson distribution. This distribution is defined about its means as:

$$y_i \simeq Poisson(\bar{y}_i) = \frac{\bar{y}_i^{y_i}}{y_i!} \exp(-\bar{y}_i), \ i = 0, ..., I-1. \tag{2}$$

The Expectation-Maximization Algorithm for PET Reconstruction Following the seminal papers of [14, 15], the *Maximum Likelihood* (ML) estimate is derived using the *Expectation-Maximization* (EM) algorithm [16]. In this case, the complete data set, denoted ψ_{ij}, is defined to be the number of emissions occurring in pixel j and recorded in the ith detector pair. The measured [incomplete] data can now be re-expressed in terms of the complete data as $\bar{y}_i = \sum_{j=0}^{J-1} \bar{\psi}_{ij}$, for all $i = 0, ..., I-1$. The alternative summation yields the total number of emissions that have occurred in pixel j and have been detected in any of the crystals, $\sum_{i=0}^{I-1} \bar{\psi}_{ij} = \lambda_j$, for all $j = 0, ..., J-1$. On this basis, the problem of PET reconstruction now becomes one of estimating the means of the ψ_{ij}.

The likelihood function of the complete data set is a measure of the chance that we would have obtained the data that we actually observed (y), if the value of the tracer distribution (λ) were known:

$$L(\bar{\psi}) = \prod_{i=0}^{I-1} \prod_{j=0}^{J-1} \frac{\bar{\psi}_{ij}^{\psi_{ij}}}{\psi_{ij}!} \exp(-\bar{\psi}_{ij}). \tag{3}$$

From the above one is able to derive the following iterative scheme [15]:

$$\lambda_j^{k+1} = \frac{\lambda_j^k}{\sum_{i=0}^{I-1} a_{ij}} \sum_{i=0}^{I-1} \frac{a_{ij} y_i}{\sum_{j'=0}^{J-1} a_{ij'} \lambda_{j'}^k}, \tag{4}$$

where k denotes the iterate number.

The Ill-Conditioned Nature of the ML Estimates Unfortunately, as the ML estimates become more refined, they also become excessively noisy. That is, the wrong form of variance is increased and a pseudo over-fitting occurs to progressively worsen the final image. The point at which this deterioration begins, the start of *over-convergence*, depends upon a number of factors and is all but intractable.

Part solutions for this problem may be found by implementing new stopping rules [17, 18], post filtering [19], or by using some form of regularisation to impose appropriate constraints on the reconstruction solution (see, for example, [7, 20]). Alternatively, the solution may be constrained to distribution models [8], which has proved popular among the image reconstruction community [9, 11, 21–24]. Applicable though these random field models are, PET reconstruction can benefit further from priors derived from Magnetic Resonance Imaging (MRI) data. This additional information typically provides boundaries within which the random field models may be applied [12, 13, 25–27]. The resulting increase in accuracy is restricted only according to the crudity of the assumptions necessary to realise the implementation.

3 The Bayesian Approach

The parameters to be estimated amount in many senses to a hypothesis of the data. In this sense, a likelihood function only tells us how well our hypothesis explains the observed data. Bayes's theorem tells us instead the probability that the hypothesis is true *given the data* [28]. The hypothesis, or prior, to be derived is a first estimation of the activity distribution of the subject, and the Bayesian paradigm allows us to maintain our faith in this estimate, letting only the observed data persuade us otherwise.

Gindi [26], and Leahy [27] achieve this with *Maximum A Posteriori* (MAP) estimates of simulated PET data incorporating *line sites* (originally proposed by [8]) from associated MRI, which are appended to the set of parameters to be estimated in each maximisation step of the procedure. The obvious extension to this approach is to weight these sites such that they find agreement to edges present the PET data [29]. Instead of line sites, Bowsher et al. [30] build a segmentation model into their reconstruction process whose repeated re-estimation is able to gauge the progress of the reconstruction. Lipinski et al. [12] use MRI priors to delineate different Markov and Gaussian energy fields acting to regularise the solution. Sastry and Carson [13] adopt a similar approach, this time coupling both the Markov and Gaussian fields in an effort to reconcile two desirable properties of the reconstruction solution. These, in the words of Leahy [31], being that: images are locally smooth; except where they are not!

The models, and the assumptions behind the two aforementioned random field priors used in [12, 13] are the following:

- *Global Homogeneity*: within each tissue compartment, the activity concentrations are Gaussian distributed with a unique mean. This is expressed using a Gaussian Random Field.
- *Local Homogeneity*: within a homogeneous activity distribution, neighbouring pixels tend to have similar values. This they model with a Markov Random Field (MRF) defined over a first order neighbourhood [32].

[13] apply these to *tissue-type* activities individually, where the tissue-type n is derived from associated MRI data. The first of the above priors does not enforce any local neighbourhood properties on the reconstruction, but instead assumes that an activity level corresponding to a tissue of a known class will not be significantly different from that of the mean activity level for that class. This is termed the *Gaussian prior*. In the case of the second prior, the piecewise smoothness assumption common to many Bayesian methods is used, and the PET image is thought to contain activities that vary little across neighbouring pixels. [13], however, are original in applying this prior in concert with the first (thus forming the so-called *Smoothness-Gaussian prior*), in an attempt to constrain local variation alongside the restriction that activity levels should remain within sensible, global bounds.

4 Developing New Priors Based on a High Resolution PET Image

The development of a good prior can be all important to the success of the Bayesian scheme. As such, a possible weakness of the above approach is the use of mean activity estimates, which, in meeting the necessary regularisation requirements, are likely to encourage homogeneous distributions. The estimation of these means is derived from knowledge of the tissue type, where, for each compartment, CSF, WM, GM, and *other*, this is given by the following activity ratios: 0:0.005:1:4, respectively (see, [6, 33–37] for similar such estimations). Under the assumption that the activity in each tissue-type is uniform, the mean activities, denoted $\bar{\lambda}^n$ for tissue-type n, are derivable from a least-squares solution to the tomographic system from a pseudo-inverse of the system matrix: $\bar{\lambda}^n = \mathbf{A}^{-1} y$ (see the original definition from equation 1).

The following proposes a new approach to defining initial activity estimates, whose emphasis is to allow inhomogeneous distributions to occur in each of the tissue classes. It is a model-based approach whose solution is an image-space PET correction. The development of a reconstruction algorithm designed to iteratively update the solution is left as the subject of the next section.

Some Assumptions on the Activity Distribution If at all avoidable, presumed homogeneous distributions should not be a part of any attempt to model (and thus, correct) the PET signal. From [38], for example, we learn that it is normally GM which shows the greatest amount of variance; CSF is likely to show some, although this is negligible; and WM should show some aspects of variability in cases of diseases, such as multiple sclerosis. Additionally, this work cites lesions, focal activations and field heterogeneities as typical factors that would violate the homogeneity assumption. Justification is also to be found in the discussion of Ma [34], where the opinion that sharp activity changes occurring at structural boundaries is refuted (i.e., those occurring under the homogeneity assumption are unlikely), and that it is more probable that the distributions can be characterised by some gradient in tracer level occurring within and possibly across structures.

For practical purposes, the increased knowledge won from simple homogeneity assumptions is applied in different forms by [3, 5, 6, 33, 37] for PET correction and redistribution, and [34] for PET simulation (a process applied to [36, 39, 40]). Improving the assumptions, however, requires a better knowledge of the emission source, where it is difficult to avoid generalisations. The method developed in the following, however, heeds the advise of [38] and attempts to do just that.

Intensity Normalisation - A High Resolution PET Prior We derive in the following a model for activity distributions based on an agreement with the original ideas of Friston et al. [38]. In fitting the underlying flow-distribution of the PET image, one is able to derive a correction for the observable PET data constrained in accordance to the MRI delineations of true regions and knowledge regarding the scanner response. The solution is a corrected "PET" image at the resolution of the MRI data, although in the context of its later usage, we will refer to this distribution as a high resolution prior.

Supposing that the MRI and PET images are in co-registration[1], we say that the original PET image (λ_j^o, reconstructed without the use of any model) can be described by an intensity transformation of the associated segmented MRI image (m_j^s) *and* a convolution that relates the differences in resolution. That is,

$$\lambda_j^o \approx h_j * \gamma_j\{m_j^s\}, \tag{5}$$

where $*$ denotes convolution, h_j is the convolution kernel that reflects the resolution mismatch in the PET and MRI data, and γ_j is the intensity transformation that we wish to derive. The MRI data are segmented into regions of GM, WM, and CSF, from which activity levels are assigned according to estimated ratios [6, 13, 33–37] (the implicit fourth compartment is background, for which zero activity is the expected level). The expansion of equation 5 in [38] is in the form of a the Taylor Series about a single GM segmentation function, and operates therefore under the premise that the PET signal in WM and CSF regions is sufficiently negligible to be absent from the model. The coefficients of the expansion are themselves expanded in terms of B basis functions such that they are non-stationary and smoothly varying about a given local [38]. In the following, only this latter aspect of the model is retained, as the segmentation is made explicit using a fuzzy algorithm [43] to derive probability maps of affinities to GM, WM and CSF; denoted below as m_j^g, m_j^w and m_j^c, respectively. The model is now defined to contain three intensity transformation functions, γ_j^g, γ_j^w and γ_j^c, operating on the GM, WM and CSF segmentations:

$$\lambda_j^o \approx h_j * [g\gamma_j^g\{m_j^g\} + w\gamma_j^w\{m_j^w\} + c\gamma_j^c\{m_j^c\}]. \tag{6}$$

where again each intensity transformation function is made up of basis functions and non-stationary coefficients (equation 7), and g, w and c are *normalised* ratio contributions. The basis functions, β_b, applied to each intensity transformation γ^η for

---

[1] It is important to note here that as the images are in co-registration, then their pixel dimensions and the physical sizes of each pixel are the same. Nonetheless, one can still talk about one data set being of a higher resolution than another in that it *exhibits* a higher resolution. In this case, although the PET data may be at the same physical resolution as the MRI data, it does not utilise this sampling density as well as the MRI, and is considered therefore, to be of a lower effective resolution [41, 42].

$\eta \in \{g, w, c\}$, are derived from,

$$\gamma_j^\eta \{m_j^\eta\} = m_j^\eta \sum_{b=0}^{B-1} u_{\eta,b} \beta_j^b, \qquad (7)$$

where $u_{\eta,b}$ denote the [unknown] coefficients of the expansion.

Deriving the High Resolution PET Prior Once a solution for each of the B vectors $\hat{u}_\eta$ is found, we are able to derive a high resolution PET prior (λ_j^p) by simply removing the convolution term:

$$\lambda_j^p = g\gamma_j^g\{m_j^g\} + w\gamma_j^w\{m_j^w\} + c\gamma_j^c\{m_j^c\}. \qquad (8)$$

This yields a PET image at the resolution of the MRI data, valid in accordance to the original assumptions made of the model. That is, it emulates a "restored" PET image, where the restoration is predominantly a deconvolution, constrained by the delineation of the GM, WM and CSF distributions in the MRI data.

The Resulting Intensity Transformation Results from the use of this transformation are shown in figure 1. As the figure shows, with only a small number of basis functions the transformation is quite convincing in its ability to map the MRI data to the PET data. For purposes of validation, figure 2 shows the transformation applied to a Monte Carlo based simulation of the acquisition process, where in this instance some notion of ground-truth is available, yet noise and other distorting effects are realistic.

In the context of the image-space correction techniques that use MRI data to compensate for the PVE in PET, the result itself constitutes a solution to the problem of PET redistribution. It is of high resolution, and of low noise. Uncertainties in the segmentations, however, are propagated to uncertainties in the solution. This effect is evident in figure 1 where regions of the resulting prior seem smeared; the algorithm is only able to average a fit. As such, the solution cannot guarantee uniqueness, and the validity of the correction method must therefore be gauged on some localised basis. Further justification for this requirement is apparent in regions where the MRI data is homogeneous and the PVE does not occur. In this instance, the prior should have no influence on the reconstruction solution.

Such heuristics can, to a good extent, be used to drive the reconstruction process, and this PET correction result may yet be iteratively improved upon. This requires a formulation of the problem within the Bayesian framework, which is the basis of the remainder of this work.

5 Applying the High Resolution PET data as a Prior

The Adopted Form of the Prior To build the above derived prior distribution into a reconstruction algorithm, the approach given here follows the general methods of [11–13]. The prior is wrapped in an energy function, designed to be minimal when the estimate for the reconstruction, λ_j, matches that of the activity estimates derived from

The Intensity Transformation with 8 Basis Functions, on 64x64 Pixel Images:

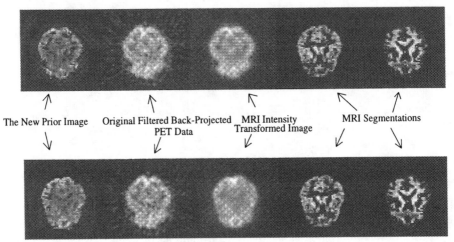

The New Prior Image   Original Filtered Back-Projected   MRI Intensity   MRI Segmentations
PET Data   Transformed Image

and the Intensity Transformation with just 4 Basis Functions, on 64x64 Pixel Images.

Fig. 1. From left to right, the figures shown above are: the high resolution PET image (the new prior - λ_j^p of equation 8); the PET image reconstructed from the sinogram data using Filtered Back-Projection (FBP); the MRI intensity transformed image (the PET model with the convolution term present: $h_j * \lambda_j^p$); the MRI GM segmentation (m_j^g); and the MRI WM segmentation (m_j^w).

the intensity transformation prior, denoted λ_j^p. This gives us a prior probability model, applied in the following form,

$$P(\lambda) = \frac{1}{Z} \exp(-U(\lambda)), \tag{9}$$

where Z is a normalisation term (the *partition function*), and $U(\lambda)$ is the energy function chosen to impose localised constraints on the reconstructed activities. The *a posteriori* probability for tissue-type activities given the sinogram data, for which a maximum is sought, is $P(\lambda|y) \propto p(y|\lambda)P(\lambda)$, where, $p(y|\lambda)$ is our likelihood function, and $P(\lambda)$ is the prior probability model. The applied prior is thus defined as $P(\lambda) \propto \exp(-U(\lambda))$, with the energy function to be minimised given as:

$$U(\lambda) = \sum_j \frac{(\lambda_j - \lambda_j^p)^2}{2\sigma_j^2}. \tag{10}$$

This takes the form of a Gaussian distribution, where in this implementation the estimation of tissue activities, λ_j^p, are those of the estimated prior distribution of equation 8. This involves the individual estimation of each pixel intensity, with an additional notion of local smoothness implicitly included as a result of restrictions on local variation set by basis functions in equation 7. The consequence is that the constraints combine in manner akin to [13]'s aforementioned smoothness-global prior. In estimating activity

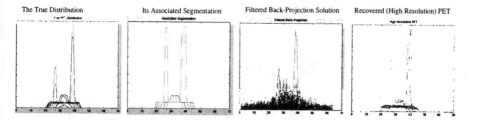

Fig. 2. From left to right, the figures shown above are: the true distribution of a PET image (Monte Carlo simulation based); its associated segmentation; the Filtered Back-Projection (FBP) reconstruction based on the simulated sinogram (including noise, scatter and random coincidences); The high resolution PET image, λ_j^p, recovered from the segmentation and the FBP image alone.

at each pixel (through the λ_j^p), however, it is the granularity of the basis functions that determines how well the assumption on piecewise smoothness is avoided.

The σ_j control the Gaussian's standard deviations, which in turn reflects the stringency of the reconstruction's coupling to the prior. This can be altered on a localised basis, allowing the algorithm to show great versatility. As $\sigma_j \to 0$, the reconstruction tends toward the prior (or the image-space correction method), and as $\sigma_j \to \infty$, it tends toward the EM solution. The estimation of these *hyperparameters* take on physical interpretations from [13] in that they relate to the allowable variation in activity levels. Here, an additional interpretation can be afforded, as these values are estimated in accordance to the likely influence of the PVE; i.e., related to the degree of correction required. This, in turn, may be estimated from the MRI data, as discussed below.

Deriving the Iterative Algorithm The likelihood function of the means of the complete data set, $L(\bar{\psi})$, was given in equation 3. Relating the complete data set to the activity distribution, $\sum_{i=0}^{I-1} \bar{\psi}_{ij} = \lambda_j$, and defining the conditional expectation value for the ψ_{ij} on the basis of a standard probability result, yields the expectation step of the EM reconstruction algorithm. This, denoted N_{ij}^k, is given as:

$$N_{ij}^k = \mathcal{E}(\psi^{k+1}|\lambda_j^k, y_i) = \frac{a_{ij}\lambda_j^k y_i}{\sum_{j=0}^{J-1} a_{ij}\lambda_j^k}, \tag{11}$$

where k denotes the iteration number, and $\mathcal{E}$ the expectation. From equation 3, the log likelihood expressed in terms of N_{ij}^k is,

$$l(\bar{\psi}^{k+1}) = \ln L(\bar{\psi}^{k+1}) = \sum_{i=0}^{I-1}\sum_{j=0}^{J-1}(N_{ij}^k \ln a_{ij}\lambda_j - a_{ij}\lambda_j - \ln N_{ij}^k!). \tag{12}$$

In the Bayesian approach, instead of maximising the likelihood function, it is necessary to maximise the *a posteriori* probability. This must include our energy function from the form of equation 9, yielding,

$$argmax_\lambda\{l(\psi) - U(\lambda)\}. \tag{13}$$

From [11, 13], we optimise by setting the derivative of $\{l(\psi) - U(\lambda)\}$ (with respect to λ) to zero:

$$\frac{\lambda_j^2}{\sigma_j^2} - \lambda_j(-\sum_{i=0}^{I-1} a_{ij} + \frac{\lambda_j^p}{\sigma_j^2}) - \sum_{i=0}^{I-1} N_{ij}^k = 0. \tag{14}$$

Taking only the positive root to the solution of this quadratic equation, yields the following iterative reconstruction scheme due to [11, 13]:

$$\lambda_j^{k+1} = \frac{1}{2}(-\sigma_j^2 \sum_i a_{ij} + \lambda_j^p) + \frac{1}{2}\sqrt{(-\sigma_j^2 \sum_i a_{ij} + \lambda_j^p)^2 + 4\sigma_j^2 \sum_i N_{ij}^k}, \tag{15}$$

where N_{ij}^k is defined according to equation 11.

Selecting the Gaussian Function's Standard Deviations (σ_j) The purpose of the algorithm in section 4 was to derive a prior estimation of activity levels for each PET pixel at the resolution of the MRI image. As such, the energy term of equation 10 should also be chosen for each pixel individually, and the standard deviation of each Gaussian field thus reflects this.

Use of an Entropy Measure Regions likely to suffer the PVE are those where the variation in structure is the greatest; outside of such regions there is no need to redistribute the data. As such, σ_j must be steered according to a local measure taken on the MRI data. This is achieved using an entropy measure about small windowed regions. For example, in areas of high entropy, where we the data exhibits less structural variation, the corresponding σ_j terms should become wider.

The application of such an entropy measure requires that its actual value and range be made explicit. It is then used to prompt variation in the σ_j terms, which are otherwise set in accordance to the paper of [13]:

$$\sigma_j = \frac{K.C.e_j}{2(\ln 2)^{\frac{1}{2}}100}, \tag{16}$$

where K defines the full-width half-maximum of the Gaussian, interpreted as being $K\%$ of some scaling constant C. In the results given in the remainder of this section, the entropy image, e_j, is simply scaled to within $\pm 0.6 + 1$ (0.4 where the prior shows greatest structure, and 1.6 corresponds to completely homogeneous regions), and C is set to the average intensity value in our prior image ($\bar{\lambda}^p$). Admittedly, these are relatively arbitrary choices, but ones that succeeded in consistently reducing the least-squares error in tests where the ground-truth data is known.

6   Results

Experiments have been performed on both artificial, Monte Carlo, and actual PET-MRI studies. In the first instance, the manufacturing of test data involved the blurring of a phantom image, the addition of Poisson noise, and finally its forward-projection to produce the sinogram data[2]. Insight into the algorithms workings was then gleened, serving primarily to put the variation according to the entropy measure to within sensible bounds.

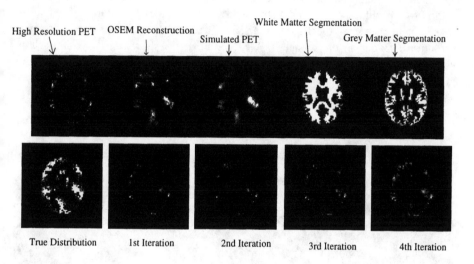

Fig. 3. In the above, the top 5 images show the intensity transformation scheme, as discussed in section 4. The first image is the high resolution PET image (λ_j^p), which in this case is derived from a PET reconstruction (the image second from the left - λ_j^o) based on the Ordered-Subsets EM (OSEM) algorithm [45]. This is a fast EM reconstruction method operating at speeds that are now clinically acceptable. The central image is the model of the OSEM distribution, and the remaining images show the WM and GM segmentations, respectively. The bottom row displays the following: the true, underlying PET distribution (simulated to exhibit the resolution of the MRI data, note); and the 1st, 2nd, 3rd and 4th iterations of the Bayesian reconstruction for $K = 30$. Only four iterations are shown, as convergence is seen to occur rapidly. The reconstruction "snaps" very quickly onto a solution, which is characteristic behaviour for methods involving such constrained priors. One must, therefore, be careful that the solution is the appropriate one.

The results of the simulated experiments shown in figure 3 show how the reconstruction solution is able to very quickly take the form of the true distribution. From this it might be thought that the algorithm is simply converging toward the prior. Evidence against this is that the least-squares error between the true distribution and the reconstruction solution improves upon that between the true distribution and the prior after only a few iterations. That is, iterative adjustments to the prior-based solution are necessary.

Although the results of figure 4 from real MRI and PET studies have no associated ground-truth, visually they are able to demonstrate two important features: structure is retained, delineating known tissue regions; and the general intensity values are, on

---

[2] This data set was taken from the *BrainWeb* Database [44].

a regional basis, the same as those of the Ordered Subset EM (OSEM) reconstructed image. The top row shows images reconstructed with the Bayesian scheme presented in this paper. In this case, the entropy measure is applied, and $K = 35$ (equation 16). The next row shows increased iterations (from the same starting estimate) of the OSEM algorithm. As 8 subsets were used, each iteration is approximately equal to 8 iterations of the EM algorithm [45]. Being a statistical technique, the OSEM algorithm uses the same system matrix as the Bayesian scheme, and should therefore also account for resolution loss due to the scanner's PSF. The next row shows the Bayesian scheme without the use of the entropy measure. The resulting reconstructions are able to very quickly demonstrate better contrast and structure in the images, especially so in regions where the PVE is likely to be of greatest influence (in regions of low entropy).

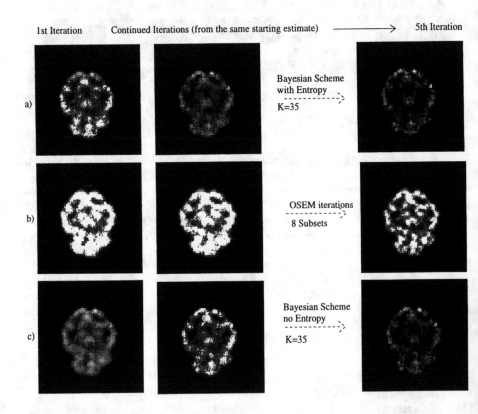

Fig. 4. The above figure shows three different reconstruction schemes each starting from the same intial estimate, an OSEM reconstruction using 8 subsets reconstruction after 4 iterations. It was also on the basis of this image that the high resolution prior (λ_j^p) was defined. The top row (a) shows increasing iterations (1,2 and 5) of our Bayesian scheme that includes the entropy measure to increase recovery from the PVE. The next row (b) shows the OSEM reconstruction scheme as it iterates further. The final row (c) shows the Bayesian scheme without the entropy measure. For both of the Bayesian methods, K of equation 16 is 35; a slightly more conservative value than that for the experiment using simulated data.

7 Discussion

In Respect of the Prior It was never envisaged that the intensity transformed MRI segmented image could be capable of replacing the reconstructed PET data. It is an artificially constructed model, and hence the distribution's use as a prior. The transform simply assigns PET values to structural objects, and the result can only constitute "a functional categorisation of anatomical structures" of the sort implied by the segmentation [38]. In a way, it does emulate a fully "restored" PET image, yet the restoration is only as good as the validity of the assumptions concerning the activity distributions. We feel, however, justified in using this method to estimate a prior distribution, as the assumptions are indeed valid, with homogeneity notable only because of its absence.

Choice of the Basis Functions As figure 1 shows, the more basis functions that can be applied, the better the fit that can be attained. However, by imposing a finite upper limit to their size, we introduce the notion of a neighbourhood to the reconstruction process. As such, it is necessary to use basis functions to capture PET activity *at its finest possible localisation of activity* (see [46]), but not beyond. Any finer than necessary would simply mean modelling the noise. What granularity captures constituent components of the activations can be approximately estimated from [47], for example. Or indeed, the approach adopted in the Statistical Parameteric Mapping methods of the Functional Imaging Laboratory in London [48]. Here, considerable smoothing is done prior to any significance testing in activation studies, and the short conclusion is that the use of 8 basis functions for 64-by-64 pixel images, 16 for 128-by-128 images, and so on, would seem about right.

Regarding the Reconstruction Algorithm The algorithm presented operates with a number of free parameters whose better selection is required to achieve improvements in performance. The appropriate selection of the σ_j in equation 16 is most critical. It basically steers the algorithm toward a normal iterative reconstruction solution at one extreme, or to an effective PET correction method at the other. For the selection of this and other such hyperparameters, there are basically two options [31]: the data-driven empirical approach; or an estimation theoretic approach.

Unfortunately, alterations to this term and the subsequent algorithmic flexibility resulting from the local variation of the σ_j of the Gaussian fields may not be appreciated by the Bayesian purists. Nonetheless, it is important in emission tomography to constrain the solution only as and when it is correct to do so. Even if a contribution cannot be afforded globally, local improvements can have a positive effect. Among Llacer's results from studies of such distributions in Bayesian reconstructions [49], was the indication that prior information applied in some areas of an imaging field has a tendency to improve the results of a reconstruction elsewhere. This may at first seem a surprising and perhaps rather dubious result, but when one considers how highly correlated an emission image is, and how its piecing together is a problem of global optimisation with constraints such as positivity and energy conservation, then increasing the certainty of the solution in one area is indeed likely to aid the solution in another.

7.1 Conclusions and Future Work

The least-squares fit of the intensity transformation gives, depending on the individual's agreement with the initial assumptions, a sensible estimate of the tracer distribution at the resolution of the MRI data. The result is a high resolution, low noise prior, whose application within the framework of a Bayesian reconstruction scheme allows for an appropriately corrected, well regularised, reconstructed PET image.

Issues of registration and segmentation errors, however, highlight the well known shortcomings in all such cross-modality reconstruction methods. This work's adoption of a fuzzy segmentation method coupled to a summation of basis functions in order to estimate the underlying activity distribution does, to a large extent, exhibit some robustness with respect to the latter of these two issues. This was seen in figure 1, but the ensuing necessary averaging of the intensity assignments does, of course, limit the effectiveness of the approach. With respect to errors in the registration of the different image sets, then the algorithm, like its counterparts, reveals its frailty. Errors need be only very slight to render the associated MRI data as all but useless.

With the fundamental aim of this work being to improve the resolution of the PET data, addressing, for example, the PVE, the approach that has been taken seeks to bridge the PET redistribution methods applied as post-processing techniques, and the model-based reconstruction methods applied at the sinogram level. Coupled in a complimentary manner, this paper has sought to demonstrate how effective this can be.

References

1. TO Videen, JS Perlmutter, MA Mintun, and ME Raichle. Regional correction of positron emission tomography for the effects of cerebral atrophy. *Journal of Cerebral Blood Flow and Metabolism*, 8:662–670, 1988.
2. CC Meltzer, RN Bryan, HH Holcomm, AW Kimball, HS Mayberg, B Sadzot, JP Leal, HN Wagner, and JJ Frost. Anatomical localization for PET using MR imaging. *Journ. Comp. Assist. Tomogr.*, 14(3):418–426, 1990.
3. HW Müller-Gärtner, JM Links, JL Prince, RN Bryan, E McVeigh, JP Leal, C Davatzikos, and JJ Frost. Measurement of radiotracer concentration in brain gray matter using positron emission tomography: MRI-based correction for partial volume effects. *Journal of Cerebral Blood Flow and Metabolism*, 12:571–583, 1996.
4. JJ Frost, CC Meltzer, and et al. MR-based correction of brain PET measurements for heterogeneous gray matter radioactivity distribution. *Neuroimage*, 2:32–32, 1995.
5. CC Meltzer, JK Zubieta, JM Links, P Brakeman, MJ Stumpf, and JJ Frost. MR-based correction of brain PET measurments for heterogeneous gray matter radioactivity distribution. *Journal of Cerebral Blood Flow and Metabolism*, 16:650–658, 1996.
6. J Yang, SC Huang, M Mega, KP Lin, AW Toga, GW Small, and ME Phelps. Investigation of partial volume correction methods for brain fdg pet studies. *IEEE Trans. Nucl. Sci.*, 43(6):3322–3327, 1996.
7. JA Fessler. Penalized weighted least-squares image reconstruction for positron emission tomography. *IEEE Trans. Med. Imaging*, 13:290–300, 1994.
8. S Geman and D Geman. Stochastic relaxation, gibbs distributions, and the bayesian restoration of images. *IEEE Trans Patt. Anal. Mach. Intel.*, 6(6):721–741, 1984.

9. S Geman and D McClure. Bayesian image analysis: An application to single photon emission tomography. In *Proceedings of the Statistical Computing Section*, pages 12–18. American Statistical Association, 1985.

10. J Qi, RM Leahy, EÜ Mumcuoüglu, SR Cherry, A Chatziioannou, and TH Farquhar. High resolution 3D bayesian image reconstruction for microPET. *1997 International Meeting in Fully 3D Image Reconstruction*, 1997.

11. E Levitan and GT Herman. A maximum a posterior probability expectation maximization algorithm for image reconstruction in emission tomography. *IEEE Trans. Med. Imag.*, 6:185–192, 1987.

12. B Lipinski, H Herzog, E Kops, W Oberschelp, and HW Müller-Gärtner. Expectation maximization reconstruction of positron emission tomography images using anatomical magnetic resonance information. *IEEE Trans on Med. Imag.*, 16(2):129–136, 1997.

13. S Sastry and RE Carson. Multimodality bayesian algorithm for image reconstruction in positron emission tomography: A tissue composition model. *IEEE Trans. Med. Imag.*, 16(6):750–761, 1997.

14. LA Shepp and Y Vardi. Maximum likelihood reconstruction in positron emission tomography. *IEEE Trans. Med. Imag.*, 1:113–122, 1982.

15. K Lange and RE Carson. EM reconstruction algorithms for emission and transmission tomography. *Journal of Computed Tomography*, 8(2):306–316, 1984.

16. AP Dempster, NM Laird, and DB Rubin. Maximum likelihood from incomplete data via the EM algorithm. *Journal of the Royal Statistical Society B*, 39:1–38, 1977.

17. E Veklerov and J Llacer. Stopping rule for the MLE algorithm based on statistical hypothesis testing. *IEEE Trans. Med. Imag.*, 6(4):313–319, 1987.

18. T Hebert. Statistical stopping criteria for iterative maximum likelihood reconstruction. *Phys. Med. Biol.*, 35:1221–1232, 1990.

19. DL Snyder, MI Miller, LJ Thomas, and DG Politte. Noise and edge artifacts in maximum likelihood reconstruction for emission tomography. *IEEE Trans. Med. Imag.*, 6:228–238, 1987.

20. AR DePierro. A modified expectation maximization algorithm for penalized likelihood estimation in emission tomography. *IEEE Trans. Med. Imag.*, 14(1):132–137, 1995.

21. S Geman and D McClure. Statisical models for tomographic image reconstruction. *Bull. Int. Statist. Inst.*, 52:5–21, 1987.

22. T Hebert and R Leahy. A generalized EM algorithm for the 3D bayesian reconstruction from poisson data using gibbs priors. *IEEE Trans. Med. Imag.*, 8(2):194–202, 1989.

23. PJ Green. Bayesian reconstruction from emission tomography data using a modified EM algorithm. *IEEE Trans. Med. Imag.*, 9:84–93, 1990.

24. EÜ Mumcuoğlu, RM Leahy, SR Cherry, and Z Zhou. Fast gradient-based methods for bayesian reconstruction of transmission and emission PET images. *IEEE Trans. Med. Imag.*, 13(4):687–701, 1994.

25. K Lange, M Bahn, and Roderick L. A theoretical study of some maximum likelihood algorithms for emission and transmission tomography. *IEEE. Trans. Med. Imag.*, 6(2):106–114, 1987.

26. G Gindi, M Lee, A Rangarajan, and IG Zubal. Bayesian reconstruction of functional images using registered anatomical images as priors. In Colchester ACF and Hawkes D, editors, *Information Processing in Medical Imaging*, pages 121–130. Springer Verlag, 1991.

27. R Leahy and X Yan. Incorporation of anatomical MR data for improved functional imaging with PET. In ACF Colchester and D Hawkes, editors, *Information Processing in Medical Imaging*, pages 105–120. Springer Verlag, 1991.

28. DS Sivia. *Data Analysis: A Bayesian Tutorial*. Oxford Science Publications, 1996.

29. X Ouyang, WH Wong, and VE Johnson. Incorporation of correlated structural images in PET image reconstruction. *IEEE Trans. Med. Imag.*, 13(2):627–640, 1994.

30. JE Bowsher, VE Johnson, TE Turkington, RJ Jaszczak, CE Floyd, and RE Coleman. Bayesian reconstruction and use of anatomical a priori information for emission tomography. *IEEE Trans. Med. Imaging*, 99:99–99, 1996.

31. RM Leahy and J Qi. Stastical approaches in quantiative positron emission tomography. Technical report, University of Southern Calafornia, 1998. Available at http://sipi.usc.edu/ jqi/.

32. R Chellappa and AK Jain. *Markov Radom Fields: Theory and Application*. Academic Press, Boston, 1993.

33. U Knorr, Y Huang, G Schlaug, RJ Seitz, and H Steinmetz. High resolution PET images through REDISTRIBUTION. In Lemke et al., editor, *Comp. Assis. Radiology*, Berlin, 1993. Springer Verlag.

34. Y Ma, M Kamber, and AC Evans. 3D simulation of PET brain images using segmented MRI data and positron tomograph characteristics. *Comp. Med. Imag. and Graph.*, 17(4):365–371, 1993.

35. Y Kosugi, M Sase, Y Suganimi, and J Nishikawa. Dissolution of PVE in positron emission tomography by an inversion recovery technique with the MR-embedded neural network model. *Neuroimage*, 2:S:35, 1995.

36. OG Rousset, Y Ma, GC Léger, AH Gjedde, and AC Evans. Correction for partial volume effects in PET using MRI-based 3D simulations of individual human brain metabolism. In *Quantification of Brain Function*, pages 113–126. Elsevier, 1993.

37. SJ Kiebel, J Ashburner, J-B Poline, and KJ Friston. A crossvalidation of SPM and AIR. *Neuroimage*, 5:271–279, 1997.

38. KJ Friston, J Ashburner, J-B Poline, CD Frith, JD Heather, and RSJ Frackowiak. Spatial registration and normalization of images. *Human Brain Mapping*, 2:165–189, 1995.

39. OG Rousset, Y Ma, M Kamber, and AC Evans. Simulations of radiotracer uptake in deep nuclei of human brain. *Journal of Comp. Med. Imag. and Graph.*, 17(4):373–379, 1993.

40. OG Rousset, Y Ma, S Marenco, DF Wong, and AC Evans. In vivo correction for partial volume accuracy and precision. *Neuroimage*, 2:33–33, 1995.

41. EJ Hoffman, S-C Huang, D Plummer, and ME Phelps. Quantitation in positron emission tomography: 6. the effect of nonuniform resolution. *Journ. Comp. Assist. Tomogr.*, 6(5):987–999, 1982.

42. JA Fessler and WL Rogers. Resolution properties of regularized image reconstruction methods. Technical Report 297, Communications and Signal Processing Laboratory, University of Michigan, 1996. Available from http://www.eecs.umich.edu/.

43. DE Gustafson and W Kessel. Fuzzy clustering with a fuzzy covariance matrix. *IEEE CDC*, 2:761–770, 1979.

44. CA Cocosco, V Kollokian, RK-S Kwan, and AC Evans. Brainweb: Online interface to a 3D MRI simulated brain database. In *Proceedings of 3rd International Conference on Functional Mapping of the Human Brain*, Copenhagen, 1997.

45. HM Hudson and RS Larkin. Accelerated image reconstruction using ordered subsets of projection data. *IEEE Trans. Med. Imag.*, 13:601–609, 1994.

46. PT Fox, JS Perlmutter, and ME Raichle. A stereotactic method of anatomical localization for positron emission tomography. *Journal of Computed Assisted Tomography*, 9:141–153, 1985.

47. KJ Worsley, AC Evans, S Marrett, and P Neelin. A three-dimensional statistical analysis for cbf activation studies in human brain. *Journal of Cerebral Blood Flow and Metabolism*, 12:900–918, 1992.

48. See http://www.fil.ion.ucl.ac.uk/.

49. J Llacer, E Veklerov, and J Nuñez. Preliminary examination of the use of case specific medical information as prior in bayesian reconstruction. In Colchester ACF and Hawkes D, editors, *Information Processing in Medical Imaging*, pages 81–93. Springer Verlag, 1991.

Markov Random Field Modelling of fMRI Data Using a Mean Field EM-algorithm

Markus Svensén, Frithjof Kruggel, and D. Yves von Cramon

Max-Plank-Institute of Cognitive Neuroscience
Postfach 500 355, D-04303, Leipzig, GERMANY
{svensen,kruggel,cramon}@cns.mpg.de

Abstract. This paper considers the use of the EM-algorithm, combined with mean field theory, for parameter estimation in Markov random field models from unlabelled data. Special attention is given to the theoretical justification for this procedure, based on recent results from the machine learning literature. With these results established, an example is given of the application of this technique for analysis of single trial functional magnetic resonance (fMR) imaging data of the human brain. The resulting model segments fMR images into regions with different 'brain response' characteristics.

1 Introduction

The purpose of this paper is two-fold: first, it reviews the theoretical underpinnings for the use of the EM-algorithm in conjunction with mean field theory for parameter estimation in Markov random field (MRF) models from unlabelled data. Second, it demonstrates the usefulness of this approach by a MRF model for single trial functional magnetic resonance imaging (fMRI) data.

Techniques for learning from unlabelled data are important in the analysis of fMRI data of the human brain, since the data generating mechanism is still far from completely understood. Obvious ethical reasons put limitations on what sort of alternative methods we can use to verify results obtained from fMRI. Other functional brain imaging techniques, which may appear as the obvious answer, suffer exactly the same problem. At the same time, the quantity and quality of fMRI data make automated analysis procedures necessary.

2 Markov Random Fields, Mean Field Theory and the EM-algorithm

In this section, we briefly review Markov random field models, the mean field theory and its connection to the EM-algorithm. Mean field theory is a since long established tool in statistical mechanics and statistical physics. It has also been extensively used in the fields of computer vision and, more recently, machine learning.

2.1 Markov Random Field Models

A MRF [24] is a set of N random variables indexed over the vertices, or sites, in an ordered lattice. The typical example is a 2-D image, where the random variables are the labels (e.g. colour) associated with the pixels. The MRF variables are not independent, but are mutually coupled; the key property of MRFs is that the distribution of the random variable associated with a site, n, given the values associated with the sites in a (typically small) *neighbourhood* of n, is independent of the rest of the sites in the MRF. This can be formalised as

$$p(x_n | x_m, n \neq m) = p(x_n | x_m \in \mathcal{N}_n) ,$$

where x_n denotes the random variable of site n and $\mathcal{N}_n$ is the set of random variables associated with the sites that are in the neighbourhood of site n.

The distribution over the MRF variables, which is assumed to be strictly positive, can be written as a Gibbs distribution,

$$p(x) = \frac{1}{Z} \exp(-E(x)) \tag{1}$$

where x is a NK-dimensional vector formed by concatenating the vectors x_n $(n = 1, \ldots, N)$, E is an *energy function* and Z is a normalisation constant,

$$Z = \sum_x \exp(-E(x)) , \tag{2}$$

where the sum runs over all possible values of x. Note that, computing Z, which is known as the *partition function*, is generally tractable only for very small MRFs, since the number of terms in the sum in (2) increase exponentially with the size of the MRF. This is due to the mutual coupling between the MRF variables. Same problem emerges if we want to compute the marginal posterior distribution over any of the individual MRF variables – e.g. for the purpose of parameter fitting – since this requires summing over all remaining variables.

The energy function E defines the properties of the MRF model and can generally be written

$$E(x, y, \Theta, \beta) = E^{\text{ext}}(x, y, \Theta) + E^{\text{int}}(x, \beta) .$$

E^{ext} denotes the energy (or potential) arising from external influence; in the context of probabilistic image modelling, this typically comes from observed data y via a model determined by parameters Θ, and corresponds to a log-likelihood term. E^{int} denotes the internal energy which, as suggested by the notation, only depends on the MRF variables x and parameter β, and corresponds to a prior distribution over x.

2.2 The Mean Field Theory

To address the computational difficulties associated with MRF models, a number of approximate methods have been proposed [24]. One popular such method is the so called *mean field* approximation, from statistical mechanics [7]. This consists of replacing $p(x|y, \Theta, \beta)$ with an approximating, computationally tractable, parameterised distribution, $q(x|m)$. As has been shown by several authors [3, 7, 31, 38], the mean field approximation can be given a formal justification as providing a computationally tractable bound on quantities of interest (e.g. the partition function). Moreover, we can optimise the variational parameter m by minimising the Kullback-Leibler (KL) divergence between $q(x|m)$ and $p(x|\Theta, \beta, y)$,

$$D(p\|q) = \sum_{x} q(x|m) \ln \frac{q(x|m)}{p(x|\Theta, \beta, y)} \quad , \tag{3}$$

which is non-negative for all probability distributions q and p and equals zero only when they are identical. The literature on statistical mechanics [7], MRFs [3, 38] and probabilistic graphical models [19, 31] provides examples of applying this methodology to different models. Section 3.3 in this paper provides an example for a multi-level logistic MRF model.

2.3 A 'Variational' View of the EM-algorithm

Traditionally, the EM-algorithm [8] is viewed as a two-step algorithm for maximum-likelihood parameter estimation from incomplete data. The first step (the E-step) consists of computing the expectation over the random variables which are missing in the data (e.g. the labels in unlabelled data), x, given the observed variables, y, and the current set of parameters, Θ. The second step (the M-step) maximises the resulting expected complete log-likelihood function with respect to its adjustable parameters, Θ. However, it can also be seen as a algorithm for minimising the variational free energy from statistical mechanics and statistical physics [35, 26], linking it to the mean field theory. From this point of view, it is natural to also consider situations where the exact distribution over the missing variables cannot be computed, but has to be replaced by an approximate distribution. This yields an algorithm which maximises a lower bound of the log-likelihood. The difference between this bound and the true log-likelihood is the KL-divergence between the exact and approximating distributions.

Following Jordan et al. [19], our objective is to maximise the log-likelihood function of the observed data $\ln p(y|\Theta)$, with respect to the parameters Θ. We now write

$$\ln p(y|\Theta) = \ln \sum_{x} p(y, x|\Theta, \beta)$$

$$= \ln \sum_{x} q(x|m) \frac{p(x|\Theta, \beta, y)p(y|\Theta)}{q(x|m)}$$

$$\geq \sum_{x} q(x|m) \ln \frac{p(x|\Theta, \beta, y)p(y|\Theta)}{q(x|m)}$$

$$= \sum_{x} q(x|m) \ln p(y|\Theta) - q(x|m) \ln \frac{q(x|m)}{p(x|\Theta, \beta, y)} \quad , \tag{4}$$

where we have used Jensen's inequality. $q(x|m)$ is an arbitrary, non-singular probability distribution, parameterised by the variational parameter m. From (4), which apart from a change of sign corresponds to the variational free energy from statistical physics, we see directly that the difference between the two sides is the KL divergence (3) between $q(x|m)$ and $p(x|\Theta, \beta, y)$. As shown by Neal and Hinton [26], maximisation of (4) with respect to m corresponds to the E-step of the EM-algorithm, whenever $q(x|m)$ is rich enough to model $p(x|\Theta, \beta, y)$ exactly. When, on the other hand, computational considerations force us to resort to simpler distribution models, we can still be certain that resulting algorithm will increase the lower bound of the log-likelihood function, unless already at (a local) maximum.

3   An Application to fMRI Data

Functional magnetic resonance imaging (fMRI) attempts to detect brain activity by localised, non-invasive measurements of the change in blood oxygenation, the so called *BOLD contrast* [27]. This is sensitive to the relative local concentrations of oxygenated hemoglobin (HbO_2) vs. deoxy-hemoglobin and provides an indirect measure of the brain's neuronal activity.

Measurements, in the form of a time-series of images, are collected under controlled conditions, where subjects are performing specific tasks, prompted by some stimulus (e.g. deciding whether a read out sentence is grammatically correct or not, perform arithmetic calculations, looking at changing scenes, etc.). We only consider fMRI experiments with a single trial (or 'event-related') design, which consist of a series of individual trials. Each trial consist one repetition of the task, followed by a period of rest during which the subject is assumed to be inactive.

When we want to model the fMRI data generating process, there are neurophysiological factors we must take into consideration. The local change in blood oxygenation as an effect of increased neuronal activity, which is called the *hemodynamic response* (HR), is delayed by 2–6 seconds from stimulus onset and dispersed by 2–3.5 seconds. This delay and dispersion vary between subjects, experimental conditions, etc. By contrast, the stimuli subjects are exposed to during data collection, which is assumed to trigger the task related activity, is normally treated as being discrete. Often it is modelled as a binary ('box-car') function, i.e. the stimuli is either present or not present.

Traditional analysis of fMRI data essentially amounts to locating so-called *activated* pixels, where the observed measurements shows significant correlation with a function representing the the task. There are different strategies for altering this function to account for the HR, ranging from simply just shifting it in time [2] to convolving it with a HR model function [13, 23, 29]. The correlation is computed for each pixel individually and the correlation scores are transformed into *Z-scores* [1]. The resulting image of Z-scores, called a *Z-map*, is then thresholded at a level chosen so that the probability of wrongly classifying a pixel as being activated is suitably low (see e.g. [13]).

We propose to model an fMR image, by which we mean a set of pixels on a regular lattice with associated time-series of measurements, as a MRF. Each pixel is assumed to belong to one out of K classes, with each class corresponding to a parametric model function for the HR. The time series associated with each pixel contains measurements collected at the corresponding location during a single trial at times $t_1, \ldots, t_D$. By choosing a MRF model, we implicitly assume that the spatial distribution of the classes will be locally smooth, so that neighbouring pixels typically belong to the same class. Thus, images will consist of one or more spatially homogeneous regions, each region associated with a parametric HR model function. The model can also be seen as a K-component mixture model [17] in the D-dimensional observable pixel space (i.e. in the temporal domain of the HR), combined with a smoothing MRF prior distribution over pixel classes (in the pixels lattice).

3.1 The Multi-level Logistic MRF model

To specify the prior pixel class distribution, we use the commonly applied multi-level logistic (MLL) model [12, 14], where we specify neighbourhoods such that each pixel only depends on its nearest neighbours (distance equal to one in the lattice of pixels). We represent the MRF variable associated with pixel n as a K-dimensional binary vector, $\boldsymbol{x}_n$. Pixel n belongs to class k if and only if the kth element of $\boldsymbol{x}_n$, denoted x_{nk}, equals 1 and all other elements equals 0. This model contains the binary MRF as a special case ($K = 2$).

We then define the energy function,

$$E^{\text{int}}(\boldsymbol{x}, \beta) = \frac{\beta}{2} \sum_n^N \sum_{\boldsymbol{x}_m \in \mathcal{N}_n} \boldsymbol{x}_m^{\text{T}} \boldsymbol{U} \boldsymbol{x}_n \ , \tag{5}$$

where $\boldsymbol{U}$ is a $K \times K$ matrix with elements along its diagonal equal to -1 and all other element equal to 1. The scalar β plays the role of a scale parameter for the prior. As β increases, so does the cost for neighbouring pixels from different classes, which in effect forces a smoother image.

3.2 Modelling the Hemodynamic Response

Several model functions for the HR have been proposed [5, 13, 23, 29]; we choose to model the HR using a Gaussian function [22], such that

$$h(t) = \eta \exp\left(-\frac{(t - \mu)^2}{\sigma}\right) + o, \tag{6}$$

where,

μ denotes the *lag*, i.e the time from the onset of the stimuli to the peak of the HR,

σ denotes the *dispersion*, which reflects the rise and decay time,

η denotes the *gain*, or amplitude, of the response, and finally
o denotes an *offset* that defines the minimum level for the HR model function, relative to some baseline level.

For numerical convenience, σ and η are expressed using auxiliary variables, z_σ and z_η, so that

$$\sigma = \exp(z_\sigma) \quad \text{and} \quad \eta = \exp(z_\eta). \tag{7}$$

This will ensure that σ and η are always positive. In case of η, this is actually a simplification, since there is evidence for localised *deactivation* in response to stimuli. We denote the parameters $\Theta = [\theta_1, \ldots, \theta_K]$, where $\theta_k = [\mu_k, z_{\sigma k}, z_{\eta k}, o_k]$.

We combine the K HR model functions with an isotropic Gaussian noise process with variance α^{-1}, common to all HR model functions. For a pixel n, which belongs to class k, the probability distribution for the D-dimensional observable trial vector y_n can then be written as

$$p(y_n | \theta_k, \alpha) = \left(\frac{\alpha}{2\pi}\right)^{D/2} \exp\left(-\frac{\alpha}{2}\|y_n - h_k\|^2\right) \tag{8}$$

where h_k is a D-dimensional vector corresponding to the HR model function, computed from (6) and (7) at times $t_1, \ldots, t_D$, using the parameter vector θ_k. Note that, this model implicitly assumes that any two random vectors y_n and y_m, $n \neq m$, are independent given the classes of the corresponding pixels.

From the negative logarithm of (8), we can derive the external energy for the MRF model

$$E^{\text{ext}}(x, y, \Theta, \alpha) = \sum_{n,k}^{N,K} \frac{\alpha}{2}\|y_n - h_k\|^2 x_{nk} , \tag{9}$$

where $\sum_{n,k}^{N,K} = \sum_n^N \sum_k^K$; this abbreviated notation will be used throughout the rest of this paper. Recall that x_{nk} is 1 if and only if pixel n belongs to class k and 0 otherwise. The term arising from the normalisation factor, $(\alpha/2\pi)^{D/2}$, has been dropped as it does depend on x.

A Prior for the HR Parameters. Given our knowledge about neurology and fMRI in general and the experimental design in particular, we have certain a-priori beliefs about what can be considered reasonable values of the HR parameters. We can express beliefs by specifying a prior distribution over the HR parameters. Here, we choose a simple independent Gaussian distribution,

$$p(\Theta) = \left(\prod_i^{4K} 2\pi V_\Theta(i,i)\right)^{-1/2} \exp\left(-\frac{1}{2}\left(\Theta - \overline{\Theta}\right)^{\text{T}} V_\Theta^{-1}\left(\Theta - \overline{\Theta}\right)\right), \tag{10}$$

where $\overline{\Theta}$ is a $4K$-element vector containing the expected values for μ_k, $z_{\sigma k}$, $z_{\eta k}$ and o_k, $k = 1, \ldots, K$, and V_Θ is a diagonal covariance matrix with the corresponding variances along its diagonal.

3.3 Mean Field Equations for the Multi-level Logistic Model

Combining (9) with (5), we get the energy function

$$E = \sum_{n,k}^{N,K} E_{nk} x_{nk} \; , \tag{11}$$

where

$$E_{nk} = \frac{\alpha}{2} \|y_n - h_k\|^2 + \frac{\beta}{2} \sum_{x_m \in \mathcal{N}_n} x_m^{\mathrm{T}} U_k \; , \tag{12}$$

where in turn U_k denotes the kth column of U. From (1), we can write the corresponding distribution over the MRF as

$$p(x|y, \Theta, \alpha, \beta) = \frac{1}{Z} \exp \left(- \sum_{n,k}^{N,K} E_{nk} x_{nk} \right) \; . \tag{13}$$

For the mean field approximation, we choose q to be a simple independent multinomial distribution, where each lattice variable, x_n, has its own variational parameter, m_n,

$$q(x|m) = \prod_{n,k}^{N,K} m_{nk}^{x_{nk}} . \tag{14}$$

m_n is a K-dimensional vector whose elements are all positive and sum to 1; the kth element of m_n, m_{nk}, represents the probability that pixel n belongs to class k. m denotes the concatenation of m_n, $n = 1, \ldots, N$.

Substituting (13) and (14) into the quotient in (3), performing some elementary algebra and then averaging with respect to $q(x|m)$, we get

$$\sum_{n,k}^{N,K} [m_{nk} \ln m_{nk} + m_{nk} E'_{nk}] + \ln Z \; ,$$

where E'_{nk} is identical to E_{nk} in (12), except that x_m has been replaced by m_m. Taking the derivative of this with respect to m_{nk}, using Lagrange multipliers, ζ_n, to ensure that $\sum_k m_{nk} = 1$ for all n (see e.g. [10]), we get

$$\ln m_{nk} + 1 + E''_{nk} + \zeta_n \; ,$$

where E''_{nk} is identical to E'_{nk}, except that the factor $\beta/2$ has been replaced by β as a consequence of neighbourhood symmetry. Setting these to zero, we can solve for ζ_n, using that $\sum_k m_{nk} = 1$, and subsequently for m_{nk}, yielding

$$m_{nk} = \frac{\exp(-E''_{nk})}{\sum_{k'} \exp(-E''_{nk'})} \; . \tag{15}$$

which are the mean field equations for the MLL model which can be solved iteratively for a fixed point solution. An alternative derivation, drawing on analogies to statistical mechanics, can be found by Zhang [36].

At the moment, it is not established under which conditions these equations converge; Zhang [37] analysed the convergence for an Ising model equivalent with a binary MRF, which was found to converge under certain conditions. In practice, convergence does not appear to be a problem – the parameter m settles rapidly, and failure to reach absolute convergence simply means that our bound on the log-likelihood will be less tight.

3.4 Parameter Estimation

Until now, we have implicitly assumed that the all the parameters are known. This is typically not the case, but given the theory in Sect. 2, we can use the EM-algorithm to estimate parameters of interest. In the E-step, we compute the mean field approximation (15) to the posterior distribution over the MRF variables. In the following M-step, we maximise the resulting expected complete log-likelihood with respect to the parameters.

Here we restrict ourselves to maximisation with respect to Θ and α. The hyperparameters for the prior distribution over HR parameters are set using knowledge about the experimental design and general HR characteristics. β is set by experimenting; experience so far suggest that the final result is not very sensitive to the exact choice of β, which was also reported in [36].

We derive our objective function from a hypothetical log-likelihood function, where the class labels, x_n, are known. As we assume that the observations at different pixels are independent given the corresponding class labels, we get the penalised log-likelihood function from (8) and (10) as

$$\ln p(\Theta) + \frac{ND}{2} \ln \alpha - \frac{\alpha}{2} \sum_{n,k}^{N,K} x_{nk} \|y_n - h_k\|^2 \ ,$$

where we have omitted terms which do not depend on Θ or α. We obtain the corresponding expected complete penalised log-likelihood function simply by replacing the x_{nk} by their correponding mean-field expectations, m_{nk}, computed from (15).

Maximisation of the resulting objective function with respect to Θ is done by numerical optimisation[1]. For α, we get an update formula in closed form

$$\alpha = \frac{ND}{\sum_{n,k}^{N,K} m_{nk} \|y_n - \tilde{h}_k\|^2} \ ,$$

where $\tilde{h}_k$ are computed using the updated parameters θ_k.

---

[1] We use the function `fsolve` from the software package Octave [11] for this purpose.

Mean Field Annealing. The parameter estimation problem is fairly difficult optimisation problem, and empirical evidence suggest that there are many poor local optima where the optimisation procedure can get stuck. To reduce the risk of this, we employ a simulated annealing scheme [6, 20], multiplying (11) by an inverse temperature factor $(1/T)$, $T \geq 1$. Setting $T > 1$ will smooth the (approximate) posterior distribution, which in effect will smooth out shallow local optima. Thus, optimisation in the high-T regime (say $T = 10$ for data normalised to zero mean and unit variance) will hopefully find a global optimum which then can be tracked by re-estimating the parameters, Θ, as T is being decreased to 1. Note that, as long as $T > 1$, α is kept fixed to 1; this annealing phase is then followed by further optimisation where both Θ and α are adapted. Annealing approaches have been used successfully with MRF models for restoration of e.g. fMRI images [9] and, in combination with mean field theory, anatomical ('non-functional') magnetic resonance images [33].

3.5 Example

In this example we use data from an fMRI experiment designed for investigating the neuronal correlates of sentence comprehension in the brain [25]. Subjects had to decide whether an aurally presented sentence contained a syntactical violation or not. The experiment employed a single trial design where each trial had a length of 24 seconds. Each trial started with a sentence being read out, which lasted 2.3–4.5 seconds. fMR images with a spatial resolution of 64×128 pixels were collected every 2 seconds, so the trial vector for each pixel consists of 12 measurements. In total, there were 76 trials, although the first 4 were not used. The data were pre-processed to correct for subject movements, remove baseline trends and filter out physiological and system noise [21].

For this example, data from the 72 selected trials were averaged, to improve the signal to noise ratio. The resulting averaged data were used to train a model with 2 HR functions and a constant 'background' function, intended to explain regions where no task related activity occur. This constant function has a single parameter, namely its value, whose maximum likelihood update is the time-averaged response at individual pixels, averaged over the posterior distribution over pixel classes. The HR functions shared a common prior given in table 1 and β was set to 1. The fitting procedure started with 20 iterations during which T was decreased linearly from 10 to 1 and α was held fixed at 1.0, followed by another 20 iterations where $T = 1$ and α was allowed to adapt.

Table 1. Hyperparameters for the prior distribution over HR parameters used in the example described in Sect. 3.5. μ and z_σ are measured in (log) time steps, while z_η and o are measured (log) relative to a normalised BOLD response

μ	V_μ	z_σ	V_{z_σ}	z_η	V_{z_η}	o	V_o
6	3	$\ln 2$	$2/2^2$	$\ln 4$	$3/4^2$	0	1

The left image in Fig. 1 shows the resulting segmentation of the functional mask obtained from our model; pixels have been assigned to the class with highest posterior class probability, computed using the mean field approximation, after the parameters had converged. Figure 2 shows the corresponding HR functions. Different types of filtering in the pre-processing [21] will cause marginal variations in these results, but the overall picture will remain the same. The right image of Fig. 1 shows a Z-map for the same data set, based on correlation with a shifted 'box-car' function, overlaid on the functional mask (see e.g. [13]); note that, only pixels with positive activation are shown, as we do not consider deactivated regions.

As can be seen the two HR functions take on different roles, one explaining regions with a relatively strong and slightly earlier response, and corresponds roughly to pixels with strong activation (high Z-scores); the other explains a weaker and slightly later response, and includes pixels with lower activation.

4 Discussion

In this paper, we have reviewed the use of the EM-algorithm combined with mean field theory for parameter estimation from unlabelled data in MRF models, and the theoretical justification for this, based on results from the machine learning literature. Furthermore, we have shown an application of this procedure for analysis of fMRI data – a learning problem of inherent unsupervised nature.

It should be clear that the overall framework is independent of the choice of HR model function, and thus other variants could be considered. Similarly, we could consider the use of a more elaborate noise model; Kruggel and von Cramon [22] discuss the use of an autoregressive (AR(1)) noise model in the spatial domain.

A limitation of the work we have presented in this paper is the remaining number of free parameters. β is currently set by experimenting. Deriving a method for updating β in the light of observed data is difficult, since the partition function for the MRF prior depends on β. Zhang [36] suggested using a mean field approximation also for the partition function, but as pointed out by Jordan et al. [19], this result in an update equation based on two different bounds, which theoretically may *decrease* the log-likelihood of the data given β. An alternative approach would be to use Monte Carlo sampling methods for the parameter fitting. Such an approach would be computationally demanding; a potential remedy could be to estimate Θ and α using mean field theory, and use Monte Carlo methods only for the updates of β, which need not be updated every iteration of the EM-algorithm. The number of HR components, K, is currently set by the user, based on empirical evidence, prior knowledge and interpretability. It would clearly be desirable to be able to estimate K from the data, but such estimation would face the same difficulties as the estimation of β, since comparing models with different values for K requires computing the corresponding partition functions. Nevertheless, methods based on minimum de-

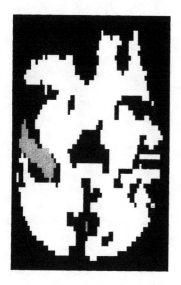

Fig. 1. A segmentation obtained using proposed method (left) and a corresponding correlation based Z-map (right), for the data described in Sect. 3.5. In the left image, pixels in the functional mask have been classified according to their maximum posterior class probabilities; the corresponding HR model functions plotted in Fig. 2; the dominating white class corresponds to the background function. In the Z-map, which is overlaid on the functional mask, pixels have been shaded according to their Z-score, where brighter pixels indicate higher activation

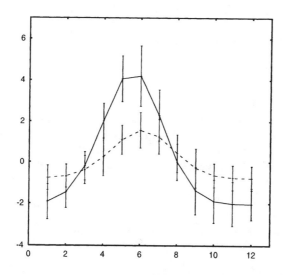

Fig. 2. The HR functions corresponding to the segmentation shown in Fig. 1. The solid line corresponds to the light grey pixels while the dashed line corresponds to dark grey pixels. The error bars corresponds to 1 standard deviation of the data from the mean given by the curve

scription length theory [30] or maximum entropy principles [34], combined with approximate methods for computing the partition function, could be considered.

The idea of deriving mean field equations by minimising the KL-divergence given a choice of approximating distribution raises the question whether other approximating distributions can be found that gives a tighter bound and remains computationally tractable. Jordan et al. [19] gives several such examples examples in the context of learning in graphical models, some of which could potentially be applied to MRF models.

An obvious limitation of the mean field approximation discussed in Sect. 3.3 is that it is unimodal, i.e. the spatial distribution of pixel class labels is centred around a single configuration. This might be a a reasonable approximation when modelling averaged data from a single experiment, as in Sect. 3.5, but if we want to investigate between-trial variance within one experiment or even the (dis)similarities between trials from different experiments, it is clearly insufficient. Jaakkola and Jordan [16] proposed the use of a mixture of fully factorised mean field distributions, and Bishop et al. [4] empirically demonstrated the usefulness of this approach in the context of sigmoid belief networks. A future direction of research will be to extend the approach presented in this paper to the use of such mixture distributions, and investigate the usefulness of this for the purpose of fMRI data modelling.

For fMRI data, it is also natural to consider modelling structure in the time domain, since an experiment consists a sequence of images corresponding to the sequence of trials. Theory for such a model could be built on the existing theory for hidden Markov models (HMM) [28], which has recently been subject to substantial development in the context of graphical models and machine learning [15, 19, 32]. This strand of research will be pursued as an extension of the mixture model discussed in the previous paragraph.

References

[1] M. Abramowitz and I. A. Stegun. *Handbook of mathematical functions with formulas, graphs and mathematical tables, 10th printing*. National Bureau of Standards Applied Mathematics Series 55, Washington, 1972.

[2] P. A. Bandettini, A. Jesmanowicz, E. C. Wong, and J. S. Hyde. Processing strategies for time-course data sets in functional MRI of the human brain. *Magnetic Resonance in Medicine*, 30:161–173, 1993.

[3] G. L. Bilbro, W. E. Snyder, and R. C. Mann. Mean-field approximation minimizes relative entropy. *Journal of the Optical Society of America A*, 8(2):290–294, 1991.

[4] C. M. Bishop, N. D. Lawrence, T. S. Jaakkola, and M. I. Jordan. Approximating posterior distributions in belief networks using mixtures. In M. I. Jordan, M. J. Kearns, and S. A. Solla, editors, *Advances in Neural Information Processing Systems*, volume 10. MIT Press, 1998.

[5] E. Bullmore, M. Brammer, S. C. R. Williams, S. Rabe-Hesketh, N. Janot, A. David, J. Mellers, R. Howard, and P. Sham. Statistical methods of estimation and inference for functional MR image analysis. *Magnetic Resonance in Medicine*, 35:261–277, 1996.

[6] V. Cerný. A thermodynamical approach to the travelling salesman problem: An efficient simulation algorithm. *Journal of Optimization Theory and Applications*, 45:41–51, 1985.

[7] D. Chandler. *Introduction to Modern Statistical Mechanics*. Oxford University Press, New York, 1987.

[8] A. P. Dempster, N. M. Laird, and D. B. Rubin. Maximum likelihood from incomplete data via the EM algorithm. *Statistical Methodology, Journals of the Royal Statistical Society, Series B*, 39(1):1–38, 1977.

[9] X. Descombes, F. Kruggel, and D. Y. von Cramon. fMRI signal restoration using an edge preserving spatio-temporal Markov random field. *NeuroImage*, 8:340–348, 1998.

[10] L. C. W. Dixon. *Nonlinear Optimisation*. English Universities Press, London, 1972.

[11] J. W. Eaton et al. GNU Octave. Available on the Internet at URL: http://www.che.wisc.edu/octave/, 1998. version 2.0.

[12] H. Elliot, H. Derin, R. Christi, and D. Geman. Application of the Gibbs distribution to image segmentation. In *Proceedings of the International Conference on Acoustics, Speech and Signal Processing*, pages 32.5.1–32.5.4, San Diego, 1984. IEEE.

[13] K. J. Friston, P. Jezzard, and R. Turner. Analysis of functional MRI time-series. *Human Brain Mapping*, 1:153–171, 1994.

[14] S. Geman and D. Geman. Stochastic relaxation, Gibbs distributions, and the Bayesian segmentation of images. *IEEE Transactions on Pattern Analysis and Machine Intelligence*, 6(6):721–741, 1984.

[15] Z. Gharamani and M. I. Jordan. Factorial hidden Markov models. *Machine Learning*, 29:245–273, 1997.

[16] T. S. Jaakkola and M. I. Jordan. Improving the mean field approximation via the use of mixture distributions. In Jordan [18]. Proceedings of the NATO Advanced Study Institute, Erice, Italy, 1996.

[17] A. Jepson and M. Black. Mixture models for image representation. Technical report, Department of Computer Science, University of Toronto, March 1996. PRECARN ARK Project Technical Report ARK96-PUB-54.

[18] M. I. Jordan, editor. *Learning in Graphical Models*. Adaptive Computation and Machine Learning. MIT Press, 1998. Proceedings of the NATO Advanced Study Institute, Erice, Italy, 1996.

[19] M. I. Jordan, Z. Ghahramani, T. S. Jaakkola, and L. K. Saul. An introduction to variational methods for graphical models. In Jordan [18]. Proceedings of the NATO Advanced Study Institute, Erice, Italy, 1996.

[20] S. Kirkpatrick and C. D. and M. P. Vecchi Gellatt, Jr. Optimization by simulated annealing. *Science*, 220:671–680, 1983.

[21] F. Kruggel, X. Descombes, and D. Y. von Cramon. Preprocessing of fMR data sets. In *Workshop on Biomedical Image Analysis*, pages 211–220, Santa Barbera, 1998. IEEE Computing Society.

[22] F. Kruggel and D. Y. von Cramon. Modelling the hemodynamic response in single trial fMRI experiments. *Magnetic Resonance in Medicine*, 1998. Under review.

[23] N. Lange and S. L. Zeger. Non-linear time-series analysis for human brain mapping by magnetic resonance imaging. *Applied Statistics, Journals of the Royal Statistical Society, Series C*, 46:1–29, 1997.

[24] S. Z. Li. *Markov Random Field Modeling in Computer Vision*. Springer-Verlag, Tokyo, 1995.

[25] M. Meyer, A. D. Friederici, D. Y. von Cramon, F. Kruggel, and C. J. Wiggins. Auditory sentence comprehension: Different BOLD patterns modulated by task demands as revealed by a 'single-trial' fMRI-study. *NeuroImage*, 7(4):S181, 1998.

[26] R. M. Neal and G. E. Hinton. A view of the EM algorithm that justifies incremental, sparse and other variants. In Jordan [18]. Proceedings of the NATO Advanced Study Institute, Erice, Italy, 1996.

[27] S. Ogawa, T. M. Lee, A. R. Kay, and D. W. Tank. Brain magnetic resonance imaging with contrast dependent blood oxygenation. *Proceedings of the National Acadademy of Sciences, USA*, 87:9868–9872, 1990.

[28] L. R. Rabiner. A tutorial on hidden Markov models and selected applications in speech recognition. *Proceedings of the IEEE*, 77(2):257–285, 1989.

[29] J. C. Rajapakse, F. Kruggel, J. M. Maisog, and D. Y. von Cramon. Modeling hemodynamic response for analysis of functional MRI time-series. *Human Brain Mapping*, 6:283–300, 1998.

[30] J. Rissanen. Modelling by shortest data description. *Automatica*, 14:465–471, 1978.

[31] L. K. Saul and M. I. Jordan. Exploiting tractable substructures in intractable networks. In D. S. Touretzky, M. C. Mozer, and M. E. Hasselmo, editors, *Advances in Neural Information Processing Systems*, volume 8, pages 486–492. MIT Press, 1996.

[32] P. Smyth, D. Heckerman, and M. I. Jordan. Probabilistic independence networks for hidden Markov probability models. *Neural Computation*, 9:227–270, 1997.

[33] W. Snyder, A. Logenthiran, P. Santago, K. Link, G. Bilbro, and S. Rajala. Segmentation of magnetic resonance images using mean field annealing. *Image and Vision Computing*, 10(6):218–226, 1992.

[34] N. Wu. *The Maximum Entropy Method*. Springer Series in Information Sciences. Springer-Verlag, 1997.

[35] A. L. Yuille, P. Stolorz, and J. Utans. Statistical physics mixtures of distributions, and the EM algorithm. *Neural Computation*, 6:334–340, 1994.

[36] J. Zhang. The mean field theory in EM procedures for Markov random fields. *IEEE Transactions on Signal Processing*, 40(10):2570–2583, 1992.

[37] J. Zhang. The convergence of mean field procedures for MRF's. *IEEE Transactions on Image Processing*, 5(12):1662–1665, 1995.

[38] J. Zhang. The application of the Gibbs-Bogoliubov-Feynman inequality in mean field calculations for Markov random fields. *IEEE Transactions on Image Processing*, 5(7):1208–1214, 1996.

Author Index

Lecture Notes in Computer Science

For information about Vols. 1–1562
please contact your bookseller or Springer-Verlag

Vol. 1606: J. Mira, J.V. Sánchez-Andrés (Eds.), Foundations and Tools for Neural Modeling. Proceedings, Vol. I, 1999. XXIII, 865 pages. 1999.

Vol. 1607: J. Mira, J.V. Sánchez-Andrés (Eds.), Engineering Applications of Bio-Inspired Artificial Neural Networks. Proceedings, Vol. II, 1999. XXIII, 907 pages. 1999.

Vol. 1608: S. Doaitse Swierstra, P.R. Henriques, J.N. Oliveira (Eds.), Advanced Functional Programming. Proceedings, 1998. XII, 289 pages. 1999.

Vol. 1609: Z. W. Raś, A. Skowron (Eds.), Foundations of Intelligent Systems. Proceedings, 1999. XII, 676 pages. 1999. (Subseries LNAI).

Vol. 1610: G. Cornuéjols, R.E. Burkard, G.J. Woeginger (Eds.), Integer Programming and Combinatorial Optimization. Proceedings, 1999. IX, 453 pages. 1999.

Vol. 1611: I. Imam, Y. Kodratoff, A. El-Dessouki, M. Ali (Eds.), Multiple Approaches to Intelligent Systems. Proceedings, 1999. XIX, 899 pages. 1999. (Subseries LNAI).

Vol. 1612: R. Bergmann, S. Breen, M. Göker, M. Manago, S. Wess, Developing Industrial Case-Based Reasoning Applications. XX, 188 pages. 1999. (Subseries LNAI).

Vol. 1613: A. Kuba, M. Šámal, A. Todd-Pokropek (Eds.), Information Processing in Medical Imaging. Proceedings, 1999. XVII, 508 pages. 1999.

Vol. 1614: D.P. Huijsmans, A.W.M. Smeulders (Eds.), Visual Information and Information Systems. Proceedings, 1999. XVII, 827 pages. 1999.

Vol. 1615: C. Polychronopoulos, K. Joe, A. Fukuda, S. Tomita (Eds.), High Performance Computing. Proceedings, 1999. XIV, 408 pages. 1999.

Vol. 1616: P. Cointe (Ed.), Meta-Level Architectures and Reflection. Proceedings, 1999. XI, 273 pages. 1999.

Vol. 1617: N.V. Murray (Ed.), Automated Reasoning with Analytic Tableaux and Related Methods. Proceedings, 1999. X, 325 pages. 1999. (Subseries LNAI).

Vol. 1618: J. Bézivin, P.-A. Muller (Eds.), The Unified Modeling Language. Proceedings, 1998. IX, 443 pages. 1999.

Vol. 1619: M.T. Goodrich, C.C. McGeoch (Eds.), Algorithm Engineering and Experimentation. Proceedings, 1999. VIII, 349 pages. 1999.

Vol. 1620: W. Horn, Y. Shahar, G. Lindberg, S. Andreassen, J. Wyatt (Eds.), Artificial Intelligence in Medicine. Proceedings, 1999. XIII, 454 pages. 1999. (Subseries LNAI).

Vol. 1621: D. Fensel, R. Studer (Eds.), Knowledge Acquisition Modeling and Management. Proceedings, 1999. XI, 404 pages. 1999. (Subseries LNAI).

Vol. 1622: M. González Harbour, J.A. de la Puente (Eds.), Reliable Software Technologies – Ada-Europe'99. Proceedings, 1999. XIII, 451 pages. 1999.

Vol. 1625: B. Reusch (Ed.), Computational Intelligence. Proceedings, 1999. XIV, 710 pages. 1999.

Vol. 1626: M. Jarke, A. Oberweis (Eds.), Advanced Information Systems Engineering. Proceedings, 1999. XIV, 478 pages. 1999.

Vol. 1627: T. Asano, H. Imai, D.T. Lee, S.-i. Nakano, T. Tokuyama (Eds.), Computing and Combinatorics. Proceedings, 1999. XIV, 494 pages. 1999.

Col. 1628: R. Guerraoui (Ed.), ECOOP'99 - Object-Oriented Programming. Proceedings, 1999. XIII, 529 pages. 1999.

Vol. 1629: H. Leopold, N. García (Eds.), Multimedia Applications, Services and Techniques - ECMAST'99. Proceedings, 1999. XV, 574 pages. 1999.

Vol. 1631: P. Narendran, M. Rusinowitch (Eds.), Rewriting Techniques and Applications. Proceedings, 1999. XI, 397 pages. 1999.

Vol. 1632: H. Ganzinger (Ed.), Automated Deduction – Cade-16. Proceedings, 1999. XIV, 429 pages. 1999. (Subseries LNAI).

Vol. 1633: N. Halbwachs, D. Peled (Eds.), Computer Aided Verification. Proceedings, 1999. XII, 506 pages. 1999.

Vol. 1634: S. Džeroski, P. Flach (Eds.), Inductive Logic Programming. Proceedings, 1999. VIII, 303 pages. 1999. (Subseries LNAI).

Vol. 1636: L. Knudsen (Ed.), Fast Software Encryption. Proceedings, 1999. VIII, 317 pages. 1999.

Vol. 1638: A. Hunter, S. Parsons (Eds.), Symbolic and Quantitative Approaches to Reasoning and Uncertainty. Proceedings, 1999. IX, 397 pages. 1999. (Subseries LNAI).

Vol. 1639: S. Donatelli, J. Kleijn (Eds.), Application and Theory of Petri Nets 1999. Proceedings, 1999. VIII, 425 pages. 1999.

Vol. 1640: W. Tepfenhart, W. Cyre (Eds.), Conceptual Structures: Standards and Practices. Proceedings, 1999. XII, 515 pages. 1999. (Subseries LNAI).

Vol. 1643: J. Nešetřil (Ed.), Algorithms – ESA '99. Proceedings, 1999. XII, 552 pages. 1999.

Vol. 1644: J. Wiedermann, P. van Emde Boas, M. Nielsen (Eds.), Automata, Languages, and Programming. Proceedings, 1999. XIV, 720 pages. 1999.

Vol. 1645: M. Crochemore, M. Paterson (Eds.), Combinatorial Pattern Matching. Proceedings, 1999. VIII, 295 pages. 1999.

Vol. 1647: F.J. Garijo, M. Boman (Eds.), Multi-Agent System Engineering. Proceedings, 1999. X, 233 pages. 1999. (Subseries LNAI).

Vol. 1649: R.Y. Pinter, S. Tsur (Eds.), Next Generation Information Technologies and Systems. Proceedings, 1999. IX, 327 pages. 1999.

Vol. 1650: K.-D. Althoff, R. Bergmann, L.K. Branting (Eds.), Case-Based Reasoning Research and Development. Proceedings, 1999. XII, 598 pages. 1999. (Subseries LNAI).

Vol. 1651: R.H. Güting, D. Papadias, F. Lochovsky (Eds.), Advances in Spatial Databases. Proceedings, 1999. XI, 371 pages. 1999.

Vol. 1653: S. Covaci (Ed.), Active Networks. Proceedings, 1999. XIII, 346 pages. 1999.

Vol. 1654: E.R. Hancock, M. Pelillo (Eds.), Energy Minimization Methods in Computer Vision and Pattern Recognition. Proceedings, 1999. IX, 331 pages. 1999.

Vol. 1663: F. Dehne, A. Gupta. J.-R. Sack, R. Tamassia (Eds.), Algorithms and Data Structures. Proceedings, 1999. X, 367 pages. 1999.